A New Plantation World

In the era between the world wars, wealthy sportsmen and sports-women created more than seventy large estates in the coastal region of South Carolina. By retaining select features from earlier periods and adding new buildings and landscapes, wealthy sporting enthusiasts created a new type of plantation. In the process, they changed the meaning of the word "plantation," with profound implications for historical memory of slavery and contemporary views of the South. *A New Plantation World* is the first critical investigation of these "sporting plantations." By examining the process that remade former sites of slave labor into places of leisure, Daniel J. Vivian explores the changing symbolism of plantations in Jim Crow–era America.

DANIEL J. VIVIAN is an associate professor in the Department of Historic Preservation at the University of Kentucky.

Cambridge Studies on the American South

Series Editors

Mark M. Smith, *University of South Carolina, Columbia*
Peter Coclanis, *University of North Carolina at Chapel Hill*

Interdisciplinary in its scope and intent, this series builds upon and extends Cambridge University Press's long-standing commitment to studies on the American South. The series offers the best new work on the South's distinctive institutional, social, economic, and cultural history and also features works in a national, comparative, and transnational perspective.

Titles in the Series

A New Plantation World

Sporting Estates in the South Carolina
Lowcountry, 1900–1940

DANIEL J. VIVIAN

University of Kentucky

CAMBRIDGE
UNIVERSITY PRESS

CAMBRIDGE
UNIVERSITY PRESS

University Printing House, Cambridge CB2 8BS, United Kingdom

One Liberty Plaza, 20th Floor, New York, NY 10006, USA

477 Williamstown Road, Port Melbourne, VIC 3207, Australia

314-321, 3rd Floor, Plot 3, Splendor Forum, Jasola District Centre, New Delhi - 110025, India

79 Anson Road, #06-04/06, Singapore 079906

Cambridge University Press is part of the University of Cambridge.

It furthers the University's mission by disseminating knowledge in the pursuit of
education, learning and research at the highest international levels of excellence.

www.cambridge.org
Information on this title: www.cambridge.org/9781108403429
DOI: 10.1017/9781108242165

© Daniel J. Vivian 2018

First published 2018
First paperback edition 2019

A catalogue record for this publication is available from the British Library

Library of Congress Cataloging in Publication data
NAMES: Vivian, Daniel, author.
TITLE: A new plantation world : sporting estates in the South Carolina
lowcountry, 1900–1940 / Daniel J. Vivian.
OTHER TITLES: Cambridge studies on the American South.
DESCRIPTION: New York : Cambridge University Press, 2018. | Series: Cambridge
studies on the American South | Includes bibliographical references and index.
IDENTIFIERS: LCCN 2017038386 | ISBN 9781108416900 (hardback : alk. paper) |
ISBN 9781108403429 (pbk. : alk. paper)
SUBJECTS: LCSH: Plantations–South Carolina–History–20th century.
CLASSIFICATION: LCC F270 .V58 2018 | DDC 975.7/04–dc23 LC record
available at https://lccn.loc.gov/2017038386

ISBN 978-1-108-41690-0 Hardback
ISBN 978-1-108-40342-9 Paperback

Contents

Figures

Acknowledgments

This book has been a long time in the making. It began as my doctoral dissertation at Johns Hopkins University, where Mary Ryan served as my advisor and dissertation director. Ronald Walters, Michael Johnson, and Stuart Leslie served on my dissertation committee and provided incisive comments on early drafts and the final dissertation. The department of history at Johns Hopkins provided a wonderfully supportive and stimulating environment for my research, and I benefited greatly from opportunities to present papers to several seminars. Although there is not room enough here to thank everyone who commented on my work, I am grateful for suggestions and criticisms offered by Philip D. Morgan, Dorothy Ross, Jane Dailey, Joe Adelman, Sarah Adelman, Andrew Deveraux, Andrew Griffith, Kate Moran, Justin Roberts, Teresa Cribelli, David Schley, James Roberts, John Matsui, and Gabe Klehr. All proved useful in helping me develop the ideas expressed herein and preventing me from making mistakes I would have regretted.

Staff at a number of institutions did a great deal to facilitate my research. At Historic Charleston Foundation, Karen Edmonds made working with their collections relating to Mulberry Plantation a pleasure. The many hours I spent at the South Carolina Historical Society provided a wealth of material. I am grateful to Mary Jo Fairchild, Virginia Ellison, Celeste Wiley, Mike Coker, John Tucker, Karen Stokes, and Jane Aldrich for their generous assistance and invaluable suggestions. At the South Carolina Department of Archives and History, Wade Ferry and Steve Tuttle provided equally helpful assistance. Several visits to the South Caroliniana Library allowed me to gather important material. I am grateful to Robin Copp, Henry Fulmer, Beth Bilderback, Brian Cuthrell, and Allen Stokes for making my research productive and enjoyable. Grace Cordial of the Beaufort County Public Library made the several days I spent working with the Beaufort District Collection especially rewarding. Harlan Green guided me through several manuscript collections at the Addlestone Library of the College of Charleston and shared insights about the Charleston Renaissance and tourism in the lowcountry between the world wars. At the Library of Congress, Chris Baer processed countless requests on my behalf and never tired of delivering books to my desk.

Support from two institutions allowed me to conduct important research as I revised the manuscript for publication. An exploratory grant from the Hagley Museum and Library in Wilmington, Delaware, allowed me to examine collections relating to the Kinloch Hunting Club and members of the Du Pont Family. The results are evident throughout Chapter 1. I also benefited from the 2014 Victor A. Olorunsola Endowed Research Award from the University of Louisville, which made possible research on northerners' estates and activities in Beaufort, Jasper, and Colleton counties. It proved essential for understanding variations within the lowcountry and developing arguments expressed in Chapter 2.

Portions of several chapters had their origins as papers presented at professional conferences. I benefited greatly from sharing an abridged version of Chapter 4 at a meeting of the Southeast Chapter of the Society of Architectural Historians and portions of Chapter 6 at meetings of the Southern Historical Association and the Midwest Modern Language Association. Comments and critiques by Clifton Ellis, Seth Bruggeman, Lisa Tolbert, Robert St. George, and Whitney Martinko proved valuable. I am grateful for the opportunity to present my work to these organizations and feel fortunate to have participated in several engaging panels.

One of the greatest joys of any major research project is meeting scholars with similar interests. Conversations with Julia Brock, Drew Swanson, Bert Way, Matthew Lockhart, John Bryan, Jennifer Betsworth, Hayden Smith, Stephen Hoffius, and Robert Cuthbert did a great deal to expand my knowledge of sport hunting, plantations, and no small number of other topics. Julia and I quickly recognized the common threads in our research and soon began working on an edited collection that became *Leisure, Plantations, and the Making of a New South: The Sporting Plantations of the South Carolina Lowcountry and Red Hills Region, 1900–1940* (2015). Our many conversations and editing the contributors' essays sharpened my thinking about South Carolina estates and the factors responsible for their development. I spoke frequently with Stephen Hoffius and Robert Cuthbert during the early stages of my research, while they were working on *Northern Money, Southern Land: The Lowcountry Plantation Sketches of Chlotilde R. Martin* (Columbia: University of South Carolina Press, 2009). Their insights about northern sportsmen and sportswomen and individual estates helped me better understand the development of northerners' "plantations," and I have frequently referred to their book since its publication.

Visits to several lowcountry estates allowed me to better understand the architecture and landscapes that northerners created. I am grateful to John Bryan for the visit he arranged to Gravel Hill Plantation and the insights that Joe Chapman shared from his long career in land management. Robert Hortman took time out of his busy schedule to show me Medway Plantation, and Ben Miller gave me a wonderful tour of Mulberry. I greatly appreciate the opportunity to visit these estates, for they helped me write more coherently about them and their counterparts.

Sandy Wood deserves special thanks for sharing insights about his grandparents and answering my many questions about Medway Plantation and the Legendre family. He also shared historical documents from his personal collections with me, for which I am especially appreciative. I would be remiss if I did not mention his patience and the consistent interest he has shown in my work. Both have been more valuable than he knows.

Friends and family have shown extraordinary support and encouragement throughout my research and writing. I am especially grateful to Jean Lee, Bradford Vivian, Sydney Goetz, James Gurley, Chris Lee, Jennifer Maxwell, John and Mary Sherrer, Al Hester, Terri Hurley, Damien and Betsy Doyle, Gigi Price, James Jacobs, Emily Cox, Aaron Pope, and John David Myles. At the University of Louisville, Tracy K'Meyer, Glenn Crothers, Cate Fosl, and Lara Kelland provided valuable advice and encouragement while also being wonderful colleagues. Robert Weyeneth has been an exceptional teacher, mentor, and friend for more than twenty years. I am fortunate to have been his student and never cease to be amazed by his dedication to his colleagues and the profession. My friend and former coworker Andy Chandler deserves particular thanks for sparking my interest in "rich Yankees" and their plantations. Whether he remembers it or not, a passing comment about "rich New Yorkers" and their duck hunting piqued my curiosity and set me on the path to this book. Although there have been times when I lamented ever hearing about the subject, for the most part I have been immensely grateful.

No author could ask for better support than I have received from Cambridge University Press. David Moltke-Hansen expressed interest in my work early on and invited me to submit a manuscript for consideration.

I am grateful for his cogent advice, his forthright critique of the manuscript I sent him, and for securing comments from two anonymous reviewers, both of which proved exceedingly useful. The editors of the Cambridge Studies on the American South series, Peter A. Coclanis and Mark M. Smith, have been supportive and encouraging while also providing valuable guidance. Deborah Gershenowitz has been an outstanding editor throughout the process, and her assistant, Kristina Deusch, has been equally helpful. I feel fortunate to have worked with such a dedicated group of professionals.

Introduction

In January 1932, journalist James Derieux informed the readers of *Country Life* magazine of developments reshaping the coastal region of South Carolina. "The visible and noble remnants of America's most colorful agrarian civilization," he announced, "are being saved by winter residents from almost total disappearance." For years, plantations all along the coast had lain in ruins, slowly succumbing to the ravages of time. Now, however, a great many had "been bought and rejuvenated by non-residents, most of them Northerners." "Scores" had been turned into winter homes and hunting retreats. Derieux admitted that "sentimentally, such a state of affairs is not perfect," but he still thought it beneficial for all involved. "Better for the grand remains of ancient glory to be saved by outsiders," he opined, "than that they be not saved at all."[1]

The phenomenon Derieux described had long since become familiar to lowcountry people. For more than a decade, residents of the region had watched northern bankers, businessmen, and heirs of industrial fortunes spend huge sums turning old plantations into country estates. From the beginning, the newcomers' inclinations had tended toward grandeur. Large houses, elaborate gardens, and extensive landscaping quickly become common. Over time, northerners grew more ambitious. By the early 1930s, the size and opulence of some estates stunned observers. "Very few of these gentlemen have less than a thousand acres," Derieux reported. "Most of them have several thousand," he

---

[1] James C. Derieux, "The Renaissance of the Plantation," *Country Life* 41, no. 3 (Jan. 1932): 34–39.

noted. Some had even more. "Mr. Baruch," Derieux observed, "is master of twenty thousand."[2]

As the "renaissance" had grown in scale, so too had the lowcountry's popularity as a seasonal destination for the wealthy. By the early 1930s, sportsmen and sportswomen came annually to shoot ducks, deer, and other game. As part of the seasonal migration from the North, the visitors arrived in the days leading up to Thanksgiving, the traditional beginning of duck-hunting season, and stayed for weeks or months at a time. Local newspapers announced northerners' arrival and described preparations made in advance. In Georgetown, the port town on the Sampit River north of Charleston, crowds gathered to watch northerners' private Pullman coaches arrive. Throughout the winter months, news about wealthy socialites, millionaire bankers, and celebrities circulated through rural communities at dizzying speed. The "winter colonists" had become a spectacle, an annually recurring drama in a region more familiar with stagnation and isolation than the antics of a social elite.[3]

Wealth, status, and enthusiasm for recreational hunting united the men and women who wintered in the region. Any of them could have chosen to spend their time elsewhere, but they instead selected the Carolina lowcountry. Even as Florida resorts had become enormously popular, some members of the upper classes opted for the Carolina coast. According to Derieux, the winter colonists found "the hunting, climate, traditions, and scenery" virtually irresistible. They succumbed to what he called "the lure of the Low Country." All came to enjoy "a winter climate of rare luxury, and a charm that is unsurpassed."[4]

Derieux's article draws back the curtain on events that attracted widespread attention at the time but have since been all but forgotten. By the late 1930s, wealthy sportsmen and sportswomen owned seventy-six "plantations" on the Carolina coast. In locations between northern

---

[2] Derieux, "The Renaissance of the Plantation," 36.

[3] Ibid., 35–39; "Black Border Plantations Lure Wealthy Northerners," *News and Courier* (Charleston, SC), Apr. 21, 1929, A9; Chalmers S. Murray, "Attracted by Climate Northerners Become Land Barons of Carolina Coast," *News and Courier*, Dec. 15, 1929, A3; Chlotilde R. Martin, "Low-Country Plantations Stir as Air Presages Coming of New Season," *News and Courier*, Oct. 16, 1932, A6; "Arrivals at Charleston," *New York Times* (New York, NY), Nov. 17, 1932, 16; "Baruch Leads Northerners to Annual Vacation in State," *News and Courier*, Nov. 30, 1933, 2; "Northern People Basking Here as Winter Grips Their Homes," *News and Courier*, Jan. 3, 1934, 5; "Winter Residents Reopen Estates," *News and Courier*, Nov. 25, 1938, B4.

[4] Derieux, "The Renaissance of the Plantation," 36.

Georgetown County and the Savannah River, wealthy sporting enthusiasts – variously called "northerners," "Yankees," and "winter colonists" by lowcountry people – wintered at handsome estates. Alongside their holdings lay others that served somewhat similar purposes. Hunting clubs controlled tens of thousands of acres. Sportsmen also owned retreats and preserves, most with smaller acreages. At virtually every turn, well-to-do people from afar had created recreational domains for their exclusive use (see Figure 0.1).[5]

The estates that northerners created are well known to lowcountry people. For decades they have been celebrated as part of the "Second Yankee Invasion," a decades-long land-buying spree that saw wealthy outsiders acquire huge expanses of territory. Recalled as elegant retreats where monied plutocrats hunted, relaxed, and socialized in idyllic circumstances, the new estates quickly became symbols of renewal. Before the Civil War, the lowcountry had ranked among the most powerful plantation districts in the South. Its planter class had wielded exceptional economic and political strength; Charleston stood as a center of trade and culture. Throughout the region, plantations produced large quantities of staple crops. As the sectional conflict escalated, the lowcountry elite committed itself to the dream of a southern nation. For a time, it appeared as though their ambition might come true. When Confederate artillery began firing on Fort Sumter during the early morning hours of April 12, 1861, planters and merchants had cheered, confident in their cause. Initially, the course of the war provided cause for optimism. Soon, however, the tide turned against the Confederacy and suffering arrived in earnest. Union forces seized control of Port Royal Sound in November 1861 and besieged Charleston in August 1863. Raiding parties struck plantations all along the coast, leading to a breakdown of authority and new assertiveness among slaves. By the time General Robert E. Lee surrendered the Army of Northern Virginia at Appomattox in April 1865, the lowcountry lay in ruins, its productive capacities shattered and its ruling elite in disarray.[6]

---

[5] Ibid.; George C. Rogers, Jr., *The History of Georgetown County, South Carolina* (Columbia: University of South Carolina Press, 1970), chap. 21; John H. Tibbetts, "The Bird Chase," *Coastal Heritage* 15, no. 4 (spring 2001), 3–13.

[6] On the colonial and antebellum lowcountry, see especially Walter Edgar, *South Carolina: A History* (Columbia: University of South Carolina Press, 1998), chaps. 3–17; Walter J. Fraser, Jr., *Charleston! Charleston!: The History of a Southern City* (Columbia: University of South Carolina Press, 1989), chaps. 1–5; Rogers, *History of Georgetown County,*

FIGURE O.I "A New Map Showing the Principal Plantations on the South Carolina Coast," 1932. See p.5 for key. Published by Elliman and Mullally, a real estate firm specializing in "plantations, shooting properties, and townhouses," this map shows most of the estates developed by wealthy northerners. By the mid-1930s, "plantation" more often denoted a private sporting estate than a site of agricultural production in the Carolina lowcountry.

From the manuscript collections of the South Carolina Historical Society, Charleston, SC.

Name	Place	Code
Allston, Mrs. Elizabeth Deas	Fairfield	BX-3
Alston, Donald	Fairview	JS-2
Amos, C. L.	The Lodge	AV-1
Atlantic Coast Lumber Co.	Hagley	AX-7
Auld, Mrs. Isaac	The Youghal	GV-14
Baldwin, Leonard D.	Cote Bas	GT-1
Barnes, A. M.	Northampton	ES-1
Barnum, Wm. Henry	Wappaoolah	FT-5
Barnwell, Arthur	The Point	NR-1
Barnwell, G. Manigault	Roxbury	JS-3
Barringer, J. L.	J. K. Hill Tract	EV-4
Baruch, Bernard M.	Hobcaw Barony	BX-6
Beach, Wm. N.	Rice Hope	DW-2
Bingham, Harry Payne	Cotton Hall	I.P.-5
Blank, I	Paradise Island	GV-11
Bonbright, G. D. B.	Pimlico	FT-9
Bonsal Roscoe	Plum Island	IU-3
Bradley, Peter B.	West Bank	JR-2
Bradley, Peter B.	Bulo Mines	HT-12
Bradley, Peter B.	Lachicotte	GV-8
Bradley, Peter B.	Charley Wood	FV-3
Brady, Wm. G.	Cedar Grove	HT-1
Brady		KR-6
Brown, J. Thompson	The Grove	KS-3
Bruorton, H. B.	100 Acre Tract	CX-9
Cain Estate	Summerton	ES-5
Cain Estate	Somerset	ES-6
Carnegie, Thomas	Cumberland Island	NP--
Carroll, Philip A.	Means Farm	LP-13
Caspary, A. H.	Bonny Doone	JQ-2
Chadwick, E. G.	The Wedge	DW-3
Chapman, C. E.	Mulberry	FT-2
Charleston Country Club	Wappoo Links	IU-5
Cheston, Radcliffe	Friendfield	BW-8
Coe, Wm. R.	Cherokee	LP-3
Coffin, Howard	Sapeloe Island	NP--
Coker, C. W.	Springwood	BW-5
Coleman, J. A.	Hardeeville	NP--
Colonial Dames	Dorchester Fort	GS-3
Copp, William M.	Spring Island	NQ-5
Corlies, Arthur	New River	NP-4
Cram, John	Colleton Neck	NQ-3
Cram, Sargent	Colleton Neck	NQ-6
Crane, Z. Marshall	Hope	KR-10
Cuthbert, C. S.	Golding	GS-2
Cypress Gardens	Dean Hall	FT-12
Dallett, Frederick A.	South Mulberry	FT-4
Dallett, Frederick A.	Buck Hall	FT-3
DeSaussure Estate	Camp Vere	FU-8
Dingle, Edward Von S.	Middleburg	FU-5
Dixon, F. E.	Mackay's Point	MP-3
Dodge, Donald D.	Snuggedy Swamp	KR-11
Dodge, Donald D.	Wayne	KR-11
Dodge, Donald D.	Fenwick Island	KS-8
Dodge, Donald D.	Seabrook Island	KS-4
Dominick, Bayard	Gregorie Neck	MP-2
Dominick, Gayer	Bulls Island	GW-1
Doubleday, Nelson	Bray's Island	MQ-1
Drayton, Charles H.	Drayton Hall	HT-5
duPont, Felix	Combahee	LP-4
duPont, Felix	The Bluff	KP-2
duPont, Felix		KQ-7
Durant. E. W.	Tyler Tea Far,	IT-2
Elbert, R. G.	Airy Hall	KR-8
Elbert, R. G.	Savage	KR-9
Elbert, R. G.	Bolder's Island	LR-5
Ellis, Geo. A. Jr.,	Bossis & Villa	FU-3
Ellis, Wm. S.	Willbrook	AX-3
Elliott, Wm.	Newberry	LP-15
Etting, A. W., Dr.	St. Helena Island	NR-5
Etting, A. W., Dr.	Pine Island	MS-1
Emerson, I. E., Dr. (Estate)	Arcadia	BX-4
Emory, Mrs.	Clifton	BX-5
Erckmann, H. L.	Muirhead	HU-7
Federal Game Refuge	Cape Romain	EW-2
Fertig, Willie E.	Waddell Ranch	AW-6
FitzSimons, Mrs. M. P.	Dungannon	IT-4
Fleming, Mathew	Rat Hall	HU-5
Frank, Fritz J.	The Bluff	FT-10
Frelinghuysen, Sen. J. S.	Comingtee	FT-15
Frelinghuysen, Sen. J. S.	Rice Hope	FT-11
Frelinghuysen, Sen. J. S.	Fish Pond	FT-16
Gaillard, J. Palmer	Walworth	EQ-1
Gibbs, Mrs. John E.	Hyde Park	FU-2
Gibbs, Benjamin	Longwood	FU-7
Gleason, Miss Kate	Dawtaw Island	MR-3
Goelet, Robert	Springbank	AR--
Graham, Allan G.	True Blue	AX-6
Grant, C. McK. Estate	Liberty Hall	GT-4
Gregorie, Ferdinand	Oakland	GV-10
Gregorie Estate	Richfield	LP-9
Grimke, Glenn	Grimke Place	HT-9
Guggenheim, S. R.	Ladies Island	MR-2
Hadden, Howard S.	Springbank	AR--
Halsey, Alfred	Calihoy	GU-4
Hamby, A. McBee	Dirleton	AW-11
Hamlin, J. L.	Snee Farm	GV-7
Hannahan, J. R.	Millbrook	HT-6
Hartford, Mrs. E. V.	Wando	GV-2
Hass, George	Tibwin	EW-1
Hastie, Norwood	Magnolia Gardens	HT-4
Hobonny Club	Hobonny	LP-6
Hollins, J. S.	Nightingale Hall	AW-7
Hollins, Gerald V.	Dawn of Hope	JQ-3
Hollins, H. B., Jr.	Hall's Island	MQ-3
Hollins, J. K.	**Mullet Hall**	JT-3
Hollins, J. K.	Bray's Island	MQ-2
Hollins, J. K.	Red Bluff	KR-1
Hornik, M.	Hornik Tract	DW-7
Hudson, Percy K.	Nieuport	LQ-8
Hudson, Percy K.	Clay Hall	LQ-2
Hume, Dr. A.	The Oaks	HT-2
Huntington, A. M.	Brookgreen	AX-2
Huntington, Robert D.	Gravel Hill	AR--
Hutton, E. F.	Cypress	KQ-3
Hutton, E. F.	Tilly Island	KR-5
Hutton, E. F.	Laurel Spring	LQ-5
Hutton, E. F.	Dale Farm	LQ-6
Hutton, E. F.		LQ-4
Hutton, E. F.		KR-3
Hutton, E. F.		KR-4
Hutton, E. F.	Oakland	KQ-7
Hutton, Franklin L.	Prospect Hill	KR-2
Huyler, Coulter D.	Capers Island	GV-4
Huyler, Coulter D.	Dewees Island	GV-5
Jackson, Baxter N.	Springbank	AR--
Jenkins, E. J.	Brick House	KS-9
Johnson, Winthrop	Bleak Hall	KT-3
Johnstone, F. S.	Oakton	CX-10
Johnstone, F. S.	Mt. Hope	CW-3
Jones, J. S.	Exeter	FT-1
Jones, J. S., Jr.	Williman Island	LQ-9
Jouannet, Alfred	Cassina Tea Farm	HV-1
Keith, W. Winchester	Wedgefield	BW-9
Kelly, Don M.	Plantersville	AW-3
Kennerty, Wm. C.	Ashley Hall	HT-10
Kiawato Club	Seabrook	KT-4
Kidder, James H.	Cowpens Point	KT-2
Kidder, James H.	Green Point	LQ-3
Kimball, Wm. A.	Wachesaw	AX-1
King Estate	Sewee	FV-2
Kittredge, Ben R.	Dean Hall	FT-12
Kress, C. W.	Buckfield	LP-12
Kress, C. W.	Kress	LP-12
Kuser, John L.	Callawassie Island	NP-5
Lawrance, Chas. E.	White Hall	KQ-1
Lawrance, Chas. E.	Paul & Dalton	LR-1
Legare Estate	Legare's Point	HU-2
Legare Estate	Old Town	HU-3
LeJendre, Sidney J.	Medway	FT-14
Lewisfield Club	Lewisfield	ET-5
Loomis, Alfred	Hilton Head	NQ-4
Lorillard, E. C.	Bindon	LR-14
Lucas Estate	Whitehall	ES-3
Lucas, Wm.	Hopsiwee	DW-1
Lyman, Arthur	Bonny Hall	LQ-1
Manigault, Robt. S.	Halidon Hill	FU-9
Marshall, Dr. J. H.	South Chachan	ET-7
Mattredge	Palmetto	HT-7
Maybank, John F.	Lavington	KR-7
Maybank, John F.	Green Pond	KQ-6
Means, Wm. B.	Ashley	HT-8
Metcalf, Jesse	Holly Grove	AW-2
Miller, John A.	Estherville	CX-4
Miller, John A.		AW-8
Miller, John A.	Cane Island	DX-4
Mills, Paul D.	Windsor	BW-4
Mitchell, Julian	Point of Pines	KT-1
Montgomery, Robert L.	Mansfield	BW-3
Morawetz, Victor	Fenwick Hall	IU-2
Mulford, Vincent S.	Bates Hill	AW-1
Municipal Air Port	North Charleston	HT-13
McCrady, Mrs.	Peachtree	DW-5
McGee, H. A.	Campbell	JQ-1
McLaughlin	Fripp's Island	NR-4
McLaughlin	St. Phillip's Island	NR-3
Nesbit, Ralph	Waverly	AX-5
Newberry, Truman H.	Newington	GS-1
Newberry, John S.	Newington	GS-1
Norris, Dr. Henry	Litchfield	AX-4
North State Lumber Co.	Silk Hope	EU-3
Orrell, C. D.	William Island	LQ-9
Osgood, Dana	Pirate House	HU-4
Otranto Club.	Otranto	GT-8
Penny, J. C.	Stono	IU-4
People's State Bank Trustees	Clayfield	FV-1
Phillips, Mrs. C. C.	Harmony	CW-2
Pickman, D. L.		AV-2
Pickman, D. L.	Springfield	AW-5
Pinckney, Mrs. C. C.	Runnymede	HT-3
Pinckney, Charles Cotesworth	Fairfield	DW-4
Pinckney, Miss Josephine S.	El Dorado	DX-5
Pompion Hill Chapel		FU-4
Poppenheim, J. F.	Hickory Hill	GT-11
Poppenheim, J. F.	Marington	GU-1
Poppenheim, J. F.	Red Bank	GU-2
Porcher Estate	Ophir	DS-1
Porcher, Henry F.	Ophir	ES-2
Porcher, Philip, Mrs.	Elm Grove	GV-15
Porcher, Philip	Stratton Hall	GV-13
Pratt, Charles	Wiggins	LR-2
Pratt, Charles	Combahee & Cheehaw	LR-3
Pratt, Frederick	Fields Point	LR-4
Pratt, Herbert I.	Good Hope	NP-3
Pulitzer, Ralph	Burroughs hall	AR--
Ramsley, Mrs. Caroline	Cat Island	CX-2
Randolph	Lost Garden	HT-11
Rathborne, J. C	Beneventum	BW-2
Reeves, Mrs. Susan	Annandale	CX-7
Reeves, Mrs. Richard E.	Crow Island	DX-1
Rhett Estate	Jehossee	KS-5
Richardson, H. A.	Longbrow	LQ-4
Robertson, Hugh S.	Bonneau & The Hut	FT-8
Roosevelt, N. G.	Gippy	ET-4
Roosevelt, N. G.	North Chachan	ET-6
Royal, Lee	Palmetto Grove	GV-6
Rutgers, Mrs. N. G.	Elwood	FT-7
Rutgers, Mrs. N. G.	Mepkin	FT-6
Rutledge, Archibald	Hampton	DW-6
Sabin, Chas. H.	The Oaks	GT-5
Sage, Henry M.	Belle Isle	CX-1
Saint Andrews Church		HT-15
Saint James Church	Santee	DW-8
Saint James Church	Goose Creek	GT-6
Sanders, Gustave	Sanders Tract	NR-2
Sanders, Leopard	Chisholm's Island	LQ-7
Schley, Kenneth B.	Ridgeland	NP--
Seabrook, A. H.	Ravenswood	EW-3
Seibels, Edwin G.	Lake Hill	BW-1
Sherburne, J. H.	Boone Hall	GV-3
Shonnard, Horatio S.	Harrietta	DX-2
Simonds, Louis	Springfield	KS-6
Simonds, Louis	Sampson Island	KS-7
Simpson, Sumner	Rose Bank	JT-2
Sloan Estate	Cat Island	GV-1
Smith, J. J. Pringle	Middleton Gardens	HT-14
Smith, F. W.	Waterford	BX-1
Speissegger	Live Oak	IT-1
Stevens, Jos. E.	Myrtle Grove	KQ-2
Stokes, Dr. L. M.	Ashland	JQ-4
Stoney, S. G.	Parnassus	GT-2
Stoney, S. David	Ararat	GT-10
Swain, W. Mosley	Belfair	NQ-2
Taylor, Geo.	Big Island	GT-3
Thorne, Edwin	Tomotley	LP-8
Thorne, Francis	Ashley Barony	HS-5
Thorne, Landon K.	Hilton Head	NQ-4
Todd, John R.	Brewton	LP-10
Turnbull, Robt. J.	Twickenham	LP-11
Tuxbury Lumber Co.	Bushy Park	GT-9
Tyson, F. J.	Maurisana	CW-1
Vanderbilt, Geo.	Arcadia	BX-4
Vanderbilt, Geo.	Bannockburn & Oak Hill	BX-2
Vanderhorst Estate	Kiawah Island	JU-1
Venning Shelby	White Hall	GV-12
Venning	Belleview	HU-6
Waddell, D. C., Sr.	Chicora Wood	AW-4
Wagner, F. W.	Ingleside & Woodstock	GS-4
Walker, Lawrance A.	Limerick	EU-2
Ward, Frederick L.	Sandy Point	JR-4
Waring, S. Vanderhorst	The Laurels	HS-1
Wayne Estate	Porcher Bluff	GV-9
Whaley, J. Swinton	Little Edisto	KS-10
Wheeling, J. S.	North Island	CX-5
Whitney, Arthur	Willtown Bluff	KS-2
Widener, Geo. D.	Mackay's Point	MP-3
Wilcox, T. Ferdinand	The Blessing	FU-6
Williams, Mrs. G. W.	Dixie	IT-3
Williams, John S.	Castle Hill	LP-7
Williams, Russell	Hanover	ES-4
Williams, Russell	Kensington	FU-1
Winston, Owen	Wappaoolah	FT-5
Wood, Alan	Maryville	BW-7
Yawkey, Thos. A.	South Island	CX-7
Yeaman's Hall	Yeaman's Hall	GT-7

In the aftermath of the war, lowcountry planters, yeomen, and freed-people struggled to rebuild. Ultimately, planters managed to restart agricultural production, but only on a greatly reduced scale. Rice, the principal export crop grown in the region, reached a postwar peak of about 52 million pounds in 1879 and then tracked downward. By the early twentieth century it became economically marginal. Sea Island cotton, the other major staple grown in the lowcountry, fared better, but not by much. The arrival of the boll weevil in 1916 signaled its doom. Meanwhile, attempts to bring industry to the region either proved short-lived or failed to offset the agricultural downturn. Phosphate mining fueled an economic rebound during the 1870s and 1880s but declined rapidly after a devastating hurricane in August 1893. By the 1910s, only limited mining operations and several fertilizer factories on the Charleston Neck remained.[7]

chaps. 1–18; Robert M. Weir, *Colonial South Carolina: A History* (New York: KTO Press, 1983); Peter H. Wood, *Black Majority: Negroes in Colonial South Carolina from 1670 through the Stono Rebellion* (New York: W. W. Norton and Co., 1974); Charles Joyner, *Down by the Riverside: A South Carolina Slave Community* (Urbana: University of Illinois Press, 1984); William Dusinberre, *Them Dark Days: Slavery in the American Rice Swamps* (New York: Oxford University Press, 1996); S. Max Edelson, *Plantation Enterprise in Colonial South Carolina* (Cambridge: Harvard University Press, 2006); Joyce E. Chaplin, *An Anxious Pursuit: Agricultural Innovation and Modernity in the Lower South, 1730–1815* (Chapel Hill: Institute of Early American History and Culture by University of North Carolina Press, 1993); Philip D. Morgan, *Slave Counterpoint: Black Culture in the Eighteenth-Century Chesapeake and Lowcountry* (Chapel Hill: Omohundro Institute of Early American History and Culture by University of North Carolina Press, 1998); Stephanie McCurry, *Masters of Small Worlds: Yeoman House-holds, Gender Relations, and the Political Culture of the Antebellum South Carolina Low Country* (New York: Oxford University Press, 1995); Don H. Doyle, *New Men, New Cities, New South: Atlanta, Nashville, Charleston, Mobile, 1860–1910* (Chapel Hill: University of North Carolina Press, 1990), 51–61; Maurie D. McInnis, *The Politics of Taste in Antebellum Charleston* (Chapel Hill: University of North Carolina Press, 2005); Emma Hart, *Building Charleston: Town and Society in the Eighteenth-Century British Atlantic World* (Charlottesville: University of Virginia Press, 2010).

[7] On Reconstruction-era conflicts, see especially Eric Foner, *Nothing but Freedom: Emancipation and Its Legacy* (Baton Rouge: Louisiana State University Press, 1983), chap. 3; Eric Foner, *Reconstruction: America's Unfinished Revolution, 1863–1877* (New York: Harper and Row, 1988), 5–54, 70–72, 77–88, 110, 112–119, 124–175; John Scott Strickland, "'No More Mud Work': The Struggle for Control of Labor and Production in Low Country South Carolina, 1863–1880," in *The Southern Enigma: Essays on Race, Class, and Folk Culture*, ed. Walter J. Fraser, Jr., and Winfred B. Moore (Westport, CT: Greenwood Press), 42–63; John Scott Strickland, "Traditional Culture and Moral Economy," in *The Countryside in the Age of Capitalist Transformation: Essays in the Social History of Rural America*, ed. Steven Hahn and Jonathan Prude (Chapel Hill: University of North Carolina Press, 1985), 141–178; Brian Kelly, "Black Laborers, the Republican Party, and the Crisis of Reconstruction in Lowcountry South Carolina," *International*

By the World War I era, many commentators saw the coastal region as virtually lifeless. Once prosperous plantations lay ruined and abandoned. Large stretches of countryside lay virtually dormant, punctuated intermittently by timber camps and small farms but otherwise domains of flora and fauna. Historians have tended to portray the lowcountry of the early twentieth century as contemporaries saw it: economically depressed and socially and culturally isolated. As Peter A. Coclanis has written, "the area's lifeblood drained to the last. Without rice the lowcountry became what it was before rice: a desolate wasteland."[8] To be sure, economic statistics and contemporary accounts support such a view. At the same time, the emphasis on decay and stagnation overlooks important developments. Throughout the region, investors had established new enterprises, and people had begun using coastal lands for new purposes. Northerners' "plantations" figured at the center of the latter trend. By adapting formerly productive tracts for leisure, wealthy sporting enthusiasts introduced new social and economic relations, new

---

Review of Social History 51 (2006): 375–414; Julie Saville, *The Work of Reconstruction: From Slave to Wage Laborer in South Carolina, 1860–1870* (New York: Cambridge University Press, 1994); Leslie A. Schwalm, *A Hard Fight for We: Women's Transition from Slavery to Freedom in South Carolina* (Urbana: University of Illinois Press, 1997); Willie Lee Rose, *Rehearsal for Reconstruction: The Port Royal Experiment* (Indianapolis: Bobbs-Merrill, 1964); Joel Williamson, *After Slavery: The Negro in South Carolina during Reconstruction, 1861–1877* (Chapel Hill: University of North Carolina Press, 1956); Thomas C. Holt, *Black over White: Negro Political Leadership in South Carolina during Reconstruction* (Urbana: University of Illinois Press, 1977). On economic conditions, see Doyle, *New Men, New Cities, New South*, 59–61, 71–76, 79–86, 111–129; Peter A. Coclanis, *The Shadow of a Dream: Economic Life and Death in the South Carolina Low Country, 1670–1920* (New York: Oxford University Press, 1989), 128–158; James H. Tuten, *Lowcountry Time and Tide: The Fall of the South Carolina Rice Kingdom* (Columbia: University of South Carolina Press, 2010); Richard D. Porcher and Sarah Fick, *The Story of Sea Island Cotton* (Charleston: Wyrick and Co., 2005), 327–330; Edgar, *South Carolina: A History*, 381, 423, 485; Rogers, *History of Georgetown County*, chaps. 19–20; Shepherd W. McKinley, *Stinking Stones and Rocks of Gold: Phosphate, Fertilizer, and Industrialization in Postbellum South Carolina* (Gainesville: University Press of Florida); Jamie W. Moore, "The Lowcountry in Economic Transition: Charleston since 1865," *South Carolina Historical Magazine* 80, no. 2 (Apr. 1979): 156–171; James M. Clifton, "Twilight Comes to the Rice Kingdom: Postbellum Rice Culture on the South Atlantic Coast, 1820–1880," *Georgia Historical Quarterly* 62 (summer 1978): 146–152.
[8] Coclanis, *Shadow of a Dream*, 142. See also Walter B. Edgar, *History of Santee Cooper, 1934–1984* (Columbia: R. L. Bryan Co., 1984), 5, 27; Edgar, *South Carolina: A History*, 490–491; Fraser, *Charleston! Charleston!*, 326–328, 339–341; Lawrence S. Rowland and Stephen R. Wise, *The History of Beaufort County, South Carolina*, vol. 3: *Bridging the Sea Islands' Past and Present, 1893–2006* (Columbia: University of South Carolina Press, 2015), 19–25, 38–69, 207–211.

land-management practices, and new forms of activity. All demonstrated the potential of outdoor recreation at a crucial moment in time.[9]

Despite the interest northerners' estates attracted in their heyday, historians have paid them little attention. In the late 1960s, scholars such as George Brown Tindall and George C. Rogers, Jr., surveyed the origins of the new estates and the activities northerners came to the region to enjoy. Tindall saw the new "plantations" as the product of "Yankee millionaires looking for playgrounds." Noting that sportsmen sought lands where they could enjoy their favorite pastimes and avoid the chill of winter, he characterized the lowcountry as one of several southern havens where "the idle rich deported themselves in or out of season." Rogers investigated the phenomenon more thoroughly. He devoted a chapter of his *History of Georgetown County, South Carolina*, to the estate-making process. Seeing northerners' love of old plantations as inspired by yearnings for legitimacy, he highlighted the social benefits of plantation ownership. "Northern industrialists were rich and in search of status," wrote Rogers. Some married European nobles; others collected great works of art. Still others bought southern plantations. Demand for the latter, Rogers contended, typified an era "when new families were searching for old roots." "Rich Yankees" eagerly bought plantations with "historic pasts" and "appropriate settings for their gentlemanly sports."[10]

More than four decades later, Rogers's account remains the most detailed investigation of the phenomenon. Other scholars have recently shown interest in the topic and several studies have chronicled the creation of individual "plantations." These investigations offer insight into owners' motivations, northerners' activities, and the goals of architects and landscape architects who worked on owners' behalf. Memoirs by members of estate-owning families and plantation superintendents have also illuminated the rise of the new estates and the rhythms of the

---

[9] On new commercial activity, see, for example, Robert B. Cuthbert and Stephen G. Hoffius, eds., *Northern Money, Southern Land: The Lowcountry Plantation Sketches of Chlotilde R. Martin* (Columbia: University of South Carolina Press, 2009), 21–25, 28–29, 58–65; Rowland and Wise, *Bridging the Sea Islands' Past and Present*, 69–73, 211–219; Albertine Moore, "Do You Know Your South Carolina? Cypress Woods Farms," newspaper clipping cited as *News and Courier*, Dec. 31, 1945, South Carolina Vertical File Collection, Charleston County Public Library, Charleston, SC (hereafter SCVFC); T. R. W., "Do You Know Your South Carolina? Coosaw Plantation," newspaper clipping cited as *News and Courier*, May 25, 1942, SCVFC.

[10] George Brown Tindall, *The Emergence of the New South, 1913–1945* (Baton Rouge: Louisiana State University Press, 1967), 103; Rogers, *History of Georgetown County*, chap. 21.

lowcountry sporting scene. Yet even as knowledge about the new "plantations" has grown, critical analyses have remained elusive. In general, historians have tended to take the new estates for granted, as though they scarcely demand explanation. One historian has characterized them as a near-predictable use of coastal lands in an era of agricultural decline; others have seen them simply as old plantations repurposed as private shooting retreats. Still another has erroneously identified estate owners as the wealthiest echelon of the tourists who came to see Charleston as its popularity surged during the 1920s and 1930s. These interpretations are not only inaccurate but fail to explain northerners' motivations and actions. Moreover, they overlook fundamental reasons for the development of the new estates. By linking northerners to tangential events, these views mislead more than they reveal.[11]

The failure to probe the origins of northerners' plantations has left a dramatic sequence of activity unexplored. Northerners did not simply buy old plantations; they refigured architecture, landscape, and space. Owners rehabilitated select buildings, restored others, demolished some, and erected new structures. Landscapes and gardens received similar treatment. Owners unearthed, revived, and enlarged old gardens, added new

---

[11] Charles Kovacik, "South Carolina Rice Coast Landscape Changes," in *Proceedings: Tall Timbers Ecology and Management Conference*, no. 16 (Tallahassee: Tall Timbers Research Station, 1982): 56–60; Coclanis, *Shadow of a Dream*, 156; Stephanie E. Yuhl, *A Golden Haze of Memory: The Making of Historic Charleston* (Chapel Hill: University of North Carolina Press, 2005), 177–183; Cuthbert and Hoffius, eds., *Northern Money, Southern Land*; Jennifer Betsworth, "'Then Came the Peaceful Invasion of the Northerners': The Impact of Outsiders on Plantation Architecture in Georgetown County, South Carolina" (MA thesis, University of South Carolina, 2011); Angela C. Halfacre, *A Delicate Balance: Constructing a Conservation Culture in the South Carolina Lowcountry* (Columbia: University of South Carolina Press, 2012), 36–37; Rowland and Wise, *Bridging the Sea Islands' Past and Present*, 284–299; Virginia C. Beach, *Rice and Ducks: The Surprising Convergence That Saved the Carolina Lowcountry* (Charleston: Evening Post Books, 2014), 58–106. On the making of individual estates, see, for example, Virginia C. Beach, *Medway* (Charleston: Wyrick and Co., 1999); David De Long, *Auldbrass: Frank Lloyd Wright's Southern Plantation* (New York: Rizzoli, 2003); Alexander Moore, *Poco Sabo Plantation: A Place in Time* (N.p.: privately published, 2005); Lee Brockington, *Plantation between the Waters: A Brief History of Hobcaw Barony* (Charleston: History Press, 2006), chaps. 4–7. For memoirs, see Gertrude S. Legendre, *The Time of My Life* (Charleston: Wyrick, 1987); Neal Cox, *Neal Cox of Arcadia Plantation: Memoirs of a Renaissance Man* (Georgetown, SC: Alice Cox Harrelson, 2003); Frances Cheston Train, *A Carolina Plantation Remembered: In Those Days* (Charleston: History Press, 2007). For critical perspectives, see Julia Brock and Daniel Vivian, eds., *Leisure, Plantations, and the Making of a New South: The Sporting Plantations of the South Carolina Lowcountry and Red Hills Regions, 1900–1940* (Lanham, MD: Lexington Books, 2015).

plantings, and removed features they considered undesirable. Forests and fields became recreational domains, managed specifically for leisure. All told, northerners transformed the environments they acquired, altering existing remains for new purposes. Although the results varied, the overriding approach lay in retaining elements of earlier periods, adding liberally to them, and creating the appearance of a coherent whole. In this manner, owners refashioned places historically devoted to large-scale commodity production for recreation and domestic use.

Behind the process of material change lay profound cultural and intellectual transformations. Northerners and other people of the era did not retain the term "plantation" simply for the sake of convenience or out of habit. Nor did they continue using it because of respect for tradition. Laden with meaning, it anchored views of what plantations had been and what many had become. It merged past and present within a singular idiom, emphasizing continuities and marginalizing radical disjunctures. Owners and others saw the new estates as rooted in well-established practices. All who commented on northerners' activities saw evolutionary development, not a sharp break with the past. Use of terms such as "restoration" and "renewal" made this clear. Moreover, people of the era saw commonalities between northerners' estates and plantations devoted to agriculture. Contemporaries recognized the new estates as "plantations" not because of what they had been but for what they had become.[12]

Race and class played a crucial role in fostering such judgments. Onlookers saw the new estates as restoring traditional social hierarchies. With white elites as landowners and landless African Americans as laborers, the new estates seemed to reconstitute relationships that emancipation had overturned. Popular and scholarly views of the past also played a role. Most Americans of the era looked back on slavery through a haze of romance and nostalgia. In this perspective, white authority and black servitude seemed inherent to plantations, part of a seemingly "natural" social order. Those who recognized the violence and oppression of slavery found their voices marginalized by the rigidly exclusionary public culture of Jim Crow. Among historians, the patently racist interpretations of

---

[12] See, for example, Susan Lowndes Allston, "Windsor Place Spot of Beauty," *News and Courier*, Dec. 22, 1929, A3; Chlotilde R. Martin, "Lowcountry Gossip," *News and Courier*, Oct. 31, 1934, A5. For instances of use of terms such as restoration and revival, see Chlotilde R. Martin, "Tomotley, Brewton Plantation, Bindon, Castle Hill – Beaufort Restorations," *News and Courier*, Nov. 23, 1930, A12; "Interest in Reviving Old Plantations," *New York Times* (New York, NY), Dec. 22, 1933, sec. 10, p. 2.

Columbia University professor William A. Dunning and his students, the so-called Dunning School historians, held sway. In viewing Reconstruction as an ill-conceived experiment gone terribly wrong, the Dunning School rationalized white supremacy. Racial control seemed logical and necessary, essential to preventing the horrors of Reconstruction from recurring.[13]

As northerners' estates proliferated, plantations became central to popular views of the lowcountry and its past. During the 1930s, interest in the lowcountry surged. Promoters, artists, and writers depicted a region of spectacular beauty, rich traditions, and historical charm. Plantations figured at the center of such portrayals. "The shadow of the plantation has fallen across the whole history of the Low Country," declared *Life* magazine in December 1939. "Many historians," it noted, believed that "the plantation culture of the Old South reached its apogee in the Low Country."[14] *Life*'s interest stemmed in part from the fact that its owner, publishing tycoon Henry Luce, owned a showplace estate on the Cooper River near Moncks Corner, upstream from Charleston. Like other seasonal residents, Luce used his influence to promote his interests. Regardless, *Life* celebrated the lowcountry in ways that had become common. It expressed a view of the region that had become widely accepted and, to many, seemed well deserved.[15]

Life's account showed that plantations remained a part of the low-country landscape, even though they no longer produced agricultural staples. "Today the big cash crop is rich Northerners who have come South and bought the plantations where Carolina aristocrats once lived and ruled," *Life* reported. The northerners, *Life* added, came to hunt, to relax, and "to enjoy the feudal feeling of property which owning thousands of acres gives them." Although natives – "romantic Carolinians," in

---

[13] David W. Blight, *Race and Reunion: The Civil War in American Memory* (Cambridge: Belknap Press of Harvard University Press, 2001); Nina Silber, *The Romance of Reunion: Northerners and the South, 1865–1900* (Chapel Hill: University of North Carolina Press, 1993); William L. Van Deburg, *Slavery and Race in American Popular Culture* (Madison: University of Wisconsin Press, 1984), 67–86, 89–127; Thomas J. Pressly, *Americans Interpret Their Civil War* (Princeton: Princeton University Press), chaps. 4 and 5; John David Smith, *An Old Creed for the New South: Proslavery Ideology and Historiography, 1865–1918* (Westport, CT: Greenwood Books, 1985); John David Smith and J. Vincent Lowery, eds., *The Dunning School: Historians, Race, and the Meaning of Reconstruction* (Lexington: University Press of Kentucky, 2013); Foner, *Reconstruction*, xix–xxvi.

[14] "The Low Country," *Life*, Dec. 25, 1939, 38–41.

[15] On Luce's estate, see Thomas R. Waring, Jr., "Do You Know Your Charleston? Mepkin Plantation," *News and Courier*, Jan. 24, 1938, 10.

Life's terminology – still looked back fondly on "their antebellum Golden Age," they welcomed the northerners with open arms. The newcomers, *Life* reported, brought "new money to a poor but proud land." As one lowcountry woman observed, "In 1865 the Devil sent the Yankees. Today God sends them."[16]

Life's account casts vital questions in sharp relief. What inspired the new estates? Why did people call them "plantations," and how did they compare to plantations of earlier periods? And, most important, how should they be understood? Plantations, after all, had never before served principally as private recreational venues. For centuries, the term "plantation" had denoted a site of intensive agriculture. Historical examples had relied on the labor of black slaves; now, sharecroppers and tenants fulfilled that role. Reformers, critics, and politicians decried plantations as fundamental to the problems facing the South. As part of an outmoded labor system that limited economic development and kept millions in poverty, plantations signified inefficiency, underachievement, and social conflict.[17] Yet contemporary ills represented only part of the problem. With roots in slavery, plantations also embodied a sordid and shameful history. They marked the foundation of a labor system and social order that had nearly torn the nation apart. When Americans of the era looked upon plantations, they invariably encountered a record of tyranny, conflict, and strife. Why some people saw select examples as ripe for renewal is difficult to explain.

Northerners' estates pose a greater problem than historians have recognized. Conventional accounts have long attributed their origins to the plantation myth, the romantic idyll that recalled plantations as

---

[16] "The Low Country," 38.

[17] National Emergency Council, *Report on Economic Conditions of the South* (Washington, DC: Government Printing Office, 1938), 1; Natalie J. Ring, *The Problem South: Region, Empire, and the New Liberal State, 1880–1930* (Athens: University of Georgia Press, 2012), chap. 3. On plantations in the early twentieth-century South, see especially Charles S. Aiken, *The Cotton Plantation South since the Civil War* (Baltimore: Johns Hopkins University Press, 1998); Pete Daniel, *Breaking the Land: The Transformation of Cotton, Tobacco, and Rice Cultures since 1880* (Urbana: University of Illinois Press, 1985), chaps. 1, 11; Jack Temple Kirby, *Rural Worlds Lost: The American South, 1920–1960* (Baton Rouge: Louisiana State University Press, 1987), chap. 1; Jack Temple Kirby, *Mockingbird Song: Ecological Landscapes of the South* (Chapel Hill: University of North Carolina Press, 2006), 103–109; US Bureau of the Census, *Plantation Farming in the United States* (Washington, DC: Government Printing Office, 1906); William C. Holly, Ellen Winston, and T. J. Woofer, Jr., *The Plantation South, 1934–1937* (Washington, DC: Government Printing Office, 1940); US Bureau of the Census, *Special Study, Plantations, 1940* (Washington, DC: Government Printing Office, 1940).

bucolic estates. Rogers made this claim in 1970, and James H. Tuten has recently elaborated it by emphasizing the "symbolic capital" of old rice plantations. According to Tuten, as plantations lost their economic value, they continued to symbolize power, wealth, and leisure. Northern sporting enthusiasts bought old plantations, he argues, because "they subscribed to the plantation mystique." In creating their estates, northerners "adopted and put on display the plantation mystique" and thus became progenitors of "a glamorous vision of the plantation heritage."[18] Although old plantations may have stirred wistful longings for some sportsmen and sportswomen, how the plantation myth could have inspired the creation of dozens of new estates is difficult to understand. Northerners did not cite the symbolism of plantations among the reasons they purchased lowcountry lands, and none of them sought to make the mythic world of Old South lore a reality. Certainly, none sought to emulate the lives of the antebellum planter class. Instead, northerners acted with their own interests in mind. Sportsmen and sportswomen came to the lowcountry principally for mild weather and sport. These concerns formed the basis of their interest in the region and sustained their activities over time.

The problem is made more complex by recent scholarship on remembrance of slavery, the Old South, and the Civil War. For most of the twentieth century, historians saw the plantation legend as a sort of national fantasy that mislead because of its inaccuracy but did little harm.

---

[18] Rogers, *History of Georgetown County*, 485–486, 496–497; Tuten, *Lowcountry Time and Tide*, 112–113, 117. On the "plantation myth," the "myth of the Old South," and mythology in southern history in general, see George B. Tindall, "Mythology: A New Frontier in Southern History," in *The Idea of the South: Pursuit of a Central Theme*, ed. Frank E. Vandiver (Chicago: William Marsh Rice University by University of Chicago Press, 1964): 1–15; William R. Taylor, *Cavalier and Yankee: The Old South and American National Character* (New York: George Braziller, 1961); C. Vann Woodward, *Origins of the New South, 1877–1913* (Baton Rouge: Louisiana University Press, 1951), 154–158; Paul M. Gaston, *The New South Creed: A Study in Southern Myth-Making* (Baton Rouge: Louisiana State University Press, 1970); Susan-Mary Grant, *North over South: Northern Nationalism and American Identity in the Antebellum Era* (Lawrence: University Press of Kansas, 2000); James C. Cobb, *Away Down South: A History of Southern Identity* (New York: Oxford University Press, 2005), 1–98; Blight, *Race and Reunion*, chap. 7; Silber, *Romance of Reunion*; Grace Hale, *Making Whiteness: The Culture of Segregation in the South, 1890–1940* (New York: Pantheon Books, 1998); Karen Cox, *Dreaming of Dixie: How the South Was Created in American Popular Culture* (Chapel Hill: University of North Carolina Press, 2011); Edward D. C. Campbell, *The Celluloid South: Hollywood and the Southern Myth* (Knoxville: University of Tennessee Press, 1981), chaps. 3–4; Jack Temple Kirby, *Media-Made Dixie: The South in the American Imagination*, rev. ed. (Athens: University of Georgia Press, 1986), chap. 4.

C. Vann Woodward and Paul Gaston saw its primary significance as easing white southerners' transition to the New South. As Woodward wrote, "the bitter mixture of recantation and heresy could never have been swallowed so readily had it not been dissolved in the syrup of romanticism." Valorizing the Confederacy soothed the wounds of defeat and aided southerners' adjustment to radically changed circumstances.[19] In the 1960s, George Brown Tindall identified moonlight-and-magnolias fantasies as the "standard image of the Old South." Emphasizing their role in maintaining southern distinctiveness, Tindall cast such portrayals as one of the many ways popular culture misrepresented the South and its past. Although he mentioned the role of "the old-time Negro" and the conspicuous absence of non-slaveowning whites in such depictions, he said nothing about the implications. If Tindall saw a relationship between contemporary politics and Old South lore, he left it unspecified.[20]

In the 1980s and 1990s, sustained investigations of Confederate memorialization and a groundswell of interest in historical memory laid the groundwork for new interpretations of the plantation myth. By detailing the many ways in which southerners celebrated the Confederacy, Charles Reagan Wilson and Gaines M. Foster revealed the inadequacy of myth as a framework for understanding the social and political conflicts of the New South and demonstrated the cultural significance of Confederate remembrance. Foster's *Ghosts of the Confederacy* proved particularly influential because of its emphasis on public rituals and their role in solidifying social and racial hierarchies.[21] More recently, Nina Silber, David W. Blight, and Grace Hale have placed remembrance of the Old South at the center of American culture during the late nineteenth and early twentieth centuries. Silber and Blight have shown the appeal of the plantation idyll among middle- and upper-class northerners and its role in fostering reunion of North and South, and Hale has elucidated the ways in which fantasies of a romantic past legitimated Jim Crow. In short, whereas historians once saw the myth of the Old South as a signifier of regional identity, recent studies have shown its significance to the

---

[19] Woodward, *Origins of the New South*, 154–158 (quotation on 158); Gaston, *New South Creed*. See also Paul Buck, *The Road to Reunion, 1865–1900* (Boston: Little, Brown and Co., 1937), 172–175, 196–219.

[20] Tindall, "Mythology: A New Frontier in Southern History."

[21] Charles Reagan Wilson, *Baptized in Blood: The Religion of the Lost Cause, 1865–1920* (Athens: University of Georgia Press, 1980); Gaines M. Foster, *Ghosts of the Confederacy: Defeat, the Lost Cause, and the Emergence of the New South, 1865–1913* (New York: Oxford University Press, 1987).

racial divisions that became deeply entrenched in American society at the dawn of the twentieth century.[22]

No serious investigation of northerners' estates, then, can overlook their role at the center of struggles over the memory and meaning of the Civil War. In 1911, the Princeton economist Winthrop M. Daniels observed that "the embers of the great conflict in which slavery perished are still hot." Despite the passing of nearly half a century, sectional animosities remained strong.[23] As sportsmen and sportswomen began remaking old plantations, the significance of those sites in "the great conflict" remained obvious to all present. That northerners encountered little opposition is beside the point. The long-term consequences are a different matter. As scholars such as Ira Berlin, Peter Wood, and Blight have observed, a more just society awaits an honest reckoning with the historical realities of plantation slavery.[24] No understanding of the

[22] Silber, *Romance of Reunion*; Blight, *Race and Reunion*; Hale, *Making Whiteness*. Scholarship on historical memory of the Civil War is vast. For especially influential studies, see W. Fitzhugh Brundage, *The Southern Past: A Clash of Race and Memory* (Cambridge: Belknap Press of Harvard University Press, 2005; W. Fitzhugh Brundage, ed., *Where These Memories Grow: History, Memory, and Southern Identity* (Chapel Hill: University of North Carolina Press, 2000); Stuart McConnell, *Glorious Contentment: The Grand Army of the Republic, 1865–1900* (Chapel Hill: University of North Carolina Press, 1992); David W. Blight, *Beyond the Battlefield: Race, Memory, and the American Civil War* (Amherst: University of Massachusetts Press, 2002); Alice Fahs and Joan Waugh, eds., *The Memory of the Civil War in American Culture* (Chapel Hill: University of North Carolina Press, 2004); John R. Neff, *Honoring the Civil War Dead: Commemoration and the Problem of Reconciliation* (Lawrence: University Press of Kansas, 2005); Donald R. Shaffer, *After the Glory: The Struggles of Black Civil War Veterans* (Lawrence: University Press of Kansas, 2004); Cynthia J. Mills and Pamela H. Simpson, eds., *Monuments to the Lost Cause: Women, Art, and the Landscapes of Southern Memory* (Knoxville: University of Tennessee Press, 2003); Karen L. Cox, *Dixie's Daughters: The United Daughters of the Confederacy and the Preservation of Confederate Culture* (Gainesville: University Press of Florida, 2003); William A. Blair, *Cities of the Dead: Contesting the Memory of the Civil War in the South, 1865–1914* (Chapel Hill: University of North Carolina Press, 2004); Caroline E. Janney, *Burying the Dead but Not the Past: Ladies' Memorial Associations and the Lost Cause* (Chapel Hill: University of North Carolina Press, 2008); Caroline E. Janney, *Remembering the Civil War: Reunion and the Limits of Reconciliation* (Chapel Hill: University of North Carolina Press, 2013); Thomas J. Brown, *Civil War Canon: Sites of Confederate Memory in South Carolina* (Chapel Hill: University of North Carolina Press, 2015).

[23] Winthrop M. Daniels, "The Slave Plantation in Retrospect," *Atlantic Monthly* 107, no. 3 (March 11, 1911), 363.

[24] Ira Berlin, "American Slavery in History and Memory and the Search for Social Justice," *Journal of American History* 90, no. 4 (March 2004): 1251–1268; Peter Wood, "Slave Labor Camps in Early America: Overcoming Denial and Discovering the Gulag," in *Slave Labor Camps in Early America: Overcoming Denial and Discovering the Gulag*, ed. Carla Gardina Pestana and Sharon V. Salinger (Hanover, NH: University Press of New

reasons Americans remained blind to slavery's horrors for so long is possible without attention to northerners' "plantations."

This book chronicles the making of those estates. It is about the origins of a new plantation world: a place where plantations served as elegant country seats for members of an aristocratic class, not in an imagined fantasy but in a part of the South with an outsized role in slavery, secession, and the Civil War. It is about remembering and forgetting, alliances of race and class, and shifting perspectives on places called plantations. Above all, it is about the ways in which Americans of the early twentieth century viewed social and material legacies of slavery. Put simply, it is about one chapter in the nation's long struggle with racial slavery and its aftermath. Despite the historical prominence of plantations in American culture and extensive scholarly attention, some types remain virtually unknown. This book brings an unusual, highly influential variant to light.

* * *

The study that follows is organized in seven chapters. Throughout, the narrative focuses on changes in use, material form, and representation. These categories highlight the major transformations of the era and how lowcountry people and others interpreted them. Northerners employed a Janus-faced vision in creating their estates. Simultaneously retaining select remains while making extensive additions, they carried out a material dialogue with the past that refashioned surviving edifices for new purposes, created substantially new settings, and removed evidence of productive activities. Ultimately, northerners created estates that accommodated contemporary needs, gave age a conspicuous presence, and possessed a form and layout that onlookers recognized as characteristic of plantations. As lowcountry people and other observers chronicled northerners' activities, they formulated interpretations of places commonly referred to as "plantations." In the process, they came to view plantations and their role in the history of the lowcountry in new ways.[25]

England, 1999): 222–238; Blight, *Race and Reunion*, 1–5, 397; James O. Horton and Lois E. Horton, eds., *Slavery and Public History: The Tough Stuff of American Memory* (New York: New Press, 2006).

[25] The theoretical underpinnings of this framework derive from scholarship on the social construction of space, place, and landscape, particularly Henri Lefebvre, *The Production of Space*, trans. Donald Nicholson-Smith (Cambridge, MA: Blackwell, 1991); Allan Pred, "Place as Historically Contingent Process: Structuration and the Time-Geography of Becoming Places," *Annals of the Association of American Geographers* 74, no. 2 (June

Setting the new estates in a historical perspective brings the processes responsible for their creation into focus, in ways that incorporate the viewpoints of multiple actors and underscore the changing spatial and physical characteristics of places called "plantations." Although historians of the postbellum South have traditionally viewed plantations as defined by the union of staple agriculture and coerced labor, in the low-country of the early twentieth century, those factors mattered little.[26] Contemporary perspectives emphasized other qualities. The rise of the new estates compels recognition of how malleable ideas about plantations became as time passed and the South became part of a national fantasy sustained by advertising, film, fiction, and well-known stereotypes. Once plantations became objects of the imagination, expansive possibilities resulted. By the 1930s, countless novels, films, and plays celebrated the glories of a supposedly romantic past. Advertisers used images of black mammies and silver-haired gentleman to hawk consumer goods to would-be buyers, and seemingly authoritative histories portrayed slavery as a civilizing institution, a "school" for people barely removed from savagery. Amid these circumstances, many Americans viewed aging plantations as a valued part of the nation's past, and a few saw opportunities for renewal. The actions of the latter group are the focus of this study. They affirmed the myth of a romantic past, extended the significance of plantations in new directions, and changed the way Americans conceived of plantations for good.[27]

1984): 279–297; Yi-Fu Tuan, *Space and Place: The Perspective of Experience* (Minneapolis: University of Minnesota Press, 1977); Timothy Oakes, "Place and the Paradox of Modernity," *Annals of the Association of American Geographers* 87, no. 3 (Sept. 1997): 509–531; J. Nicholas Entrikin, *The Betweenness of Place: Towards a Geography of Modernity* (Baltimore: Johns Hopkins University Press, 1991); Richard H. Schein, "Normative Dimensions of Landscape," in *Everyday America: Cultural Landscape Studies after J. B. Jackson*, ed. Chris Wilson and Paul Groth (Berkeley: University of California Press, 2003), 199–218; Richard H. Schein, "The Place of Landscape: A Conceptual Framework for Interpreting an American Scene." *Annals of the Association of American Geographers* 87, no. 4 (Dec. 1997): 660–680.

26 See, for example, Aiken, *Cotton Plantation South since the Civil War*, chaps. 1–3; Kirby, *Rural Worlds Lost*, chap. 1; John M. Vlach, "The Plantation Tradition in an Urban Setting: The Case of the Aiken-Rhett House in Charleston, South Carolina," *Southern Cultures* 5, no. 4 (winter 1999): 52. Recent characterizations have emphasized the instrumentality of plantations instead of fundamental qualities. See, for example, Foner, *Nothing but Freedom*, 9; Edelson, *Plantation Enterprise in Colonial South Carolina*, 2–5.

27 Hale, *Making Whiteness*; Cox, *Dreaming of Dixie*; Campbell, *Celluloid South*, chaps. 3–4; M. M. Manring, *Slave in a Box: The Strange Career of Aunt Jemima* (Charlottesville: University Press of Virginia, 1998); Micki McElya, *Clinging to Mammy: The*

Chapter 1 examines the initial phases of northern sportsmen's activities, covering the period from the 1880s to the early 1910s. During this era, sportsmen came to the lowcountry solely to hunt and stayed for short periods of time. They established hunting clubs and retreats but took limited steps toward rehabilitating old plantations. Only after beginning to see greater potential in the lowcountry and, importantly, as women and children began wintering in the region, did a few northerners begin adapting old plantations for use as seasonal residences. These developments, however, did not disrupt the meaning of the term "plantation" or identify some plantations as having special potential. Modest steps toward creating more comfortable and commodious accommodations did nothing to reimagine the fundamental use and purpose of plantations or give new meaning to select examples. At the same time, new portrayals of lowcountry history urged people to see the region in new ways, and depictions of select plantations as vestiges of a bygone era stimulated interest in renewal. These developments laid a foundation with expansive possibilities. Although seemingly inconsequential at the time, they proved crucial to explorations that later occurred and ultimately began to reimagine select plantations in powerful, far-reaching ways.

Chapter 2 chronicles northerners' activities from the 1910s through the peak of the invasion in the 1930s. During the World War I era, small numbers of sportsmen and sportswomen began rehabilitating old plantations for new purposes. By the early 1920s, northerners devised several methods for turning old plantations into seasonal residences and recreational venues. In 1925, wealthy sportsmen and sportswomen, many of them prominent in society, inaugurated a spectacular burst of activity. Northerners began purchasing large tracts of land and immediately creating handsome estates. By the eve of World War II, seventy-six lay between the northern end of the Waccamaw Neck and the Savannah River. These "plantations" refigured the meaning of the term in American culture. They merged visions of a romantic past with new material environments, new purposes, and new symbolism. By tying upper-class pastimes to former sites of slave labor, they effected what contemporaries viewed as a "renaissance" while obscuring memory of slavery, labor, and commercial enterprise.

Faithful Slave in Twentieth-Century America (Cambridge: Harvard University Press, 2007); Marilyn Kern-Foxworth, *Aunt Jemima, Uncle Ben, and Rastus: Blacks in Advertising, Yesterday, Today, and Tomorrow* (Westport, CT: Greenwood Press, 1994), chaps. 2–4; Kenneth W. Goings, *Mammy and Uncle Mose: Black Collectables and American Stereotyping* (Bloomington: Indiana University Press, 1994).

Chapter 3 surveys the general characteristics of the new estates and their relationship to lowcountry plantations of the late antebellum era. Northerners' estates differed from their predecessors in several ways. They occupied more land, possessed different organization, displayed greater refinement, and served different purposes. Although northerners often retained historical names and select buildings and landscape features, they otherwise carried out extensive changes. Inspiration regarding architecture, landscapes, and organization came from country estates in the North. Although many owners privileged surviving buildings and landscape features and some may have embraced popular views of antebellum plantations, the model of the American country house guided northerners' efforts. Recognizing this raises powerful questions about northerners' intentions. For all the attention that new "plantations" received, the evidence indicates that northerners sought to replicate conditions typical of country estates in the North, albeit with less opulence and grandeur, even as they also came to see themselves as rehabilitating old plantations. In the end, patterns of material change and associated discourses proved crucial to shaping understandings of the new estates.

Chapters 4 and 5 examine the remaking of two plantations in close detail. Chapter 4 recounts the history of Mulberry Plantation, one of the most revered plantations in the southeastern United States. Established in the early eighteenth century, Mulberry grew to become a large and productive rice plantation. During the late antebellum era it ranked among the most prosperous in St. John's Berkeley Parish. Left in ruins by the Civil War, Mulberry enjoyed renewed prosperity in the 1870s and 1880s but later fell into decline. By about 1900 it lay in dilapidated condition. In 1915, Clarence and Adelaide Chapman of New Jersey purchased Mulberry. The Chapmans restored the main dwelling to its early nineteenth-century appearance and rehabilitated the surrounding grounds. They continued making improvements in later years. In 1932, the Chapmans created new gardens and landscaping that gave the domestic core of the plantation an elegant appearance. The Chapmans' efforts exemplify methods sportsmen and sportswomen used in creating their estates. They also highlight the relationship between the material change and historical memory. Although Mulberry drew interest from architectural historians and devotees of regional history long before the Chapmans' arrival, rehabilitation brought new attention. Specialized treatment of surviving edifices and material development led Mulberry to be viewed as exceptional. Transformative change figured at the center of the process. As renewal stimulated interest, new views of the plantation's significance and a wave of celebratory accounts developed.

Chapter 5 explores the history of Medway Plantation. During the eighteenth and early nineteenth centuries, Medway thrived as a rice plantation and site of brick manufacture. The Civil War brought its operations to a standstill, and Medway subsequently declined. By the early twentieth century it lay in desperate condition. In 1929, Gertrude and Sidney Legendre purchased Medway for a winter residence and shooting estate. They rehabilitated the main house, erected new outbuildings, and developed modest landscaping. In later years they continued adding to the domestic complex and associated grounds. The Legendres became renowned for their sociability, love of travel, and enthusiasm for recreational hunting. They also made distinctive aesthetic choices. While most northerners favored restoration of early buildings or construction of new buildings designed in historicized styles, the Legendres cultivated an aged and timeworn appearance. They left the exterior of the main house at Medway little changed and made extensive use of salvaged building materials specifically for the sake of aesthetics. The Legendres' efforts reveal fascination with the exoticism of the lowcountry, picturesque decay, and authentic remains. As sportsmen and sportswomen made plantations anew, they insisted that age remain conspicuous. The Legendres' version of Medway offers a compelling example.

Chapter 6 surveys newspaper and magazine portrayals of northern-owned estates and narratives of lowcountry history. Lowcountry newspapers such as the Charleston *News and Courier* and the Georgetown *Times* devoted extensive attention to northerners' activities. Feature articles discussed owners' backgrounds and interests, the history of select plantations, and choices regarding architecture and landscaping. Magazines such as *Country Life*, *Home and Garden*, and *Life* also offered even more detailed portraits. Through such portrayals, readers came to see plantations as symbols of present-day wealth and power. Meanwhile, new histories of the lowcountry emphasized the role of plantations across time. Although plantations had long been part of the lowcountry landscape, not until the 1930s did people explicitly identify them as central to lowcountry history. In part, new claims about the significance of plantations reflected the growing potential of tourism, but the making of new estates also played a role. The destabilizing force of northerners' activities stimulated interest in plantations of earlier eras and the role of plantations in lowcountry history. Self-appointed protectors of regional history proffered new narratives to account for the changes taking place and valorize the antebellum lowcountry before other perspectives developed.

Chapter 7 explores the activities and experiences that made up "plantation life," the experiential realm estate owners enjoyed during the heyday of the lowcountry sporting scene. Plantation life melded visions of a romantic South with upper-class sporting pursuits and the physical space of remade plantations. Although most observers saw it as emulating the fabled lifestyle of the old planter class, plantation life grew out of contemporary yearnings for status and distinction. Virtuous forms of leisure demonstrated commitment to physical and mental strength. Environmental and social conditions supplied owners and guests with characteristically "southern" experiences. Vestiges of a majestic past and throngs of "plantation Negroes" imparted exoticism and a sense of racial privilege. In short, plantation life affirmed northerners' individual and collective sense of self while layering meaning onto familiar sporting pursuits. It also reshaped the symbolism of plantations. In expressing northerners' views of their estates and interpreting them for other audiences, plantation life revealed the significance of plantations created specifically for upper-class leisure and recreation and the challenges inherent in "virtuous" recreation.

The Epilogue chronicles the end of the Second Yankee Invasion and the decline of plantation life. Changes set in motion by the New Deal, new modes of upper-class behavior, and World War II constrained northerners' activities, leading to an end of estate-making and the decline of plantation life. The Epilogue surveys these developments and considers the legacy of northerners' plantations. Today, more than seventy-five years after the creation of the last example, the events of the era between the world wars continue to obscure important dimensions of the region's history. Conceptions of lowcountry plantations owe significant debts to northerners' estates and the perspectives that developed with them. Although integrating northerners' plantations into narratives of regional history offers potential for fuller, better-informed understandings of the lowcountry and its past, so far, historical writing and popular discourses have ignored important shifts in use, purpose, and meaning. The result is a flattening of crucial distinctions, presumptions of undue continuity, and the loss of histories vital to understanding the Jim Crow South.

Recovering the history of the new estates casts fundamental questions about the history of the early twentieth-century United States in fresh perspectives. It underscores the extraordinary influence of the myth of a majestic southern past, the appeal that mythology held among northern elites, and the beliefs that developed as rose-tinted perspectives on the past all but pushed competing views of history aside. Northerners' "plantations" reveal important dimensions of the lowcountry's rise as a leisure destination, the narratives that became tied to select plantations as interest in the region surged, and the circumstances under which those ties developed. They especially highlight reasons for the lowcountry's prominence in memory of the antebellum era. By the 1930s, many Americans saw the region as central to the nation's beginnings, one of a few places where the slaveholding "civilization" of the Old South had reached an advanced stage of development, and where, decades later, remnants of the colonial and antebellum eras remained plainly visible.[28] Fascination with the antebellum era partly reflected contemporary practices. Upper-class enthusiasm for "plantations" and the experiences they afforded became the lens through which commentators viewed the lowcountry past. Celebrations of pre–Civil War wealth and power owed as much to northerners' activities as historical knowledge and patterns of remembrance.

To see the new estates simply as underscoring the influence of the plantation legend, however, misjudges their significance. Northerners did not romanticize old plantations or draw inspiration from popular myths so much as they engaged in parallel processes that instantiated confidence in the renowned grandeur of the South of yesteryear. Although historians have long emphasized the remarkable range of representations of the South and its past that saturated American culture from the late 1880s through the civil rights era, scholars have paid comparatively little attention to perspectives on authentic vestiges of the colonial and antebellum eras. Studies of early historic preservation efforts in the former Confederacy have concentrated on urban settings, Civil War battlefields, and sites associated with well-known military and political figures.[29]

---

[28] See, for example, Edward Hungerford, "Charleston of the Real South," *Travel* 11, no. 6 (Oct. 1913): 32–35, 57–58; Mildred Cram, *Old Seaport Towns of the South* (New York: Dodd, Mead and Co., 1917), 114–129; "Charleston, South Carolina," *Fortune* 7, no. 3 (March 1933): 79–81, 83; Dixon Wecter, *The Saga of American Society: A Record of Social Aspiration, 1607–1937* (New York: Charles Scribners' Sons, 1937), 7; DuBose Heyward, "Charleston: Where Mellow Past and Present Meet," *National Geographic* 75, no. 3 (March 1939): 273–312; "The Low Country."

[29] Yuhl, *A Golden Haze of Memory*, chap. 1; Charles B. Hosmer, Jr., *Presence of the Past: A History of the Preservation Movement in the United States before Williamsburg* (New York: G. P. Putnam's Sons, 1965), chap. 3; Robert R. Weyeneth, *Historic Preservation for a Living City: Historic Charleston Foundation, 1947–1997* (Columbia: University of South Carolina Press, 2000); Melvin I. Urofsky, *The Levy Family and Monticello, 1834–1923: Saving Thomas Jefferson's House* (Charlottesville: Thomas Jefferson

Investigations of plantation landscapes have emphasized master-slave relations, slaves' agency, and human interaction with natural environments while paying little attention to symbolism.[30] Although scholars of landscape have long emphasized the evocative power of material remains and their tendency to legitimate particular visions of history, how aging plantations shaped popular views of the past remains little examined. How some plantations came to be seen as "historic" and what meanings people associated with them has yet to be understood.[31]

Northerners' estates cast light on this lacunae. Although unusual in several respects, at bottom, their development reflected growing fascination with authentic remains of the pre–Civil War South. To varying degrees, northerners acted to "preserve" valued remains for posterity, for self-definition, and as part of the deepening interest in the American past that unfolded between the world wars. Recognition of a majestic past as an embodied presence formed the basis for a range of approaches that balanced reverence with evolutionary development. Restoration, renewal, imitation, emulation – all proceed from that point. As northerners created

Foundation, 2001); Marc Leepson, *Saving Monticello: The Levy Family's Epic Quest to Rescue the House That Jefferson Built* (New York: Free Press, 2001); James M. Lindgren, *Preserving the Old Dominion: Historic Preservation and Virginia Traditionalism* (Charlottesville: University Press of Virginia, 1993); Patricia West, *Domesticating History: The Political Origins of America's House Museums* (Washington, DC: Smithsonian Institution Press, 1999), chaps. 1 and 3; Timothy B. Smith, *This Great Battlefield of Shiloh: History, Memory, and the Establishment of a Civil War National Military Park* (Knoxville: University of Tennessee Press, 2004); Timothy B. Smith, *The Chickamauga Memorial: The Establishment of America's First Civil War National Military Park* (Knoxville: University of Tennessee Press, 2009); Timothy B. Smith, *The Golden Age of Battlefield Preservation: The Decade of the 1890s and the Establishment of America's First Five Military Parks* (Knoxville: University of Tennessee Press, 2008), chaps. 3, 5, and 7.

[30] John Michael Vlach, *Back of the Big House: The Architecture of Plantation Slavery* (Chapel Hill: University of North Carolina Press, 1993); Clifton Ellis and Rebecca Ginsburg, eds., *Cabin, Quarter, Plantation: Architecture and Landscapes of North American Slavery* (New Haven: Yale University Press, 2010); Mart A. Stewart, *"What Nature Suffers to Groe": Life, Labor, and Landscape on the Georgia Coast, 1680–1920* (Athens: University of Georgia Press, 1996); Mart A. Stewart, "Rice, Water, and Power: Landscapes of Domination and Resistance in the Lowcountry, 1790–1880," *Environmental History Review* 15, no. 3 (autumn 1991): 47–64; Drew A. Swanson, *Remaking Wormsloe Plantation: The Environmental History of a Lowcountry Landscape* (Athens: University of Georgia Press, 2012).

[31] David Lowenthal, *The Past Is a Foreign Country* (Cambridge: Cambridge University Press, 1985), 125–182, 238–259; Barbara Bender, "Introduction: Landscape – Meaning and Action," in *Landscape: Politics and Perspectives*, ed. Barbara Bender (Providence, RI: Berg, 1993): 1–11; Andrew Jones, *Memory and Material Culture* (Cambridge: Cambridge University Press, 2007).

their estates, they conserved select vestiges of earlier periods while fashioning environments suited to their needs. Like better-known undertakings such as John D. Rockefeller's restoration of Colonial Williamsburg and Henry and Clara Ford's creation of Greenfield Village, northerners crafted an ennobling, empowering heritage that provided meaning and inspiration. That heritage, of course, proved highly exclusionary. Incomplete, rife with biases, and hardly impartial, it reflected all the prejudices of the era. Yet as a body of beliefs that affirmed contemporary social and political orders, it had immense power. Moreover, material presence gave it unmatched authority and durability. As enthusiasm for the romances faded and other views gained strength, northerners' "plantations" and associated narratives remained, their significance reinforced by the interlocking relationship they shared.[32]

Lest northerners' "plantations" be taken as unique, a survey of contemporary examples reveals their place at the apex of several intersecting trends. The "quail plantations" of south Georgia and north Florida offer the closest parallels. Between the 1880s and 1930s, wealthy sportsmen and sportswomen developed thirty-eight estates totaling more than 250,000 acres in the Red Hills region, between Thomasville, Georgia, and Tallahassee, Florida, and near Albany, Georgia. As in the lowcountry, large plantations had dominated the area before the Civil War. Northerners restored early houses and erected new Southern Colonial–style mansions. Owners resided in northern cities and belonged to several overlapping social networks, much like their counterparts in the lowcountry.[33] Meanwhile, members of the Carnegie Family and automobile executives such as Henry Ford and Howard Coffin created lavish estates along the Georgia coast, some with restored mansions, others with

---

[32] Michael Kammen, *Mystic Chords of Memory: The Transformation of Tradition in American Culture* (New York: Alfred A. Knopf, 1991), 299–327, 351–374; Lowenthal, *The Past Is a Foreign Country*, 275–278, 290–295, 301–319, 344–362; Anders Greenspan, *Creating Colonial Williamsburg: The Restoration of Virginia's Eighteenth-Century Capital*, 2nd ed. (Chapel Hill: University of North Carolina Press, 2009); Richard Handler and Eric Gable, *The New History in an Old Museum: Creating the Past at Colonial Williamsburg* (Durham: Duke University Press, 1997), 31–37, 62–65, 70–77.

[33] Albert G. Way, *Conserving Southern Longleaf: Herbert Stoddard and the Rise of Ecological Land Management* (Athens: University of Georgia Press, 2011), 25–27; Julia Brock, "Land, Labor, and Leisure: Northern Tourism in the Red Hills Region, 1890–1950" (PhD diss., University of California at Santa Barbara, 2012); William R. Brueckheimer, "The Quail Plantations of the Thomasville-Tallahassee-Albany Regions," *Journal of Southwest Georgia History* 3 (fall 1965): 44–63; Clifton Paisley, *From Cotton to Quail: An Agricultural Chronicle of Leon County, Florida, 1860–1967* (Gainesville: University of Florida Press, 1968), 74–98.

dwellings designed in historical styles.[34] In Virginia, wealthy men and women restored a handful of old plantations, motivated by what one historian has described as "a timeless image of a wealthy, leisured, cultivated and aristocratic way of life."[35] Elsewhere, native southerners and other Americans launched efforts to preserve and restore valued sites and buildings.[36] In short, northerners' lowcountry "plantations" stood apart because of their opulence, their number, and their celebrity, but in no way did they constitute a *sui generis* phenomenon. They demonstrated the growing importance of leisure, cultivation of a noble past, and continued positioning of the South as an exotic other, simultaneously denigrated and desired, lauded and lamented, and imagined and idealized like no other part of the nation.

* * *

A few words about methodology and language are in order. The geographic area considered in this study encompasses the territory where northern sportsmen and sportswomen created sporting estates that incorporated remains of old plantations and exhibited significant aesthetic pretention. This territory ranged from the northern reaches of Georgetown County south to the Savannah River and as far west as Williamsburg, Berkeley, Colleton, Hampton, and Jasper counties. It approximated, but did not match, the region commonly known as "the lowcountry" during the early twentieth century. It is important to recognize that no clear definition of "the lowcountry" exists or ever has. Scholars of the colonial and antebellum eras have often viewed the lowcountry as encompassing the entire coastal plain, or close to it, in keeping with contemporary usage. During the early twentieth century, most commentators recognized the lowcountry as extending between fifty and eighty miles inland but tended to emphasize areas closer to the coast

---

[34] Mary R. Bullard, *Cumberland Island: A History* (Athens: University of Georgia Press, 2003), 179–235, 241–258; Stewart, *"What Nature Suffers to Groe,"* 220; Franklin Long and Lucy Long, *The Henry Ford Era at Richmond Hill, Georgia* (Darien, GA: Darien Graphics, 1998); Maxwell Taylor Courson, "Howard Earle Coffin, King of the Georgia Coast," *Georgia Historical Quarterly* 83, no. 2 (summer 1999): 322–341.

[35] Marilyn Harper, "'What it Ought to Have Been': Three Case Studies of Early Restoration Work in Virginia" (MA thesis, George Washington University, 1989) (quotation on 3).

[36] See, for example, Swanson, *Remaking Wormsloe Plantation*, chap. 4; Brundage, *Southern Past*, 194–226; Steven Hoelscher, "The White-Pillared Past: Landscapes of Memory and Race in the American South," in *Landscape and Race in the United States*, ed. Richard H. Schein (New York: Routledge, 2006), 39–73; Jonathan Daniels, *A Southerner Discovers the South* (New York: Macmillan Co., 1938), 219.

where tidal rice culture and Sea Island cotton production had historically taken place. This latter definition is used herein, for it captures the way people of the era thought about and spoke of the region.[37]

The scope of this study is necessarily selective, given the scale of northerners' activities, the number of estates they created, and the discourses that developed. The broad sweep of the first three chapters is contrasted by the comparatively narrow focus of Chapters 4 and 5. Although Mulberry and Medway plantations represent only a fraction of the estates northerners created, the patterns illustrated appeared throughout the lowcountry. Each estate illustrates an approach that northerners used broadly, the general contours of northerners' activities, and the close relationship between material change and remembrance. Chapters 6 and 7 return to a region-wide perspective in order to account for representations of northerners' estates and northerners' experiences. In this fashion, the study assesses the overall significance of northerners' plantations while also examining estate-making practices in detail.

As for terms relating to northerners' "plantations," the language used throughout the study mirrors the vocabulary employed during the 1920s and 1930s. The terms "estate" and "plantation" are used interchangeably, in a manner consistent with contemporary practices. "Sporting plantations" and "shooting plantations" are also, for they too figured in the vernacular of the era. "Northerners" is used in referring to sportsmen and sportswomen who wintered in the lowcountry, most of whom, although not all, hailed from above the Mason-Dixon Line. "Yankees"

---

[37] On the difficulty of defining the lowcountry, see Charles F. Kovacik and John J. Winberry, *South Carolina: A Geography* (Boulder: Westview Press, 1987), 7, 18–26; Jane Lareau and Richard Dwight Porcher, *Lowcountry: The Natural Landscape* (Greensboro, NC: Legacy Publications, 1988), 7, 25, 27. For historical definitions, see, for example, Weir, *Colonial South Carolina*, 44–46; William Gilmore Simms, *The Geography of South Carolina* (Charleston: Babcock and Co., 1843), 9–10. Robert Mills's 1825 *Atlas of the State of South Carolina* does not mention the lowcountry. See Robert Mills, *Atlas of the State of South Carolina* (Baltimore: F. Lucas, Jr., 1825). For examples of late nineteenth- and early twentieth-century usage, see Harry Hammond, *South Carolina: Resources and Population, Institutions and Industries* (Charleston, SC: Walker, Evans and Cogswell, 1883), 4; Herbert Ravenel Sass, "The Low-Country," in *The Carolina Low-Country*, ed. Augustine T. Smythe et al. (New York: Macmillan Co., 1931), 6–10; William Watts Ball, *The State That Forgot* (Indianapolis: Bobbs-Merrill Co., 1932), p. 15; "The Low Country," 38. On historians' use of the term, see, for example, Rachel N. Klein, *Unification of a Slave State: The Rise of a Planter Class in the South Carolina Backcountry, 1760–1808* (Chapel Hill: Institute of Early American History and Culture by University of North Carolina Press, 1990), 7; McCurry, *Masters of Small Worlds*, 23; Coclanis, *Shadow of a Dream*, 28–33, 268 n. 5; Joyner, *Down by the Riverside*, 11.

and "winter colonists" are also used for the same purpose. Contemporaries used all of these terms in referring to estate owners and their guests. Although some readers may object to the apparent imprecision, this terminology is consistent with contemporary practices and illustrates how lowcountry people and other observers thought of the newcomers and their estates.

The seventy-six estates that constitute the focus of this study shared several unifying characteristics. They possessed significant material refinement, served as settings for upper-class recreation and social intercourse, intermingled colonial and antebellum-era remains with new buildings and landscapes, and figured in discourses about the resurgence of the lowcountry and the role of plantations in its history and present-day landscape. These qualities proved fundamental to the new type of plantation that developed during the era and differentiated such plantations from other holdings. They also distinguished northerners' lowcountry estates from similar properties within the region and in other parts of the South. Some of the hunting clubs, preserves, and retreats northerners created in the lowcountry possessed one or more of these qualities but lacked others. Moreover, compared with the sporting estates of the Red Hills region and the recreational plantations of the Georgia coast, lowcountry estates derived significance from the discourses surrounding their creation. In part, fervent attention to the combination of a valued past and new significance distinguished northerners' lowcountry estates as a group. Discourses concerning the individual and collective history of plantations and the role of new estates in a resurgent lowcountry became the primary lens through which contemporaries saw the new estates and has remained influential since.

Discovering the Lowcountry

Northern Sportsmen in Paradise, 1880–1915

In the closing decades of the nineteenth century, sportsmen from northern cities began traveling to coastal South Carolina to hunt ducks, deer, and other game during the fall and winter months. As wildlife populations declined across the North and Midwest, sportsmen sought out new territory. The lowcountry attracted attention early on. Ease of access from northern cities, plentiful wildlife, and inexpensive land made the region a prime destination for dedicated sport hunters. Beginning in the mid-1880s, sportsmen moved aggressively to take advantage of these circumstances. They formed hunting clubs that bought and leased large acreages and also bought and leased land individually, in pairs, and in small groups. In most cases, sportsmen received assistance from white southerners. Northerners purchased old plantations and parcels comprised mainly of forest and fields. No matter what they acquired, sportsmen used land under their control solely for hunting. They concentrated on a pastime that middle- and upper-class men saw as vital to their sense of self and physical and moral well-being. Although the lowcountry offered myriad possibilities, sportsmen focused exclusively on that one pursuit.

Sportsmen's activities became the leading edge of a phenomenon that lowcountry people soon termed the "Second Yankee Invasion." Barely two decades after General William Tecumseh Sherman's famously destructive march through the Carolinas during the closing phases of the Civil War, northerners again returned to the lowcountry with rifles in hand. This time they arrived in smaller numbers and only with ambitions of exploiting the region's recreational potential, not subjugating a rebel army. Initially, their activities gained little notice. Only when sportsmen became part of a larger pattern did they attract significant attention. Yet despite their modest profile, sportsmen became crucial to the long-term development of the region. They took plantation lands out of agricultural use and adapted them for recreation. They established social and business relationships with white southerners and secured African American labor for their use. Sportsmen also developed needed infrastructure and instituted systems that made it possible for them to own and manage large landholdings from afar. In short, sportsmen secured large tracts of territory and ensured their ability to return year after year.

Sportsmen's arrival in the lowcountry coincided with a series of natural disasters and economic decline. Between 1885 and 1911, seven major hurricanes and a severe earthquake struck the region. Together, these events caused extensive devastation and upended the lowcountry economy. Agricultural conditions began to deteriorate in the 1880s and continued to decline into the 1920s. Rice production collapsed during the 1890s and the first decade of the twentieth century. By 1910, only a few plantations remained in commercial production. Growers of Sea Island cotton did somewhat better but nonetheless struggled to adapt to new circumstances. For a time, phosphate mining offered cause for hope. During the 1870s and 1880s, the industry provided jobs for thousands of African Americans and profits for landowners, mining companies, and shippers. Yet in August 1893, a devastating hurricane damaged mining operations all along the coast, plunging the industry into a downward spiral. Although some mining continued in later years, phosphates ceased to be a leading force in the regional economy.[1]

As economic conditions deteriorated, evidence of decay and stagnation became ubiquitous. Throughout the region, thousands of acres lay idle and abandoned. Once-grand plantations lay in varying states of distress.

---

[1] Walter J. Fraser, *Lowcountry Hurricanes: Three Centuries of Storms at Sea and Ashore* (Athens: University of Georgia Press, 2006), 155–218; Walter Edgar, *South Carolina: A History* (Columbia: University of South Carolina Press, 1998), 426; Peter A. Coclanis, *A Shadow of a Dream: Economic Life and Death in the South Carolina Low Country, 1670–1920* (New York: Oxford University Press, 1989), 136–142, 154–155; James H. Tuten, *Lowcountry Time and Tide: The Fall of the South Carolina Rice Kingdom* (Columbia: University of South Carolina Press, 2010), chaps. 2–3; Richard D. Porcher and Sarah Fick, *The Story of Sea Island Cotton* (Charleston: Wyrick and Co., 2005), 326–333; Charles F. Kovacik and Robert E. Mason, "Changes in the South Carolina Sea Island Cotton Industry," *Southeastern Geographer* 25, no. 2 (November 1985): 91–96; Shepherd W. McKinley, *Stinking Stones and Rocks of Gold: Phosphate, Fertilizer, and Industrialization in Postbellum South Carolina* (Gainesville: University Press of Florida, 2014); Lawrence S. Rowland and Stephen R. Wise, *The History of Beaufort County, South Carolina*, vol. 3: *Bridging the Sea Islands' Past and Present, 1893–2006* (Columbia: University of South Carolina Press, 2015), 1–25.

Others already lay in ruin. Material decay evidenced the downfall of a once-powerful plantation empire. As the years passed and conditions worsened, many observers saw the countryside as frozen in time. Visitors routinely commented on aging houses, flooded fields, and scenes of desolation. Journalists and antiquarians fashioned narratives of loss and decline to explain the changes under way. By the early twentieth century, lament became a dominant theme of such accounts, old plantations silent memorials to a fast-disappearing way of life.[2]

Had state-level politics taken a different course, other narratives might have had greater influence. During Reconstruction, the nation had trained its watch on the lowcountry, eager to see if a biracial democracy would develop. Large black majorities made the coastal region a Republican stronghold. In 1868, 1870, 1872, and 1874, voters sent black candidates to the US House of Representatives, both houses of the state legislature, and a host of county-level offices. From 1868 to 1876, Republicans controlled the governor's office and the legislature. Over time, however, corruption, infighting, and rampant violence undermined Republican strength. In 1876, a fiercely contested election resulted in an armed stand-off between rival Democratic and Republican administrations, with each side claiming victory. Four months later, the negotiations that made Rutherford B. Hayes president also made Confederate General Wade Hampton III governor of South Carolina and returned white Democrats to power. South Carolina's "redemption" from Republican rule coincided with the withdrawal of federal troops from the South and African Americans' decreased ability to tell their version of history in public space.[3]

---

[2] Edward King, *The Great South* (Hartford, CT: American Publishing Co., 1875), 451; Belton O'Neil Townsend, "South Carolina Society," *Atlantic Monthly* 39, no. 237 (June 1877): 678; Coyne Fletcher, "In the Lowlands of South Carolina," *Frank Leslie's Popular Monthly*, March 1891, 284–286; E. Eldon Deane, "An Autumn Trip to South Carolina," in *The Georgian Period: A Collection of Papers Dealing with "Colonial" or XVIII-Century Architecture in the United States*, ed. William Rotch Ware (New York: American Architect, 1908), III: 52–53; Joseph Ioor Waring, "Homes of Long Ago in the Carolina Lowlands," *News and Courier* (Charleston, SC), May 14, 1911, 19.

[3] Edgar, *South Carolina: A History*, chap. 17; Stephen R. Wise and Lawrence S. Rowland, *The History of Beaufort County, South Carolina*, vol. 2: *Rebellion, Reconstruction, and Redemption, 1861–1893* (Columbia: University of South Carolina Press, 2015), chaps. 18, 19, and 22; George C. Rogers, Jr., *The History of Georgetown County, South Carolina* (Columbia: University of South Carolina Press, 1970), 441–462; Eric Foner, *Reconstruction: America's Unfinished Revolution, 1863–1877* (New York: Harper and Row, 1988), 332–333, 352–379, 535–553, 570–582; Thomas Holt, *Black over White: Negro Political Leadership in South Carolina during Reconstruction* (Urbana: University of Illinois Press,

In the years that followed, blacks retained significant political power in large portions of the lowcountry, Beaufort and Georgetown counties especially. In 1895, however, Democrats secured passage of a new state constitution that disenfranchised more than 100,000 African Americans. Thereafter, lowcountry blacks retreated from public life, determined to preserve their autonomy and gains made as landowners and independent farmers. Beaufort County became sharply divided between a large black population on the coastal islands, where Reconstruction-era land sales had given African American exceptional strength, and whites on the mainland. Elsewhere, rural blacks sustained themselves through a combination of farming, wage labor, and hunting and fishing. Throughout the region, sheer numbers told of slavery's legacy. In 1900, African Americans made up between 68 and 92 percent of the rural population. As northerners explored the lowcountry's recreational potential, lowcountry whites became the sole narrators of the region's history. Interpretation of the Civil War and its aftermath fell mainly to descendants of the old planter class.[4]

BEGINNINGS: THE PINELAND CLUB
AND ITS PROGENY, 1880–1910

Northern sportsmen's interest in coastal South Carolina grew out of three intersecting trends. One was the rapid growth of recreational hunting after the Civil War. Sport hunting had a limited following in the United States before about 1850 but grew rapidly thereafter, mainly because of burgeoning interest among middle- and upper-class men in industrializing cities. Clerks, salesmen, managers, and professionals took up sport hunting for the sake of camaraderie and adventure, to maintain contact with nature, and in pursuit of vigorous masculinity. A small sporting press promoted the growth of the pastime by disseminating information about proper "sporting" practices and the mental and physical benefits of hunting, and improvements in firearms technology also made certain types of hunting more accessible. Whereas shooting wild animals for pleasure had

---

1977); William J. Cooper, Jr., *The Conservative Regime: South Carolina, 1877–1890* (Baltimore: Johns Hopkins University Press, 1968), 21–23.

[4] Edgar, *South Carolina: A History*, 443–448; Wise and Rowland, *Rebellion, Reconstruction, and Redemption*, 574–586; Rowland and Wise, *Bridging the Sea Islands' Past and Present*, 74–110, 169–178; Rogers, *History of Georgetown County*, 474–484; Cooper, *The Conservative Regime*, 84–115.

once been the province of a small elite, during the late nineteenth century, it developed a large and devoted following.[5]

Growing numbers of sport hunters pressured wildlife populations. By the era of the Civil War, the seemingly inexhaustible numbers of animals that had greeted early European settlers to North America had begun a spiraling decline. Not only did recreational hunting take a toll, but intensifying market hunting did also. Urbanization created unprecedented demand for fresh meat. Commercial hunters responded by killing huge numbers of buffalo, deer, elk, and waterfowl. Railroad cars filled with animals packed on ice traveled from the West to eastern cities, bound for urban markets and restaurants. As game populations declined, competition between sport and market hunters escalated. By the 1870s, the two groups fought constantly, vying for access to good shooting grounds.[6]

Sportsmen responded by waging a vigorous campaign to protect their interests. They portrayed their brand of hunting as more virtuous than others and derided men who hunted for substance or market. Sportsmen viewed themselves as heirs to the sporting traditions of the English gentry. By emphasizing the challenge of the hunt over actual results, they cast themselves as responsible stewards of wildlife. Men who hunted to feed themselves or for income killed indiscriminately, they argued; sportsmen gave wildlife a fair chance. Sport and conservation thus went hand and hand. Sportsmen also lobbied for laws establishing closed hunting seasons, bag limits, and regulations on the sale, storage, and transportation of game. As a result of their efforts, states across the nation passed a spate of game laws, most of which benefited sportsmen at the expense of other hunters.[7]

At the same time, sportsmen bought and leased land for their exclusive use. Taking control of land outright protected it from development and

[5] James A. Tober, *Who Owns the Wildlife?: The Political Economy of Conservation in Nineteenth-Century America* (Westport, CT: Greenwood Press, 1981), 43–49; Daniel Justin Herman, *Hunting and the American Imagination* (Washington, DC: Smithsonian Institution Press, 2001), chaps. 10–13; Thomas R. Dunlap, *Saving America's Wildlife* (Princeton: Princeton University Press, 1988), 5–13; Peter J. Schmitt, *Back to Nature: The Arcadian Myth in Urban America* (New York: Oxford University Press, 1969), 7–14.

[6] Tober, *Who Owns the Wildlife?*, 69–81; C. John Sullivan, *Waterfowling on the Chesapeake, 1819–1936* (Baltimore: Johns Hopkins University Press, 2005), 72–73; Herman, *Hunting and the American Imagination*, 245–247; Jack Baum, "A History of Market Hunting in the Currituck Sound Area, Part 1," *Wildlife in North Carolina* 32, no. 11 (Nov. 1968): 13–15; Jack Baum, "A History of Market Hunting in the Currituck Sound Area, Part 2," *Wildlife in North Carolina* 32, no. 12 (Dec. 1968): 4–8, 31.

[7] Tober, *Who Owns the Wildlife?*, 43–50; Herman, *Hunting and the American Imagination*, 152–158, 173–187, 240.

ensured access to wildlife. Hunting clubs became the principal instrument of this strategy. By pooling members' resources, clubs secured good hunting grounds, instituted habitat management, and reserved wildlife for sportsmen's pleasure.[8]

Between the mid-1880s and the early 1910s, northern sportsmen established at least eight large clubs in coastal South Carolina. In 1884, two northern bankers, Harry Hollins and Edward Dennison, founded the Pineland Club in lower Beaufort County. With fifteen members and extensive resources, it quickly took control of about 13,000 acres. In 1891, the Chelsea Club, an eight-member organization, secured 20,000 acres nearby, southeast of Ridgeland. Soon thereafter, several sportsmen founded the Okeetee Club, a twenty-member organization that amassed about 35,000 acres. Meanwhile, in neighboring Hampton County, another group of sportsmen established the Palachucola Club, a ten-member organization that took control of about 12,000 acres. By the beginning of the twentieth century, sportsmen had established a cluster of large clubs in the densely forested lowlands of the Savannah River Valley, between the towns of Okatie and Robertsville.[9]

During the same era, northerners also established clubs near Georgetown, where extensive rice fields and swamps provided outstanding nesting grounds for migrating waterfowl. During the 1890s, several well-publicized visits by President Grover Cleveland brought the Georgetown area national attention. Cleveland, an avid sportsmen, hunted near Georgetown on eight occasions between 1894 and 1907.[10] As newspapers such as the *New York Times* and the *Washington Post* reported on his outings, other sportsmen secured nearby territory for their use. The Annandale Club, an organization formed of a "syndicate" of men from Darlington, South Carolina, and

---

[8] Tober, *Who Owns the Wildlife?*, 49–52; Herman, *Hunting and the American Imagination*, 238–240, 250–252.

[9] Chlotilde R. Martin, "Pineland, Mother of Hunting Clubs," *News and Courier*, Feb. 1, 1931, B1; Chlotilde R. Martin, "Chelsea, Where Northerners Hunt," *News and Courier*, June 7, 1931, 2; Chlotilde R. Martin, "Okeetee Club, 42,000-Acre Preserve," *News and Courier*, May 3, 1931, B6; Chlotilde R. Martin, "Palachucola Club in Hampton," *News and Courier*, Feb. 22, 1931, A6; Robert B. Cuthbert and Stephen G. Hoffius, eds., *Northern Money, Southern Land: The Lowcountry Plantation Sketches of Chlotilde R. Martin* (Columbia: University of South Carolina Press, 2009), 136–137, 154–156, 159–163, 190–192.

[10] Rogers, *History of Georgetown County, South Carolina*, 487; Matthew A. Lockhart, "From Rice Fields to Duck Marshes: Sport Hunters and Environmental Change on the South Carolina Coast, 1890–1950" (PhD diss., University of South Carolina, 2017), 192–196, 199–201.

"several wealthy men from New York and Philadelphia," already owned 4,600 acres on the Waccamaw River. In 1898, a group of sportsmen founded the Santee Gun Club on lands in the Santee Delta and northern Charleston County. It quickly grew to become one of the largest and best-managed clubs on the East Coast. Over time, it amassed 20,000 acres. In 1912, the Kinloch Club, an organization made up of Du Pont Company executives and close associates, acquired 6,700 acres immediately to the west. By the early 1920s its holdings totaled nearly 8,000 acres.[11]

Northerners also bought and leased land in Berkeley and Williamsburg counties, due north of Charleston. About 1905, two Charleston lumber merchants, Robert L. Montigue and Robert P. Tucker, took "some friends from the North" quail hunting near the town of St. Stephen. Impressed by the number of coveys they found, the group decided to establish a hunting club. Within two years, the Oakland Club purchased more than 27,000 acres and secured long-term leases to another 33,000. Several smaller clubs followed. By World War I, Williamsburg and western Berkeley counties became a leading destination for hunting quail, deer, and wild turkey.[12]

The clubs that northern sportsmen established shared several characteristics. First, they maintained all-male memberships. Although small numbers of upper-class women hunted for recreation, sport hunting remained a predominantly male pastime, full of rituals and behaviors facilitated by the absence of women and children. Sportsmen relished the fraternity of camp life and the raw equality of wilderness, where skill and character signified status more than birth or occupation. Shared pursuit of wild animals and the excitement of the hunt forged bonds among men whose

[11] Chalmers S. Murray, "Annandale, Secluded in Georgetown," *News and Courier*, May 31, 1931, B3; Annandale Plantation (Georgetown Co.) File, South Carolina Vertical File Collection, Charleston County Public Library, Charleston, SC (hereafter SCVFC); J. M. L., Jr., "Do You Know Your Lowcountry? Santee Gun Club," *News and Courier*, Nov. 15, 1937, 10; Henry H. Carter, *Early History of the Santee Club* (Boston [?]: n.p., 1934); Matthew A. Lockhart, "'Rice Planters in Their Own Right': Northern Sportsmen and Waterfowl Management on the Santee River Plantations during the Baiting Era, 1905–1935," in *Leisure, Plantations, and the Making of a New South: The Sporting Plantations of the South Carolina Lowcountry and Red Hills Region, 1900–1940*, ed. Julia Brock and Daniel Vivian (Lanham, MD: Lexington Books, 2015): 108–109; Chalmers S. Murray, "Kinloch, Exclusive Santee Club," *News and Courier*, July 12, 1931, B7; President's Report, May 17, 1913, folder "Kinloch Gun Club – Annual Reports of President and Treasurer, 1912–21," box 17, Miscellaneous Papers of Eugene du Pont, Hagley Library, Wilmington, DE (hereafter MPEDP).

[12] J. K. C., "Do You Know Your Charleston? Oakland Club," *News and Courier*, Jan. 28, 1935, 10; Laura C. Hemingway, "Counties of Coast Returned to 'Happy Hunting Ground,'" *News and Courier*, Dec. 2, 1934, C2.

professional and personal lives tended to be routine. Hunting demanded skill, self-control, and restraint. Moreover, it thrived off competition. As the historian Nicholas Proctor has observed, "triumphs and defeats in the field meant little unless observed by others." By taking to the field in small groups, sportsmen made hunting a test of manhood and mastery and a forum for building trust and respect.[13]

Northerners also organized their clubs as joint-stock corporations, a measure that maintained exclusivity and protected members' interests. Clubs operated according to formal bylaws, which specified procedures for admitting new members, annual dues, bag limits, pay for hunting guides and servants, and a host of other rules. Operating in this fashion ensured stability and upheld social boundaries. Since belonging to a hunting club demarcated elite status, members insisted on admitting only men of appropriate character and social standing. Members purchased shares to join. Only when a member died or opted to leave did openings occur, and new members had to be elected by existing members. In this manner, sportsmen protected the social significance of sport hunting.[14]

Hunting clubs did more than simply acquire land. All of the clubs that northerners established carried out extensive land management activities, and many instituted baiting in order to provide members with favorable hunting conditions. By planting feed crops for game, clubs attracted wildlife and therefore benefited members while disadvantaging other hunters. The Pineland Club, for example, planted between two and three hundred acres of grain annually. The Palachucola Club planted feed crops on "a small portion" of its lands. Waterfowling clubs planted large acreages in rice. During the early 1910s, the Kinloch Club planted 125 acres annually. The Santee Club planted about ten acres of rice in 1906 and steadily increased its baiting operations over time. By 1920, the club planted about 100 acres each year.[15]

[13] Herman, *Hunting and the American Imagination*, 238–239, 243–253; Nicholas W. Proctor, *Bathed in Blood: Hunting and Mastery in the Old South* (Charlottesville: University of Virginia Press, 2002), 61–118.

[14] *Oakland Club, St. Stephens P.O., Berkeley County, South Carolina* (N.p.: n.p., 1908); *Santee Club* (N.p.: n.p. 1919), copy in folder "Misc. Leaflets," box 38, MPEDP. See also Charles Hallock, *Hallock's American Club List and Sportsman's Glossary* (New York: Forest and Stream Publishing Co., 1878), 7–9.

[15] Martin, "Pineland, Mother of Hunting Clubs"; Martin, "Palachucola Club in Hampton"; Report of House Committee, March 31, 1914, p. 2, folder "Kinloch Gun Club – Annual Reports of President and Treasurer, 1912–21," MPEDP; Lockhart, "Rice Planters in Their Own Right," 113–114: Carter, *Early History of the Santee Club*, 15; Hemingway, "Williamsburg Boasts Many Game Preserves for Hunt"; N. L. Willett,

Clubs also developed essential infrastructure. Initially, sportsmen relied on existing buildings wherever possible. By using old houses for lodgings and barns and sheds for storage and to shelter dogs and horses, they satisfied basic needs easily and at low cost. Over time, clubs built new lodges, stables, and kennels. The Pineland Club, for example, initially used a small frame building as a clubhouse. About 1900, members erected a complex of Shingle-style buildings consisting of a two-story lodge, an annex, and five log cabins. The Palachucola Club lodged in an old house until it burned in 1916. Members immediately built a cypress-shingled complex in its place. In 1923, the Kinloch Club erected a fifteen-room lodge that contained a club room, nine bedrooms, and dining and kitchen facilities. Clubs also rehabilitated tidal rice fields, built and installed hunting blinds, and maintained canals and roads. By ensuring members' comfort and access to good hunting grounds, clubs facilitated sportsmen's aims. They made it possible for members and their guests to concentrate on their preferred pastime during their stays in the region.[16]

Sportsmen generally hired white southerners, usually former planters or men with experience supervising agricultural operations, to manage their lowcountry holdings. In 1905, for example, the Santee Club hired Ludwig A. Beckman as its superintendent. Beckman had previously planted rice on Blackwood Plantation in the Santee Delta. When he sold Blackwood to Eben Jordan, a club member, for use as a private hunting domain, Beckman put his knowledge of local conditions and his relationships with laborers, landowners, and merchants to use on behalf of his new employer. John Edwin Fripp followed a similar path. A Beaufort County planter, Fripp became superintendent of the Chelsea Club after selling his plantation. In Berkeley County, John B. Gasden, an avid hunter and the former superintendent of the Charleston light-house district, became manager of the Oakland Club. Superintendents often brought stability to club operations. When the Charleston *News and Courier* surveyed northern-owned hunting clubs in the early 1930s,

Game Preserves and Game of Beaufort, Colleton and Jasper Counties, South Carolina: Hunters' Paradise, Manly Sports (Beaufort, SC: Charleston and Western Carolina Railroad Co., 1927), 8.

[16] Martin, "Pineland, Mother of Hunting Clubs"; Martin, "Palachucola Club in Hampton"; Carter, *Early History of the Santee Club*, 3, 10, 18–19; Murray, "Kinloch, Exclusive Santee Gun Club"; Report of House Committee, 1917, 6–7, folder "Annual Reports of President and Treasurer, 1912–21," box 17, MPEDP.

reporters identified three superintendents who had been on the job for two decades or more.[17]

Northerners had little difficulty hiring lowcountry men as superintendents and generally found lowcountry people accepting of their activities. To be sure, not all lowcountry residents welcomed the newcomers. Hollins and Denison encountered hostility during their initial foray in Beaufort County, and during the 1910s, landowners and politicians in Hampton, Jasper, Colleton, and Beaufort counties attacked the "hunting club evil." Ownership of large tracts by wealthy outsiders, they believed, limited agricultural productivity. County leaders considered enacting a special tax on large preserves and taking other measures to curtail northerners' activities but ultimately did nothing, in part because the Okeetee Club agreed to work with local authorities to promote economic development. Despite occasional friction, lowcountry people acquiesced in northerners' activities and often lent support. Although sectional rhetoric and laments about the loss of ancestral lands suffused discussions of northerners' clubs and activities, lowcountry people put up little resistance. In many cases, they provided assistance out of self-interest.[18]

Three factors explain lowcountry people's general stance toward northern sportsmen. First, Gilded Age sporting culture fostered bonds of mutual respect and assistance between northern and southern sportsmen. Northerners benefited from several operatives who provided assistance and material support. In Beaufort County, John K. Garnett helped sportsmen establish the Pineland, Okeetee, Chelsea, and Palachucola clubs. A member of a prominent family, Garnett acted as a local agent of sorts by brokering real estate deals, hiring superintendents and laborers, and cultivating amicable relations with locals. Garnett sold his personal residence to the Palachucola Club and served at least one term as president of the Pineland Club. The Okeetee Club also named its clubhouse in honor of his father.[19]

---

[17] Lockhart, "Rice Planters in Their Own Right," 113; Scott E. Giltner, *Hunting and Fishing in the New South: Black Labor and White Leisure after the Civil War* (Baltimore: Johns Hopkins University Press, 2008), 27, 28–29, 133; Martin, "Pineland, Mother of Hunting Clubs"; Martin, "Okeetee Club, 42,000-Acre Preserve"; "Do You Know Your Charleston? Oakland Club."

[18] Martin, "Pineland, Mother of Hunting Clubs"; Rowland and Wise, *Bridging the Sea Islands' Past and Present*, 284–285.

[19] Martin, "Pineland, Mother of Hunting Clubs"; Martin, "Okeetee Club, 42,000-Acre Preserve"; Martin, "Palachucola Club in Hampton"; John E. Davis, "The Plantation Broker," *South Carolina Wildlife* 50, no. 6 (Nov.–Dec. 2003): 7–13.

Hugh R. Garden played a similar role with the Santee Club. A native South Carolinian, Garden practiced law in New York City. In 1898 he founded the Santee Club. The organization's initial members included five men from New York and five from South Carolina, making it a truly intersectional venture.[20] Edward Porter Alexander also proved influential. A former Confederate general and railroad executive, Alexander owned South Island, a barrier island below Georgetown that he developed as a residence and a hunting and fishing preserve. Alexander introduced Hollins and Dennison to the territory where they established the Pineland Club and hosted Grover Cleveland on two of his visits to the Georgetown area. He also sold seed to the Santee Club when it began planting rice to attract waterfowl.[21]

The shared identity of sportsman undergirded relations between northern and southern men. As a class-specific persona, "sportsman" denoted more than a recreational preference; it identified men who considered themselves gentlemen and subscribed to the behaviors, values, and ethics associated with the term. Gentlemen maintained a strict code of conduct designed to maintain personal reputation and honor. Concern for sporting practices and fairness in the field formed part of an outlook that extended into business and personal affairs. Moreover, sportsmen shared a common approach to wildlife conservation and land use. As wildlife populations declined, sportsmen saw themselves engaged in a struggle to protect game and good shooting grounds from market and subsistence hunters. Common devotion to sport hunting transcended sectional divisions and formed the basis of friendships that quickly became crucial to northerners' activities.[22]

If shared commitment to sport made upper-class southerners receptive to northerners' activities, economic interests had broader appeal. Northerners' clubs spent huge sums of money. The Chelsea Club spent between $40,000 and $50,000 on labor and supplies each year. The Santee Club's annual expenditures totaled about $30,000. Throughout the 1910s, the Kinloch

[20] Henry H. Carter, *Early History of the Santee Club* ([Boston?]: n.p., [1934?]), 3–4; J. M. L., Jr., "Do You Know Your Lowcountry? Santee Gun Club," *News and Courier*, Nov. 15, 1937, 10; Lockhart, "Rice Planters in Their Own Right," 108–109. On Garden, see "Hugh R. Garden," *National Magazine: A Monthly Journal of American History*, May 1892, 84–89.

[21] Michael Golay, *To Gettysburg and Beyond: The Parallel Lives of Joshua Lawrence Chamberlain and Edward Porter Alexander* (New York: Crown Publishers, 1994), 286–287; Martin, "Pineland, Mother of Hunting Clubs"; Lockhart, "Rice Planters in Their Own Right," 113.

[22] Herman, *Hunting and the American Imagination*, chaps. 10–13; Proctor, *Bathed in Blood*, chap. 5; Giltner, *Hunting and Fishing in the New South*, chaps. 3–5.

Club spent between $6,000 and $10,000 annually. In a region suffering severe agricultural decline, sportsmen's activities proved vital. They benefited merchants, laborers, contractors, and municipal coffers.[23]

Northerners' expenditures fell into several categories. One encompassed land acquisition costs and related fees. Land purchases put money in sellers' hands, and a small cadre of real estate brokers became known for selling hunting tracts to monied people from afar. Although northerners generally obtained land at low prices, sometimes for as little as $1 or $2 per acre, landowners eagerly dispensed of their holdings. The depressed agricultural economy left most planters ready to sell when the opportunity arose.[24]

Second, construction of new buildings and repairs and improvements to existing structures led clubs to purchase building materials and hire laborers. Large African American populations provided sportsmen with a ready supply of skilled and unskilled labor. Throughout the lowcountry, blacks grew vegetables and rice for their own consumption, raised poultry, grew cotton for market, hunted and fished, and engaged in limited wage labor. Unusually high rates of land ownership and social networks rooted in slavery gave lowcountry blacks exceptional autonomy. Historically, lowcountry slaves had labored under the task system, an arrangement that organized work by volume rather than time. Whereas gang labor prevailed on short-staple cotton plantations, in the lowcountry, masters assigned slaves a specific amount of work to perform each day or week. Once a slave completed his or her assigned task, they were free to do as they wished. Slaves generally used their time to hunt and fish, to tend to garden plots, and to assist other slaves with their assigned tasks. The task system forged solidarity among slaves and made African Americans resistant to direct oversight. In the postbellum era, its legacies included community strength and distinctive attitudes toward labor and working conditions.[25]

[23] Martin, "Chelsea, Where Northerners Hunt"; "Do You Know Your Lowcountry? Santee Gun Club"; House Committee Reports, 1914–1918, folder "Annual Reports of President and Treasurer, 1912–21," box 17, MPEDP.

[24] "Real Estate Change Hands," *Beaufort Gazette* (Beaufort, SC), Apr. 7, 1927, 1; "Big Land Deal Made Last Week," *Beaufort Gazette*, Aug. 4, 1927, 1; Hoffius and Cuthbert, eds., *Northern Money, Southern Land*, 162; Davis, "The Plantation Broker," 7–11.

[25] Rogers, *History of Georgetown County*, 445–446; John Scott Strickland, "Traditional Culture and Moral Economy: Social and Economic Change in the South Carolina Low Country, 1865–1910," in *The Countryside in the Age of Capitalist Transformation: Essays in the Social History of Rural America*, ed. Steven Hahn and Jonathan Prude (Chapel Hill: University of North Carolina Press, 1985): 141–178; John Scott Strickland,

When northerners purchased lowcountry lands, they continued the leases of African American tenants. The Chelsea Club, for example, had "several hundred negroes" living on its property. At the Kinloch Club, tenants lived in sixty-four houses scattered across its acreage. Retaining tenants offset operating costs and provided clubs with a ready source of part-time labor. Clubs employed African American laborers in several capacities. Building maintenance and construction created strong demand for laborers. In its early years of operation, for example, the Kinloch Club refurbished and adapted an old plantation house for use as a clubhouse, retrofitted outbuildings for use as storehouses and a kennel, rebuilt a stable, turned two dwellings into residences for domestic laborers, and repaired another plantation house to provide "extra sleeping quarters" when needed. Subsequent improvements included repairs to a rice mill and tenant houses. Materials for these activities totaled more than $5,200 alone. Although demand for construction labor proved most intensive in clubs' early years of operation, when members sought to have clubhouses outfitted, stables and kennels built, and other basic facilities readied, it never ceased. Clubs made improvements continually and storms, fires, and other mishaps necessitated repairs from time to time.[26]

A third category of expenditures pertained to land management and related activities. Planting feed crops, building and maintaining fences, and rehabilitating and maintaining tidal rice fields required laborers and supplies. Throughout the lowcountry, clubs relied on small numbers of white workers and large numbers of African Americans for these tasks.

"'No More Mud Work': The Struggle for the Control of Labor and Production in Low Country South Carolina, 1863–1880," in *The Southern Enigma: Essays on Race, Class, and Folk Culture*, ed. Walter J. Fraser and Winfred B. Moore (Westport, CT: Greenwood Press, 1983): 43–62; Mart A. Stewart, *"What Nature Suffers to Groe": Life, Labor, and Landscape on the Georgia Coast, 1680–1920* (Athens: University of Georgia Press, 1996), 193–196, 238–242. On the task system, see Philip D. Morgan, "Work and Culture: The Task System of Lowcountry Blacks, 1700 to 1880," *William and Mary Quarterly* 39, no. 4 (Oct. 1982): 563–599; Philip D. Morgan, *Slave Counterpoint: Black Culture in the Eighteenth-Century Chesapeake and Lowcountry* (Chapel Hill: Omohundro Institute of Early American History and Culture by the University of North Carolina Press, 1998), 179–187; Charles Joyner, *Down by the Riverside: A South Carolina Slave Community* (Urbana: University of Illinois Press, 1984), 43–45, 128–130; S. Max Edelson, *Plantation Enterprise in Colonial South Carolina* (Cambridge: Harvard University Press, 2006), 83–89, 157–161.

[26] Martin, "Chelsea, Where Northerners Hunt"; Kinloch Gun Club Property, March 21, 1913, p. 1, and President's Reports and Reports of House Committee, 1913–1918, folder "Kinloch Gun Club – Annual Reports of President and Treasurer, 1912–21," box 17, MPEDP.

Land-management activities took place year-round but increased during the spring planting season and again in the fall, in advance of sportsmen's arrival. Clubs generally used African American workers for such tasks but hired white crews for some skilled jobs such as rebuilding the earthen banks of tidal fields and harvesting timber.[27]

Labor and supplies related to annual hunting seasons constituted a fourth category of expenditures. Club members expected to have domestic help; dogs, horses, and equipment well maintained and ready for use; and hunting guides and assistants at their disposal. Superintendents oversaw preparations for the annual hunting season and ensured smooth operation during members' stays. The number of people employed by clubs varied. Most employed black females as domestic workers and cooks and black men as stewards, hunting guides, and general laborers. Domestic labor included cooking, laundry, housekeeping, and general cleaning. Hunting labor consisted mainly of guiding; caring for dogs, horses, and wagons; porting equipment; cleaning and dressing game; and standing watch to guard against poachers and forest fires. Labor associated with sportsmen's activities tended to be menial and monotonous but compared favorably with most of the other varieties of employment available. Agricultural labor, work in timber camps, and phosphate mining all imposed greater physical demands and harsher conditions. Moreover, some forms of hunting labor offered African Americans opportunities to earn tips and recognition for their skill in the field. As Scott E. Giltner has shown, sportsmen praised guides for their knowledge of the lowcountry landscape and animal behavior. Guiding thus provided rare opportunities for African Americans to receive recognition from whites and earn material benefits for abilities that they alone possessed.[28]

A final reason that lowcountry whites accommodated northern sportsmen's activities has to do with racial control. Even as rice and Sea Island

---

[27] Martin, "Okeetee Club, 42,000-Acre Preserve"; "Do You Know Your Lowcountry, Santee Gun Club"; Hemingway, "Williamsburg Boasts Many Game Preserves for Hunt"; "Do You Know Your Charleston? Oakland Club"; President's Reports and Reports of House Committee, 1913–1918, folder "Kinloch Gun Club – Annual Reports of President and Treasurer, 1912–21," box 17, and Payroll sheets for periods ending Oct. 15, Oct. 31, Nov. 30, and Dec. 15, 1920, folder "Kinloch Gun Club – Payrolls, 1918–21," box 32, MPEDP. See also Giltner, *Hunting and Fishing in the New South*, 129–130.

[28] Giltner, *Hunting and Fishing in the New South*, 81–82, 88–101. See also Hayden R. Smith, "Knowledge of the Hunt: African American Hunting Guides in the South Carolina Lowcountry at the Turn of the Twentieth Century," in *Leisure, Plantations, and the Making of a New South*, 131–148.

cotton declined, most whites believed agriculture would eventually rebound or that some other productive use for coastal lands would be found. The immense prosperity that coastal lands had historically produced gave residents reason to believe that staple crops would figure in the lowcountry's future, and various schemes to develop industry offered other possibilities. In this context, maintaining a large force of rural laborers became a priority. Although lowcountry whites did not seek to limit black mobility as aggressively as their counterparts in cotton-producing areas, they nonetheless took steps to retain large numbers of African Americans. Initially, landowners appear to have expected that northerners' presence would be short-lived. As soon as agricultural prices rose or new industries arrived, they believed, increased land values would compel sportsmen to sell. Of course, neither occurred, and, over time, landowners and others came to see hunting clubs as assets rather than a poor alternative to productive activities. Hence the shift from tentative acquiescence to endorsement. In the 1890s, most landowners viewed hunting clubs as acceptable only because no better options existed. By the 1930s, civic leaders, landowners, and commercial interests celebrated northerners as central to an economy increasingly dependent on leisure, having long since realized that agriculture no longer produced reasonable returns.[29]

Overall, the South Carolina coast offered northern sportsmen unusually accommodating circumstances. Economic and social conditions proved conducive to northerners' activities and the development of landholdings devoted to recreation. Although conflicts occurred, they proved less severe and fewer in number than might have been expected. In recent years, historians have uncovered instances of fierce resistance to elite hunter's efforts to secure access to wildlife. In *Crimes against Nature*, for example, Karl Jacoby recounts acts of vandalism, trespassing, poaching, and even murder by rural residents of the Adirondacks who opposed sport hunters' efforts to establish private game preserves and end customary access to wildlife. According to Jacoby, such acts of "environmental banditry" form part of a "hidden history of conservation" that forces reconsideration of Progressive Era struggles over natural resources.[30] Although Jacoby's

[29] Tuten, *Lowcountry Time and Tide*, chaps. 2–4; Giltner, *Hunting and Fishing in the New South*, chap. 5. On support for sportsmen during the 1930s, see, for example, Burnet R. Maybank, "Justice for Mr. Hutton," *New York Times*, Feb. 2, 1933, 16.

[30] Karl Jacoby, *Crimes against Nature: Squatters, Poachers, Thieves, and the Hidden History of American Conservation* (Berkeley: University of California Press, 2001), 15–28, 39–47, 58–78. On resistance to game laws in the lowcountry, see Giltner, *Hunting and Fishing in the New South*, 165.

study casts light on an important phenomenon, in the lowcountry, no strong opposition to northern sportsmen's activities occurred. Most residents viewed northerners favorably or at least saw their activities as innocuous. African Americans found part-time and seasonal labor at hunting clubs a valuable supplement to their other activities, and small numbers of whites also benefited. Moreover, common devotion to sport hunting facilitated relationships between northern and southern men that quickly assumed a modest degree of reciprocity. As the lowcountry economy deteriorated, sportsmen from afar found prime shooting grounds and men like them who relished the pursuit of wild game. The combination of depressed land prices, abundant wildlife, and landowners willing to sell and lease large tracts facilitated the establishment of hunting clubs in select parts of the lowcountry. By the early 1910s, northerners had taken control of more than 170,000 acres and developed much of the basic infrastructure needed for their activities. These measures marked a beginning. Significant in their own right, they also created a foundation for future growth.

DECAY AND DECLINE: THE LOWCOUNTRY LANDSCAPE AT THE END OF THE CENTURY

Northern sportsmen encountered more than large game populations and large expanses of undeveloped land during their visits to the lowcountry. They also discovered the decaying remains of a massive plantation complex. After the Civil War, ruin became a pervasive feature of the lowcountry landscape. Plantations lay in varying states of disarray, with huge expanses of once-productive fields flooded and overgrown. Physical decay marked the decline of a plantation empire. In 1860, South Carolina had produced 63 percent of all rice grown in the United States and more than 15.5 million pounds of Sea Island cotton. By 1900, the state produced less than 20 percent of all US rice and only 2.6 million pounds of the latter crop. No longer an agricultural powerhouse, the lowcountry had become a backwater, economically moribund, impoverished, and adrift of the American mainstream.[31]

In most cases, sportsmen's yearnings for wildlife led them to wilderness. As sport hunting grew in popularity, travel to remote destinations

---

[31] Coclanis, *Shadow of a Dream*, 137–156; Tuten, *Lowcountry Time and Tide*, 45–46; Kovacik and Mason, "Changes in the South Carolina Sea Island Cotton Industry," 91–97; Porcher and Fick, *Story of Sea Island Cotton*, 115–116, 119–120, 324–330.

became common. Sportsmen in northern cities found good shooting grounds in locations within a day's travel by train. Western Pennsylvania, the Adirondacks, the Catskills, and parts of Maine and New Hampshire became popular, and especially ambitious hunters struck out for the Dakotas and the mountain West. No matter what the destination, all sought undeveloped land and wildlife. Sportsmen viewed unspoiled nature as vital to the challenge and rejuvenating influence of hunting and thus spared no effort in seeking it out.[32]

In many ways, the Carolina coast offered conditions similar to those that sportsmen found elsewhere. Extensive forests, coastal marshes, and grassy savannahs provided outstanding possibilities for outdoor recreation, and abundant wildlife consistently awed visitors. Few regions offered comparable species diversity and as varied a landscape. Moreover, a location along the Atlantic Flyway, one of the four migratory bird corridors in North America, brought huge numbers of waterfowl to the region during the fall and winter. Countless observers commented on the way the sky turned black when ducks alighted from coastal marshes en masse, and sportsmen spoke of "100 duck days" – a reference to the number of birds a good morning's shoot could attain (see Figure 1.1).[33] Yet if nature's bounty made itself plainly apparent, so too did the influence of humankind. Throughout the region, material remains of past eras exerted a powerful presence. Aging plantation complexes, flooded rice fields, and deteriorating slave streets provided a vivid record of lowcountry history. Sportsmen's activities brought them face to face with remnants of an exceptionally sophisticated agricultural complex. As sportsmen moved through the landscape, they developed an awareness of the lowcountry past and judged its bearing on the present.

[32] Herman, *Hunting and the American Imagination*, chaps. 14 and 15; Tober, *Who Owns the Wildlife?*, 71–73; Giltner, *Hunting and Fishing in the New South*; Sullivan, *Waterfowling on the Chesapeake*, 3–26.

[33] Charles F. Kovacik and John J. Winberry, *South Carolina: The Making of a Landscape* (Columbia: University of South Carolina Press, 1989), 18–29, 45–48. On wildlife populations, see J. Motte Alston, *Rice Planter and Sportsman: The Recollections of J. Motte Alston, 1821–1909* (1953; reprint, Columbia: University of South Carolina Press, 1999), 58, 75–77; A. S. Salley, Jr., *The Happy Hunting Ground: Personal Experience in the Low-Country of South Carolina* (Columbia: The State Co., 1926), vii, ix; James Henry Rice, *Glories of the Carolina Coast* (Columbia: R. L. Bryan Co., 1925), 55–59, 69–71, 73–126. On "100 duck days," see Lee Brockington, *Plantation between the Waters: A Brief History of Hobcaw Barony* (Charleston: History Press, 2006), 40, 57. On the Atlantic Flyway, see Ann Vileisis, *Discovering the Unknown Landscape: A History of America's Wetlands* (Washington, DC: Island Press, 1997), 162–163.

FIGURE 1.1 Bernard and Annie Baruch pose with friends after a hunt at Hobcaw Barony, circa 1910. Shooting conditions in the lowcountry astonished even experienced sportsmen. "Hundred duck days" – named for the number of ducks killed – thrilled hunters and novices alike.
Courtesy of the Belle W. Baruch Foundation, Georgetown, SC.

None of the northerners who hunted in the lowcountry at the turn of the twentieth century is known to have recorded their impressions of the lowcountry landscape. Most of the accounts they left are concerned solely with hunting. Still, fragmentary references offer insight into how newcomers viewed the environments they encountered. William Whitson hunted along the Cooper River in 1899. When he arrived by train he found his host waiting for him with "a number of old-fashioned South Carolina negroes." "These old negroes," Whitson noted, spoke a strange dialect that no outsider could understand. Language, physical appearance, and demeanor made them "a distinct species." For lodgings, Whitson and his compatriots stayed in "an old-time mansion" that he judged to have been "magnificent" in its day. Upon inquiring, he learned that the building had stood for more than 200 years. Nearby he encountered more evidence of the "ancient civilization" of the South Carolina coast: "Old Goose Creek Church," long the place of workshop for the planters of the neighborhood. With the "new order of things brought about by the results of the War of the Rebellion," the building remained shuttered, save for one day a year when members and their descendants gathered to worship. For Whitson, hunting along the Cooper provided an object lesson in the costs of the Civil War and its aftermath.[34]

Twenty-five years later, D. J. Hart hunted near Georgetown. He characterized the area as "a country that before the war was rich in agricultural wealth and teeming with waving fields of rice and cotton." Now, he noted, "little farming is done." Woodland covered much of the territory formerly devoted to agriculture. Hart described conditions that made for good hunting but also told of how dramatically circumstances had changed.[35]

Whitson's and Hart's comments echoed a small but growing body of literature that portrayed the South Carolina coast as a ruined land, a once-majestic civilization toppled by war and social upheaval. Beginning in the 1870s, journalists called attention to a small number of plantations whose anachronistic appearance and historical associations made them objects of intrigue. Writing in *Harper's* in 1875, Constance Woolson labeled more than a dozen along the Ashley and Cooper rivers near Charleston as vestiges of the colonial and Revolutionary eras.[36] Fifteen years later, Coyne Fletcher described the "sadly changed" scene near Georgetown, where once "handsome dwellings, barns and mills" lay "dilapidated or utterly destroyed." "The famous rice lands" that had historically produced vast wealth lay in ruins, the result of planters' "loss of regular labor." With the prosperity of the region gone, rice production stood at a fraction of its prewar output. Although planters and merchants continued to hope for a recovery, Fletcher viewed the prospects of that happening as dim. Anyone who tried to revive rice production, she opined, would have to rebuild mile after mile of tidal fields virtually from scratch.[37]

The narrative underlying such portrayals emphasized the lowcountry's role in the nation's beginnings, the effects of the Civil War, and the contrast between the antebellum era and the present. Most commentators viewed the aftermath of the Civil War as a tragedy, an unfortunate consequence of the great struggle over slavery. Although few bemoaned slavery's demise, most viewed the fate of the planter class as a grave loss. One commentator after another extolled the culture, education, and refinement that

---

[34] Whitson, "A South Carolina Hunt."
[35] D. J. Hart, "Wild Turkey Hunting in South Carolina: The Ways and Habits of Meleagris Gallapavo," *Field and Stream*, Dec. 1915, 778.
[36] Constance Woolson, "Up the Ashley and the Cooper," *Harper's New Monthly Magazine* 52, no. 307 (Dec. 1875): 1–24.
[37] Fletcher, "In the Lowlands of South Carolina," 286.

had characterized the lowcountry elite in its heyday. Fletcher recalled "a chivalric people" known for "courtesy of manners," "free hospitality," and "eloquence." "The modes and customs of South Carolina to-day," she lamented, offered "but a meager idea of its former grandeur."[38] Others used even more florid language. One account recalled "fine old houses," "retinues of thoroughly trained servants," and a vibrant social life of balls, dances, and visiting.[39] No matter what their choice of phrasing, writers of the era mourned the downfall of the old planters. For them, the diminished circumstances of the present symbolized the injustice of war, not the deserved results of a failed rebellion.

When journalists mentioned the conflicts of the Civil War era, they usually sounded the theme of sectional reconciliation. In a feature on "old Charleston and the nearby rice plantations," for example, Leila Mechlin described an exchange between a northern visitor and "an old gentleman of Charleston." According to Mechlin, the Charlestonian, with "amazing magnanimity and graciousness," admitted that the suffering of the postwar era had been "hard" but suggested it might have been necessary. "Those plantation homes were once hot-beds of sedition," he noted. Destroying them, the Charlestonian opined, "was perhaps the only remedy." How many lowcountry whites would have agreed with the Charlestonian's assessment is an open question; many remained deeply embittered. Regardless, Mechlin's story illustrates a major current of the postbellum era. As the years passed, public discourse increasingly emphasized white unity across sectional lines and downplayed the fierce conflicts of the past.[40]

That writers such as Mechlin applied reconciliationist tropes to the lowcountry illustrates the strength of reunion and its racial underpinnings. As sectional tensions had escalated during the 1850s, many Americans viewed the coastal region with distain, shocked by the radicalism of its politicians and lowcountry planters' commitment to slavery. Soon after Union troops stormed ashore at Port Royal in November 1861, the *New York Times* celebrated the arrival of "twenty thousand arms soldiers of democracy" in "the *heart* of the rebellion," the region where a "small but powerful ... class of oligarch-planters ... had led the way in all the troubles of Slavedom and the Union." When Brigadier General

[38] Ibid., 281.
[39] C. R. S. Horton, "French Santee, South Carolina," in *The Georgian Period*, III: 65–66.
[40] Leila Mechlin, "A Glimpse of Old Charleston and the Nearby Rice Plantations," *American Magazine of Art* 14, no. 9 (Sept. 1923): 480.

William T. Sherman led his army north from Savannah in January 1865, his soldiers relished the chance to punish the state where, as one wrote, the "infamous conspiracy" had begun. Yet once the fighting ended, vitriol and animosity quickly dissipated. By the mid-1870s, public expressions of distain for the lowcountry became rare, and in the years that followed they virtually disappeared. By the turn of the twentieth century, most commentators viewed Charleston as a city of faded glory and romantic decay. As white supremacy healed divisions between North and South, public discourse praised the city most closely associated with secession for its quaintness and Old World charm.[41]

When Fletcher mentioned the "the famous rice lands," she referred to an extraordinary agricultural landscape. Even in Fletcher's time, the infrastructure that had made the lowcountry's historical prosperity possible remained a sight to behold. Physical deterioration did nothing to obscure the tale of toil, struggle, and suffering it embodied. Tidal rice cultivation relied on the natural rise and fall of the ocean tides to irrigate and drain fields in river floodplains. During the late eighteenth century, colonists likely grew rice on low, moist lands where water collected during heavy rains. Experimentation with small embankments increased planters' skill at controlling water. Soon, planters developed inland fields bounded by dikes, ditches, and drains. By damming small streams, they created reservoirs of fresh water that could be used to irrigate fields during droughts.[42]

Rice became Carolina's leading staple during the 1720s. By the early 1740s, the colony exported an average of 35 million pounds annually. As colonists intensified production, slave imports rose dramatically. By 1750, the colony's population numbered approximately 39,000 blacks

---

[41] "Southern South Carolina – Its Social and Political Character," *New York Times* (New York, NY), Nov. 23, 1861, p. 4 (first and second quotations); George W. Pepper, *Personal Recollections of Sherman's Campaigns in Georgia and the Carolinas* (Zanesville, OH: Hugh Dunne, 1866), 298 (third quotation); Anne S. Rubin, *Through the Heart of Dixie: Sherman's March and American Memory* (Chapel Hill: University of North Carolina Press, 2014), 29–30, 36, 111; George Marshall Allen, "Charleston: A Typical City of the South," *Magazine of Travel* 1, no. 2 (Feb. 1895): 99–119; William D. Howells, "In Charleston," *Harper's Magazine* 131, no. 785 (Oct. 1915): 747–757; Edward Hungerford, "Charleston of the Real South," *Travel* 11, no. 6 (Oct. 1913): 32–34, 57–58.

[42] Edelson, *Plantation Enterprise in Colonial South Carolina*, 72–75, 103–109. See also Hayden R. Smith, "Reserving Water: Environmental and Technological Relationships with Colonial South Carolina Inland Rice Plantations," in *Rice: Global Networks and New Histories*, ed. Francesca Bray et al. (New York: Cambridge University Press, 2015): 189–211.

and 25,000 whites. Although colonists also had success growing indigo, rice remained Carolina's principal export, the main source of the colony's wealth.[43]

During the mid-eighteenth century, planters began creating tidal fields, which increased output and protected crops from storms and freshets. Building tidal fields required extraordinary effort. Masters sent slaves into river swamp in the middle of winter to construct large embankments, canals, and sluice gates. By cordoning off swampy terrain, planters claimed fertile land for cultivation. Associated infrastructure provided exceptional control of water. Contemporaries likened the construction of tidal fields to the building of the pyramids, a judgment historians have determined to be no exaggeration. By the early nineteenth century, huge embankments lined portions of coastal rivers all along the Carolina coast and into Georgia. The historian Philip D. Morgan has estimated that slaves moved at least 500 cubic yards of river swamp for every acre of rice field constructed. By the end of the eighteenth century, banks along the eastern branch of the Cooper River, a stretch of slightly more than ten miles, contained more than six million cubic feet of earth.[44]

---

[43] Edelson, *Plantation Enterprise in Colonial South Carolina*, 77; Coclanis, *Shadow of a Dream*, 82–83; Wood, *Black Majority*, chap. 5; Morgan, *Slave Counterpoint*, 58–61, 159–164; Edgar, *South Carolina: A History*, 139–150.

[44] Morgan, *Slave Counterpoint*, 155–157; Edelson, *Plantation Enterprise in Colonial South Carolina*, 105–113; Joyce E. Chaplin, *An Anxious Pursuit: Agricultural Innovation and Modernity in the Lower South, 1730–1815* (Chapel Hill: University of North Carolina Press for the Institute of Early American History and Culture, 1993), chap. 7. Recent debates have focused on enslaved Africans' role in the origins and development of rice culture. See Judith Carney, *Black Rice: The African Origins of Rice Cultivation in the Americas* (Cambridge, MA: Harvard University Press, 2001); Judith Carney, "Landscapes of Technology Transfer: Rice Cultivation and African Continuities," *Technology and Culture* 37, no. 1 (Jan. 1996): 5–35; Judith Carney, "The African Antecedents of Uncle Ben in U.S. Rice History," *Journal of Historical Geography* 29, no. 1 (2003): 1–21; David Eltis, Philip Morgan, and David Richardson, "Agency and Diaspora in Atlantic History: Reassessing the African Contributions to Rice Cultivation in the Americas," *American Historical Review* 112, no. 5 (2007): 1329–1358; David Eltis, Philip Morgan, and David Richardson, "Black, Brown, or White? Color-Coding American Commercial Rice Production," *American Historical Review* 115, no. 1 (Feb. 2010): 164–171; S. Max Edelson, "Beyond 'Black Rice': Reconstructing Material and Cultural Contexts for Early Plantation Agriculture," *American Historical Review* 112, no. 5 (Feb. 2010): 125–135; Stanley B. Alpern, "Did Enslaved Africans Spark South Carolina's Eighteenth-Century Rice Boom?," in *African Ethnobotany in the Americas*, ed. Robert A. Voeks and John Rashford (New York: Springer, 2013): 35–66; Edda L. Fields-Black, "Atlantic Rice and Rice Farmers: Rising from Debate, Engaging New Sources, Methods, and Modes of Inquiry, and Asking New Questions," *Atlantic Studies* 12, no. 3 (Sept. 2015): 276–295.

In the heyday of rice production, visitors marveled at the scale of large plantations and the order planters had imposed upon the landscape. Many plantations "are of great extent, sometimes covering from one to two thousand acres," noted T. Addison Richards in 1859. "The inhabitants make up a large community of themselves alone," he added. Like other commentators, Richards observed that the number of buildings on a typical plantation gave it the appearance of "a large and busy village or town."[45] No one who visited the South Atlantic coast during the late antebellum years could overlook the confluence of slave labor, hydraulic engineering, and commercial enterprise that anchored the region's prosperity. As production declined and decay mounted, the sheer power of the lowcountry rice complex remained plainly apparent. The awe and lament expressed in turn-of-the century views simultaneously referenced past prosperity and the dramatically changed circumstances of the present.[46]

Just as the lowcountry landscape informed sportsmen about the regional past, so did their interactions with African Americans. Gullah, the "brogue" to which Whitson referred, offered an object lesson in historical demography and distinguished the coastal region from other parts of the South. Gullah developed as a pidgin language during the slave trade. Pidgin refers to any speech evolved from multiple languages that is spoken by persons for whom it is not the primary tongue. As men, women, and children kidnapped from different African nations found themselves in baracoons on the western coast of Africa and stowed in the holds of slave ships, pidginized forms of English became their principal means of communication. The slave trade distributed such languages throughout the Atlantic World, and the social conditions of early Carolina, where large numbers of Africans from a limited number of ethnic groups labored under the direction of English-speaking planters and overseers, fueled development of a distinctive form. Gullah ultimately became a fully developed Creole language – a primary tongue – for generations of enslaved Carolinians. Especially after the Revolutionary War, when social conditions stabilized and importation of Africans ended, Gullah developed the distinctive vocabulary and syntax that later

---

[45] T. Addison Richards, "The Rice Lands of the South," *Harper's New Monthly Magazine* 19, no. 114 (Nov. 1859): 730. See also King, *The Great South*, 433–437; Hugh Starnes, "The Rice-Fields of Carolina," *Southern Bivouac* 2, no. 6 (Nov. 1886): 329, 333–340.

[46] Fletcher, "In the Lowlands of South Carolina"; Deane, "An Autumn Trip to South Carolina," 52–53.

fascinated visitors to the Carolina coast and became an object of scholarly study during the early twentieth century.[47]

At the end of the nineteenth century, whites saw Gullah as denoting more than simply the strange dialect spoken by African Americans throughout coastal South Carolina and Georgia. Most saw "Gullah" as also referring to the black population of the same region, which seemed more "African" than other southern blacks. As one early student wrote, lowcountry blacks constituted a "highly specialized group" with the same "characteristics" as other Negroes but in "sharply heightened and accentuated form." Gullah Negroes, he continued,

are almost unbelievably primitive, childlike, excitedly religious, ignorant, humorous, shrewd, shiftless beyond expression, superstitious, unstable, likeable, loyal, full of a homely untutored philosophy, gifted with picturesque speech, humble, kindly, generous, good-natured, and entirely without malice.

According to the prevailing wisdom, the size and stability of the slave populations of coastal Georgia and South Carolina accounted for the strength of the Gullah language and culture. After the Civil War, continuing isolation left lowcountry blacks "untouched by civilization."[48] Contemporaries saw a primitive people free of "modern" influences. As sportsmen's familiarity with the lowcountry grew, knowledge of "Gullah negroes" became central to their views of the region.

How northern sportsmen learned about Gullah and what they knew about its origins is largely a matter of speculation. At the time, only a few avocational linguists and anthropologists had published essays on the subject, so sportsmen's knowledge likely came from firsthand observation and information provided by lowcountry whites.[49] Whatever the

[47] Wood, *Black Majority*, chap. 6. See also Margaret Washington Creel, *"A Peculiar People": Slave Religion and Community Culture among the Gullahs* (New York: New York University Press, 1988), 15–25; William S. Pollitzer, *The Gullah People and Their African Heritage* (Athens: University of Georgia Press, 1999); Lorenzo Dow Turner, *Africanisms in the Gullah Dialect* (Chicago: University of Chicago Press, 1949); Michael Montgomery, ed., *The Crucible of Carolina: Essays in the Development of Gullah Language and Culture* (Athens: University of Georgia Press, 1994); Morgan, *Slave Counterpoint*, 567–571; Joyner, *Down by the Riverside*, chap. 7.

[48] Reed Smith, *Gullah: Dedicated to the Memory of Ambrose E. Gonzales* (Columbia: University of South Carolina, 1926), 7–21 (quotations on 11).

[49] On early interest in Gullah, see Charles Colcock Jones, *Negro Myths from the Georgia Coast Told in the Vernacular* (Boston: Houghton, Mifflin, and Co., 1888); A. H. M. Christensen, *Afro-American Folk Lore: Told Round Cabin Fires on the Sea Islands of South Carolina* (Boston: J. G. Cupples Co., 1892); John Bennett, "Gullah: A Negro Patois," *South Atlantic Quarterly* 7, no. 4 (Oct. 1908): 332–347; John Bennett, "Gullah: A Negro Patois, Part II," *South Atlantic Quarterly* 8, no. 1 (Jan. 1909): 39–52.

case, sportsmen's interactions with lowcountry blacks fostered a sense of exoticism and difference while foregrounding the social and environmental conditions of the rural lowcountry. Hearing Gullah spoken and observing lowcountry blacks called attention to differences within the South, the history of the lowcountry, and qualities that made the region distinctive. No matter what their impressions, sportsmen could not help but consider the conditions responsible for the numbers of dark-complected, seemingly primitive people living on old plantations throughout the countryside. Strange, curious, and quaintly picturesque, lowcountry blacks figured among the qualities that set the coastal region apart.

The lowcountry, then, offered more than plentiful wildlife and undeveloped land. More than simply the basic resources needed for sportsmen's activities, the lowcountry afforded uncommon possibilities. Exactly what those possibilities might yield and whether sportsmen and other visitors would show more than passing interest in them remained uncertain. Familiarity marked a starting point. What came next depended on factors in and outside the region, some related to sportsmen's activities, some not.

RETREATS AND PRESERVES

Alongside northerners' hunting clubs, sportsmen from outside the region also bought and leased lands individually, in pairs, and in small groups. These efforts created holdings that contemporaries referred to as hunting retreats and preserves. By providing access to wildlife and essential infrastructure, they served roughly the same purpose as hunting clubs. In size, northern-owned retreats and preserves varied from a few hundred acres to as many as ten thousand or more. Thus, some approximated the size of small clubs, but most occupied less territory. Conditions varied. Some retreats and preserves supplied hunting comparable to the best-managed clubs, but most fell into a lesser category. Maintaining large acreages for recreation required resources, and clubs generally achieved greater success than individual owners. Still, retreats and preserves usually offered good-quality hunting. Large game populations and limited competition from market and subsistence hunters generally yielded favorable conditions.

Northern-owned retreats and preserves distinguished themselves partly through the buildings that owners used for lodgings, sheltering dogs and horses, and storage. In some cases, northerners acquired old plantations where buildings remained standing. Sportsmen tended to rehabilitate such structures in rudimentary fashion. They used planter's dwellings as residences and outbuildings as stables, kennels, and storehouses. In most cases, northerners put limited effort into renewal. Making buildings weathertight and comfortable for short stays satisfied their needs. In other instances, northerners acquired tracts without buildings or where surviving structures could not be easily repaired. At these locations they generally erected new buildings. Virtually all displayed rustic styling of one kind or another. Shingle-style lodges proved especially popular, and sportsmen also built small cottages and bungalows. These buildings broke sharply with regional traditions. Whereas plantation architecture had traditionally exhibited strong classical influences, northerners' buildings took cues from contemporary trends in domestic design. They reflected the activities that northerners came to the lowcountry to enjoy.[50]

Northerners began buying and renting land for retreats soon after the founding of the Pineland Club. In 1889, W. P. Clyde, owner of the Clyde Steamship Company, began buying land on Hilton Head Island. Through purchases of small tracts, mostly at prices of a dollar or two per acre, he assembled a 9,000-acre retreat. Clyde used Honey Horn Plantation, the only plantation on the island with an extant planter's dwelling, as his residence.[51] In 1891, J. Donald Cameron, a US senator from Pennsylvania and former Grant administration official, bought Coffin Point, a 298-acre estate on St. Helena Island. He later enlarged it to nearly a thousand acres.[52] Also in 1891, Harry Hollins, one of the cofounders of the Pineland Club, purchased Good Hope Plantation, a tract immediately adjacent to the club's property. He built a group of "log cabin camp buildings" that he and his friends used while hunting in the area.[53]

During the first decade of the twentieth century, northerners continued acquiring similar properties and using them in more or less the

---

[50] Mark Gelernter, *A History of American Architecture: Buildings in their Cultural and Technological Context* (Hanover, NH: University Press of New England, 1999), 181; Leland M. Roth, *Shingle Styles: Innovation and Tradition in American Architecture, 1874 to 1982* (New York: Henry N. Abrams, 1999), 9, 12.

[51] Virginia C. Holmgren, *Hilton Head: A Sea Island Chronicle* (Hilton Head Island, SC: Hilton Head Publishing Co., 1959), 119–120; Michael N. Danielson and Patricia R. F. Danielson, *Profits and Politics in Paradise: The Development of Hilton Head Island* (Columbia: University of South Carolina Press, 1995), 11; Cuthbert and Hoffius, eds., *Northern Money, Southern Land,* 77.

[52] Cuthbert and Hoffius, eds., *Northern Money, Southern Land,* 41–48.

[53] Martin, "Pineland, Mother of Hunting Clubs"; Grace Fox Perry, *Moving Finger of Jasper* ([Ridgeland, SC?]: n.p., 1962), 148–150; "Real Estate Change Hands," *Beaufort Gazette,* Apr. 7, 1927, 1.

same way – as private hunting retreats. At the same time, others began charting a different course. The shift took place quietly and went largely unnoticed. Rather than enacting dramatic changes, a few northerners introduced variations on established themes. Still, their efforts marked the beginning of a new and important trend. It laid the groundwork for what would ultimately become an entirely new sphere of activity and a new variety of recreational landholdings.

Several estates offer illustrative examples. In 1906, Isaac Emerson, a Baltimore pharmaceutical manufacturer, acquired Prospect Hill Plantation on the Waccamaw River. Prospect Hill possessed one of largest and most refined antebellum mansions in Georgetown County. Emerson renovated and enlarged the main house and began developing gardens on a broad slope leading away from the façade. Within a few years he had created the most elaborate estate on the Carolina coast.[54] Immediately adjacent, Bernard M. Baruch, a wealthy financier with South Carolina roots, acquired three different parcels totaling about 14,500 acres. These became an estate he called "Hobcaw Barony" after the colonial land grant that had encompassed the lower end of the Waccamaw Neck. Baruch used the house at Friendfield Plantation, a Queen Anne–style dwelling built about 1890, as his residence.[55] On the western branch of the Cooper River, Benjamin Kittredge and his wife Elizabeth purchased Dean Hall Plantation.[56] Far to the south in Beaufort County, Robert H. McCurdy and his wife assumed ownership of Tomotley Plantation, a 5,600-tract on the Pocotaglio River (see Figure 1.2). They built a rambling bungalow amid a complex of nearly twenty surviving outbuildings and began

[54] Chalmers S. Murray, "Arcadia, Where Lafayette Stopped," *News and Courier*, June 21, 1931, B5; Alberta Morel Lachicotte, *Georgetown Rice Plantations* (Columbia: The State Printing Co., 1955), 18–25; Suzanne Cameron Linder and Marta Leslie Thacker, *Historical Atlas of the Rice Plantations of Georgetown County and the Santee River* (Columbia: South Carolina Department of Archives and History for the Historic Ricefields Association, Inc., 2001), 57–60.

[55] Hobcaw Barony, National Register of Historic Places Nomination, South Carolina Department of Archives and History, Columbia, SC; Brockington, *Plantation between the Waters*, 40. See also Margaret L. Coit, *Mr. Baruch* (Boston: Houghton Mifflin, 1957), chap. 12; James Grant, *Bernard M. Baruch: The Adventures of a Wall Street Legend* (New York: Simon and Schuster, 1983), chap. 6. Baruch purchased the lands that became the core of Hobcaw Barony in 1905–1907.

[56] Chlotilde R. Martin, "The Cypress Gardens, in Berkeley," *News and Courier*, undated newspaper clipping, Dean Hall Plantation Vertical File, South Carolina Historical Society, Charleston, SC (hereafter SCHS).

FIGURE 1.2 Main house at Tomotley Plantation, circa 1930. Few of the houses northerners built before World War I exhibited significant pretention. Most followed popular trends in domestic design. The bungalow that Robert McCurdy and his wife erected at Tomotley Plantation is a good example. Although unusually large, it would have fit comfortably in many upscale residential neighborhoods of the era.
From John R. Todd and Francis M. Hutson, *Prince Williams Parish and Plantations* (Richmond: Garrett and Massie, 1935), 128.

spending their winters hunting, fishing, and boating on the waters upstream from Beaufort.[57]

These men and women set themselves apart in several ways. Unlike Clyde, Cameron, and Hollins, they used their plantations as more than just hunting retreats. Although hunting remained the focus of their activities, they also engaged in other pursuits, including horseback riding, fishing, boating, and carriage rides. Thus, they made their estates venues for multiple forms of recreation.[58] Second, new social arrangements

[57] Chlotilde R. Martin, "Tomotley, Brewton Plantation, Bindon, Castle Hill – Beaufort Restorations," *News and Courier*, Nov. 23, 1930, A12.
[58] This discussion is based on analysis of the several plantations named here and others that northerners acquired before World War I (as identified and discussed in greater detail in

appeared. These estates not only differed in purpose, they also possessed a different social profile. Virtually to a rule, hunting clubs excluded women. Clubs provided middle- and upper-class men with opportunities to relax, socialize, and recreate in settings without women and children, away from the constraining influences of the home. Privately held retreats and preserves served similar roles. Like clubs, they tended to be homosocial spaces.[59]

By contrast, old plantations adapted for new uses quickly became heterosocial domains. Men and women socialized together and recreated in similar fashion. Owners brought their wives and children with them on at least some visits, and entertaining in mixed company became common. At times, an exclusively male atmosphere prevailed. Owners sometimes hosted friends and business associates in small groups and took to the field in the same manner. In general, however, heterosociality rapidly became the norm. At a minimum, these estates offered greater flexibility. Most became settings for more inclusive forms of socializing and recreation.

The advent of heterosocial leisure and longer stays introduced new practical and aesthetic demands. Rustic hunting lodges and old plantation houses satisfied dedicated sportsmen who came to the lowcountry for a few days at a time. "Roughing it," after all, formed part of the experience of manly sport. Women and children established new imperatives. Northerners moved to secure more comfortable and commodious dwellings, to install modern amenities, and to beautify their surroundings. Owners put up new paint and developed landscaping

Chapter 2). Source limitations make it difficult to determine exactly when heterosocial activity, new modes of recreation, and substantial material changes arrived. Baruch renovated and enlarged the main house and began recreating with his family at Hobcaw Barony immediately. The Kittredges did the same. Exactly when the McCurdys made substantial improvements is less clear. Some generalization is therefore unavoidable. Nonetheless, the available evidence shows a shift in behavior began around 1910, with consequences for recreational activity and the material environments of select plantations. On Baruch's use of Hobcaw, see especially Brockington, *Plantation between the Waters*. The Kittredges' activities during the period are apparent from correspondence in the Kittredge Family Papers, SCHS.

[59] Most hunting clubs excluded women, even as guests. Of the several clubs northerners established, only the Okeetee Club allowed women. See Martin, "Pineland, Mother of Hunting Clubs," B6. The Oakland Club allowed members to bring women as guests. See Oakland Club, *Oakland Club, St. Stephens P.O.*, 22. On the homosocial character of hunting clubs, see "Heard at the Clubs," *New York Times Sunday Magazine*, Feb. 22, 1903, 7; Dunlap, *Saving America's Wildlife*, 13.

reminiscent of suburban houses. In short, they began striving for greater comfort and refinement.[60]

Around 1910, then, a few northerners began a trend toward conscious elaboration and embellishment. It developed in combination with new forms of recreation, new social arrangements, and new ambitions. Northerners proceeded tentatively, hesitantly, and without clear direction. Acting principally with practical considerations in mind, they sought greater comfort and accommodation of recreational practices. Neither of these goals required dramatic alterations. At best they encouraged modest steps beyond the practices of clubs, retreats, and preserves. Nonetheless, rehabilitation of old plantations established a new realm of activity and set important precedents. Its full significance would not become clear for at least a decade, yet when it did, its role in creating a broader, more varied, more inclusive sphere of activity would be plainly apparent.

NORTHERN SPORTSMEN AND THE LIMITS OF GENTLEMANLY RECREATION, 1880–1915

The clubs, retreats, and preserves that northern sportsmen established marked a crucial step in the lowcountry's rise as a recreational destination. Between the mid-1880s and mid-1910s, sportsmen explored the region's recreational potential and secured large tracts of land for their use. They introduced new managerial practices and administrative arrangements to a region where large acreages had never before served exclusively as recreational domains. In later years, they continued to expand their activities and develop landholdings for recreation. None of the northerners who traveled to the lowcountry in the years around the beginning of the twentieth century acted with intentions of establishing a foundation for later growth. In many ways, however, they did exactly that. Northerners' efforts to secure access to wildlife became a start toward other forms of activity.

Northerners did not introduce recreational hunting to the lowcountry. White southerners had hunted for sport since the colonial era and continued to do so as sportsmen from outside the region began acquiring

---

[60] Immediately upon purchasing the 10,000-acre tract that formed the core of Hobcaw Barony, Baruch, for example, modernized and enlarged the house at Friendfield Plantation by installing modern bathrooms and electricity and adding a sun porch. See Brockington, *Plantation between the Waters*, 52. The Kittredges' rehabilitation of Dean Hall also presumed familial occupancy from the beginning.

land. Members of the two groups shared a great deal. In outlook, social standing, and the style and manner in which they hunted, northern and southern sportsmen exhibited strong similarities. The crucial difference lay in the circumstances that brought northerners to the lowcountry and their immediate actions. Northerners' interest in the region grew out of wildlife scarcity close to home. Buying and leasing land secured access to wildlife before populations dwindled. By contrast, lowcountry sportsmen had yet to see reserving land for recreation as necessary. Lowcountry sportsmen had formed hunting clubs since the Revolutionary era, but they served mainly social purposes and did not buy and lease land for members' use. Northerners' clubs reflected late nineteenth-century struggles for wildlife. During the early twentieth century, as declines in wildlife populations became more apparent, lowcountry sportsmen also formed clubs that held and managed land. Before then, however, they felt no such need. As members of the planter class and merchants and professionals with close ties to large-scale landowners, lowcountry sportsmen had ample access to wildlife.[61]

Northern and lowcountry sportsmen's interests became more closely aligned over time. As wildlife populations dwindled, both groups took steps to limit African American hunting. In 1915, the South Carolina General Assembly adopted legislation requiring hunters to purchase licenses and obtain written permission from landowners before hunting or fishing on their lands. The law also gave game wardens the power to enforce it. The "Ziegler Bill," as the act became known, directly targeted African American hunting, in that the majority of the counties affected had large black populations. African American hunting in several coastal counties subsequently plummeted. The scale of black landownership in the lowcountry, blacks' familiarity with the landscape, and vegetation

---

[61] On early lowcountry hunting clubs, see J. H. Easterby, "The St. Thomas Hunting Club, 1785–1801: Its Rules, Excerpts from Its Minutes, and a List of Members," *South Carolina Historical and Genealogical Magazine* 46, no. 3 (July 1945): 123–131; J. H. Easterby, "The St. Thomas Hunting Club, 1785–1801 (Continued)," *South Carolina Historical and Genealogical Magazine* 46, no. 4 (Oct. 1945): 209–213; Robert Wilson, *An Address Delivered before the St. John's Hunting Club, at Indianfield Plantation, St. John's, Berkeley, July 4, 1907* (Charleston: Walker, Evans, and Cogswell Co., 1907); Francis Marion Kirk, *A History of the St. John's Hunting Club* (N.p.: St. John's Hunting Club, 1950). On clubs formed by white southerners during the early twentieth century, see Linder, *Historical Atlas of the Rice Plantations of the ACE River Basin*, 244, 309–310. On southern hunting in general, see Marks, *Southern Hunting in Black and White*, chaps. 2–3; Proctor, *Bathed in Blood*; Herman, *Hunting and the American Imagination*, 21–25, 68, 149–150.

and topography in some areas militated against sportsmen's and game wardens' efforts to limit African American hunting. Moreover, by renting land to tenants, northerners' clubs allowed their black residents to take small numbers of animals for their own consumption. Preventing black hunting completely thus proved impossible. Still, the overall trend favored large landowners. Northern and lowcountry sportsmen complained continually about poachers, but illegal game-taking proved more of an annoyance than a serious threat.[62]

Other developments aided elite hunters' effort to limit access to wildlife. The advent of fence laws in lowcountry counties during the early twentieth century also curtailed hunting by landless whites and blacks, and vigorous efforts by the Audubon Society of South Carolina to promote wildlife conservation also helped. James Henry Rice, Jr., the society's secretary, spoke to farmers' societies and other groups throughout the state, imploring landowners to do their part. How much Rice's efforts accomplished is unclear. Although they encouraged landowners to conserve wildlife habitat, more stringent enforcement of game laws surely did more to limit illegal hunting. Still, Rice's efforts likely had some effect. If nothing else, his condemnations of "game hogs," "pothunters," and "vagrants" reinforced the view of sport hunters as responsible protectors of wildlife and poor whites and African Americans as the reason for declining numbers of animals.[63]

For northern sportsmen, the clubs and other holdings they established achieved important goals. By taking plantation lands out of commercial use and buying and leasing large tracts of idle and undeveloped land, sportsmen created recreational domains suited to their needs. They amassed acreages that provided unencumbered access to wildlife and pioneered land management systems that made possible ongoing use. By hiring superintendents and laborers, maintaining land and buildings, and cultivating favorable hunting conditions, sportsmen protected their investments and ensured their ability to return year after year.

Sportsmen's clubhouses, lodges, and support buildings created an extensive recreational infrastructure that made possible stays of varying length, offered comfortable accommodations, and provided direct access to game lands (see Figure 1.3). Whether sportsmen used old plantation houses or built new structures, they located themselves in the field, close

---

[62] Giltner, *Hunting and Fishing in the New South*, 138–141, 209 n. 3.
[63] Rowland and Wise, *Bridging the Sea Islands' Past and Present*, 220–221; Giltner, *Hunting and Fishing in the New South*, 150–153.

FIGURE 1.3 Santee Gun Club lodge, early twentieth century. Sportsmen favored simple, unornamented buildings for their hunting clubs, retreats, and preserves. Lodge-like buildings with rustic accents and Shingle-style buildings predominated. Both had strong associations with outdoor recreation.
From the William Cain Family Papers, South Carolina Historical Society, Charleston, SC.

to swamplands, savannahs, and forests. Immediate access to hunting grounds made their priorities clear. Initially, sportsmen showed no interest in other forms of recreation. During the first decade of the twentieth century, a few broadened the scope of their activities, but most adhered to established practices. Before the World War I era, few northerners showed interest in anything but hunting.

Sportsmen's clubs, retreats, and preserves instituted practices that slowed the decline in game populations, even if only marginally. Not only did sportsmen initiate habitat management, they adopted bag limits before required by law. Clubs specified limits as part of their basic rules, and sportsmen who owned retreats and preserves generally adhered to similar limits. For the 1916 season, for example, the Pineland Club restricted themselves to twenty-five quail and twenty woodcock per day and five turkeys each season. The Oakland Club limited members to fifteen partridges and woodcock daily and three deer and five turkeys per season. Self-discipline thus played a role in sportsmen's efforts to conserve wildlife. The effectiveness of these efforts is uncertain. Clearly, they failed to stem long-term declines, but they likely slowed the process. Probably the most important result lay in sportsmen's self-definition. By following "sporting" practices, northern sportsmen fostered alliances with white southerners and encouraged development of a conservation ethos.[64]

Arguably the single most important contribution that northerners made to the long-term development of the lowcountry lay in awareness. The sportsmen who came in these years introduced the region and its possibilities to other men like themselves. Simply by word of mouth and by bringing other sportsmen to hunt as guests, northerners played a crucial role in establishing the lowcountry as a recreational destination. The small numbers of men who traveled to the region each year gave the lowcountry a favorable reputation in elite sporting circles. Moreover, each person who hunted as a guest at a club or retreat became a potential convert. An introduction to the lowcountry's outstanding gamelands meant firsthand knowledge of a region where land could be had at low prices, where labor could be obtained easily and at modest costs, and where better hunting conditions could hardly be imagined. It meant an introduction to spectacular coastal scenery, mild weather, and a countryside where moss-draped live oaks and grand houses stirred intrigue at every turn. How many sportsmen became seriously interested in the lowcountry after hunting as guests is difficult to say, but some did. Some became members of hunting clubs themselves. Some bought retreats of their own. Some returned year after year to hunt with friends and business associates. Some bought old plantations and began turning them into comfortable estates. Over time, as northern sportsmen's interest and investments in the region grew, the lowcountry took on greater potential.[65]

[64] Martin, "Pineland, Mother of Hunting Clubs"; Martin, "Palachucola Club in Hampton"; Hemingway, "Williamsburg Boasts Many Game Preserves for Hunt"; Willett, *Game Preserves and Game of Beaufort, Colleton and Jasper Counties*, 8; addendum dated Nov. 22, 1916, included in *Articles of Agreement and Rules of the Pineland Club*, copy at Maryland Historical Society, Baltimore, MD; Oakland Club, *Oakland Club, St. Stephens P.O., Berkeley County, South Carolina*, 18–20; General Description of the Richfield Property, March 21, 1912, p. 3, folder "Annual Reports of the President and Treasurer, 1912–21," box 17, MPEDP.
[65] On men who joined clubs or purchased land after hunting in the lowcountry, see, for example, Martin, "Pineland, Mother of Hunting Clubs"; Martin, "Okeetee Club, 42,000-Acre Preserve"; "Esterville, Transformed Coastal Estate," *News and Courier*,

When viewed against the broad contours of the Second Yankee Invasion, the significance of sportsmen's early activities lies in part in what northerners did not do. Before the late 1910s, ideas about plantations mattered little. None of the large clubs saw plantations as relevant to their activities except in purely practical terms, nor did owners of retreats or preserves. Even northerners who adapted old plantations for seasonal use showed no signs of thinking about plantations imaginatively. Northerners recognized the historical importance of plantations and knew that plantation agriculture had once dominated the lowcountry, but the cultural significance of plantations appears to have elicited few thoughts. Moreover, nothing suggests that sportsmen saw some plantations as having untapped potential. Plantations throughout the region continued to operate as agricultural enterprises, others lay idle, and only traces of others remained. What the future held remained unclear. Most observers saw promise in new commercial possibilities but what enterprises, if any, could operate successfully on former rice-growing lands remained a mystery. As landowners and entrepreneurs searched for possibilities, agriculture continued to decline.[66]

For the moment, uncertainty concerning the status of old plantations mattered less than what it revealed about the scope of northerners' activities. Clubs, retreats, and preserves did not require significant intellectual labor. Simple in their conception and execution, they existed for a single purpose: a recreational pastime that upper-class hunters had enjoyed for decades. Northern sportsmen brought their brand of hunting to the lowcountry fully formed and saw no reason to modify it in any way. Clubs and retreats replicated well-established models. In terms of organization, purpose, and management, they differed little from similar institutions in the Northeast and Midwest. Moreover, although northerners amassed landholdings that included old plantations, the term "plantation" held no special significance in the context of their activities. Plantations that clubs acquired became part of larger landholdings that served principally as hunting domains. Rather than seeing plantations as holding possibilities, clubs viewed them in utilitarian terms. The same held true for owners of retreats and preserves.

Feb. 15, 1931, B7; Martin, "Gravel Hill, R. P. Huntington Home." The common spelling of "Esterville" is "Estherville." The latter is used herein, except in the notes, where the title of the *News and Courier* article is used without modification.

[66] Tuten, *Lowcountry Time and Tide*, chap. 4.

By the early 1910s, then, northern sportsmen had established recreational holdings throughout the lowcountry. In later years, their clubs would continue to grow and develop and others like them would appear on the scene. The number of retreats and preserves would also increase. At the same time, however, northerners would begin buying dilapidated plantations and exploring their possibilities. These properties would quickly become a focus for new forms of activity. Plantations undergoing adaptation would become touchstones for the imagination and venues for practices vaguely reminiscent of a romantic Old South. They would become the basis of a larger, more vibrant social scene that thrived off the lowcountry's developing role as a winter haven for wealthy, powerful people. Plantations adapted for new roles would become sites where privileged, socially prominent people enjoyed popular field sports and warm weather in idyllic surroundings. As new estates developed, so, too, would a view of the lowcountry as more than simply a recreational destination. Increasingly, people far and wide would view it as a region with an extraordinary past, rich traditions, and a distinctive heritage. In that vision lay the seeds of a region where plantations symbolized an aristocratic heritage, the possibilities of an especially alluring part of the South, and the ascendency of a new elite.

2

Creating Plantations for Sport and Leisure

Estate-Making in the Carolina Lowcountry, 1915–1940

Change came swift and fast to the lowcountry during World War I and the years that followed. After decades of decline, new patterns of activity took hold. New transportation connections, new roads and bridges, and efforts to capture the "tourist trade" marked the beginnings of new vitality. Throughout the period, the better part of the lowcountry remained a quiet expanse of old plantations, small farms, and jungle-like swamps. Yet in small and important ways, a new and dramatically different region began taking shape. Developments in the 1920s marked the beginnings of profound social and economic changes, many of which had lasting effects.

The growing popularity of Florida resorts and the land boom that swept the Sunshine State in the 1920s sparked the beginning of the transformation. The surge of real estate speculation, resort development, and population growth that made Florida the talk of the nation stirred the interest of businessmen and civic boosters all along the eastern seaboard. "The Florida boom is advertising the South," the *Georgetown Times* declared in 1925. "The coastal section from Conway, S.C., to Beaufort, S.C., is bound to be discovered," the newspaper predicted. Soon, it added, "development will take place at a rapid rate."[1] Comments of this kind exhibited a combination of optimism, hyperbole, and zealous boosterism,

---

[1] *Georgetown Times* (Georgetown, SC) quoted in "Feels Coastal Region Is Aided," *News and Courier* (Charleston, SC), Sept. 28, 1925, 2. On the Florida land boom in general, see Frederick Lewis Allen, *Only Yesterday* (New York: Harper and Brothers, 1931), chap. 11; Charleston W. Tebeau, *A History of Florida* (Coral Gables: University of Miami Press, 1971), 382–387.

but they nonetheless indicated new potential. After decades of stag-nation, new activity of almost any kind provided cause for hope. As Florida thrived and travel to the southeastern states surged, communities all along the coast looked anew at long-discussed possibilities. Ambitions quickly soared.

In Charleston, business leaders took notice of the stream of visitors that arrived every March and April, when the season at Florida resorts closed. Recognizing an untapped opportunity, the Charleston Chamber of Commerce and Mayor Thomas P. Stoney launched a campaign to "sell the city ... to the outside world." Merchants and citizens threw their support behind the effort, which grew to include new tourist hotels, beautification programs, and advertisements in major newspapers and magazines. Tourism quickly became a major industry. By 1929, the city attracted more than 47,000 visitors annually.[2]

The tourism campaign took place against a backdrop of shifting eco-nomic fortunes, continued weakness in agriculture, and surging invest-ments from afar. During World War I, Charleston experienced significant growth. The expansion of the Charleston Navy Yard provided the better part of the stimulus. When war broke out in Europe in 1914, the yard employed 1,240 civilian personnel and had an annual payroll of almost $884,000. After the United States entered the war in April 1917, the yard became the headquarters of the Sixth Naval District. The number of US Navy personnel stationed in Charleston grew dramatically and civilian employment rose to 5,000. By late 1918, the combined annual payroll for military and civilian personnel exceeded $9,000,000. Laborers rushed to the city in search of jobs. Housing ran scarce and city officials struggled to cope with increased demand for municipal services. While war raged in Europe, Charleston's prosperity seemed secure.[3]

Federal dollars continued to pour into the city through 1921. New factories, a new port terminal, increased freight traffic, and new suburban

---

[2] Walter J. Fraser, Jr., *Charleston! Charleston!: The History of a Southern City* (Columbia: University of South Carolina Press, 1989), 373–374; "Sell Charleston to Charlestonians Plea of Speakers," *News and Courier*, Nov. 20, 1924, 2. On efforts to generate favorable publicity, see, for example, "Much Publicity for Charleston," *News and Courier*, Jan. 4, 1926, 7.

[3] Fraser, *Charleston! Charleston!*, 359–361; Fritz P. Hamer, *Charleston Reborn: A Southern City, Its Navy Yard, and World War II* (Charleston: History Press, 2005); John Hammond Moore, "Charleston in World War I: Seeds of Change," *South Carolina Historical Magazine* 86, no. 1 (Jan. 1985): 39–49.

housing tracts buoyed the local economy. In the long run, however, the economic boom proved impossible to sustain. As operations at the Navy Yard returned to a peacetime footing, the local economy began to lag. New downturns in agriculture added to the city's woes. Municipal leaders found themselves facing a familiar problem: how to revive a chronically moribund economy.[4]

Historians have often portrayed the tourism campaign as a response to the postwar slowdown. According to this interpretation, civic leaders and business interests turned to tourism as a last-ditch effort to pull the city out of a steadily deepening decline. As Stephanie E. Yuhl has argued, the World War I–era boom "did not significantly alter the city's dormant economy." Civic leaders turned to tourism, she writes, in a bid to "rouse Charleston from its complacent sleep."[5] Although some officials spoke of tourism as though no other options existed and economic conditions inspired a sense of urgency, Charlestonians hardly acted out of desperation. Throughout the 1920s, a steady drumbeat of new development provided cause for optimism.

"The development of the coast is at hand," proclaimed the *News and Courier* in February 1926. New resorts and tourist hotels, real estate sales, and extensive highway construction all pointed to greater prosperity.[6] The *Georgetown Times* predicted that the Carolina coast would "soon rival Florida." Charleston, it contended, would "double in size and population."[7] Other commentators offered more restrained assessments

---

[4] Fraser, *Charleston! Charleston!*, 370.

[5] Stephanie E. Yuhl, *A Golden Haze of Memory: The Making of Historic Charleston* (Chapel Hill: University of North Carolina Press, 2005), 4.

[6] "Looking to the Coast" (editorial), *News and Courier*, Feb. 26, 1926, 4. On bridge and highway construction, see especially "Our Highway Progress" (editorial), *News and Courier*, Jan. 2, 1926, 4; "Coastal Highway Work Expected to Begin Part of Huge Program," *News and Courier*, May 17, 1926, 1; "$7,000,000 Road Project Started," *News and Courier*, Apr. 26, 1927, 12. For developments in the Georgetown area, see George C. Rogers, Jr., *The History of Georgetown County, South Carolina* (Columbia: University of South Carolina Press, 1970), 506–507; "Much Road Work for Georgetown," *News and Courier*, Jan. 3, 1925, 2; "Yauhannah Bridge about Completed," *News and Courier*, Nov. 11, 1926, 3. On resort development, see "Modern Hotel and Golf Course for the Isle of Palms," *News and Courier*, Sept. 20, 1925, 2; "Isle of Palms Hotel Started," *News and Courier*, Apr. 24, 1927, 4; "Folly Beach Control Acquired by Citizens and Southern Company," *News and Courier*, Sept. 17, 1925, 1, 4; "Folly Beach Plans Draw Attention," *News and Courier*, Oct. 4, 1925, 24; Walter B. Edgar, *South Carolina: A History* (Columbia: University of South Carolina Press, 1998), 493.

[7] *Georgetown Times* quoted in "Feels Coastal Region Is Aided."

but nonetheless saw similar trends. As the Columbia *State* opined, "The regeneration of the 'Low Country' ... has begun."[8]

For well-to-do sporting enthusiasts, the rush of development created new possibilities. Improved transportation connections made travel to the region easier than before. Completion of the Atlantic Coast Highway, new bridges across the Savannah, the Cooper, the Combahee, and the Pee Dee rivers, and faster, more reliable train service facilitated access to the lowcountry.[9] Road construction within the region made remote areas more accessible while heightening contrasts between long-isolated districts and the "modern" world.[10] Meanwhile, popular imagery infused northerners' activities with meaning. As people near and far increasingly recognized "the lowcountry" as a distinctive part of the South, they viewed northerners against the backdrop of history. In the eyes of many, sportsmen and sportswomen showed strong similarities to the antebellum planter class. Although no one overlooked the differences between the two groups, the parallels seemed too striking to ignore. As lowcountry people and others watched the "winter colonists" with interest, they saw the newcomers stepping into familiar roles.

Amid these circumstances, northerners' activities moved into new territory. Efforts to rehabilitate old plantations grew modestly during World War I and the early 1920s. Then, in the middle of the decade, they suddenly exploded. Wealthy sportsmen and sportswomen began creating large and opulent estates with clear resolve. Their efforts became bolder, more surefooted, and more ambitious. Simultaneously, northerners' activities took on new dimensions. With larger numbers of sportsmen and sportswomen wintering in the lowcountry and the advent of a lively social scene, conspicuous display became the norm. Estate-owners took steps to

---

[8] *The State* (Columbia, SC) quoted in "Regeneration of the Low Country," *News and Courier*, Sept. 28, 1925, 4.

[9] Fraser, *Charleston! Charleston!*, 368, 370, 376. On bridge construction, see "The Friendship Tour" (editorial), *News and Courier*, Sept. 8, 1925, 4; "Charleston Is Saying, Howdy'e" (advertisement), *News and Courier*, Sept. 8, 1925, 7; "Many Present at Big Celebration," *News and Courier*, Oct. 8, 1925, 10; "Bridge Opening Fete," *News and Courier*, March 13, 1927, 2. On bridge and highway construction near Beaufort, see Lawrence S. Rowland and Stephen R. Wise, *The History of Beaufort County, South Carolina*, vol. 3: *Bridging the Sea Islands' Past and Present, 1893–2006* (Columbia: University of South Carolina Press, 2015), 225–237. On railroad improvements, see "Seaboard to Open Short Line Soon," *News and Courier*, Dec. 30, 1924, 10; "Better Service by Coast Line," *News and Courier*, Oct. 27, 1927, 11.

[10] An especially good example is "Edisto Park Site Is like a Jungle," *News and Courier*, Dec. 2, 1934, B5.

set themselves apart. Social and recreational activities became exclusive
undertakings, much like comparable gatherings in the North. Meanwhile,
northerners' "plantations" became central to portrayals of the region. As
the lowcountry became a focus of national attention and tourist travel
swelled, newspapers and magazines highlighted handsome new sporting
estates. Northerners' activities increasingly symbolized the region's resurgence and growing prominence.

MAKING OLD PLANTATIONS ANEW

During World War I, the remaking of old plantations took on new significance. Although many sportsmen continued hunting at preserves and
retreats, others turned their attention to rehabilitating old plantations.
Men and women who had already created large estates continued making
improvements, and others followed in their footsteps. By the early 1920s,
estate-making assumed a modest degree of coherence. Northerners made
no attempt to coordinate their efforts and whether they paid much attention to each other's estates is unclear. Nonetheless, a discernible pattern
began to emerge and, with it, a distinct set of practices. In turn, boundaries between sporting estates and domains devoted exclusively to hunting
grew more pronounced.

Several estates illustrate the trend. At Arcadia, Isaac Emerson continued developing the grounds surrounding his mansion and adding
outbuildings in the rear. During the 1910s, he added new barns, a kennel,
and portions of a garden that would ultimately stretch away from the
main house in an impressive sequence of hedgerows and flowerbeds. By
the end of the decade he created one of the most elaborate estates in the
lowcountry. Baruch, Emerson's neighbor, developed his estate in similar
fashion. He moved at a slower pace and showed less interest in aesthetic
pretention but nonetheless continued making improvements. The most
significant included landscaping of the grounds around the main house
and the dock where visitors alighted. Developing these spaces provided
visual cues for guests and the beginnings of a stronger organizational
scheme. Less is known about the activities of the McCurdys and the
Kittredges. The extent to which they continued "improving" their plantations is unclear. Certainly they continued upkeep and maintenance, and
they may also have developed select areas. Firm evidence, however, is
elusive. Even so, their efforts placed them in a special category. By
rehabilitating and adapting old plantations for new uses, both couples
took a course different from that followed by most of their peers.

Other northerners followed in a similar vein. Throughout the 1910s, sportsmen and sportswomen continued buying old plantations and remaking them for seasonal residency and recreation. Georgetown County quickly emerged as the major center of activity. Between 1910 and the early 1920s, sportsmen acquired at least seven plantations along the Black, the Pee Dee, and the Santee rivers. In 1911, Dudley L. Pickman of Boston purchased Springfield Plantation, a 4,600-acre estate on the Pee Dee. The following year Charles W. Tuttle acquired Mansfield Plantation on the Black River. In 1916, Dr. Emerson W. Hitchcock bought an adjoining plantation, Beneventum. Shortly thereafter, Jacquelin S. Holliday, the head of a steel and iron wholesaling firm in Indianapolis, and his wife, Florence, purchased Nightingale Hall, an estate of about 1,200 acres on the Pee Dee. In 1918, John A. Miller, president of the Pennsylvania-Dixie Cement Company, purchased Estherville Plantation on the edge of Winyah Bay.[11] By the end of the decade, these men and women and a handful of others had joined Baruch and Emerson in the countryside around Georgetown.

Northerners also acquired plantations farther south. On the Cooper River in Berkeley County, Clarence E. Chapman, a New York stockbroker, and his wife, Adelaide, purchased Mulberry Plantation in 1915. Two years later, Frederick A. Dallett, heir to a Philadelphia shipping fortune, and his wife, Katherine, purchased South Mulberry, a historically related tract lying immediately adjacent. These estates took their place alongside the Kittredges' Dean Hall and a handful of other properties northerners owned along the river.[12] In Beaufort County, R. J. Turnbull, a New York attorney, became the owner of Twickenham Plantation.[13] In newly created Jasper County, Julian B. Clark and his wife, Louise,

[11] Chalmers S. Murray, "Springfield on the Peedee," *News and Courier*, July 27, 1931, 3; "Mansfield, Ancestral Parker Home," *News and Courier*, March 1, 1931, A9; "Polo Player Buys Georgetown Estate," *News and Courier*, May 4, 1931, 3; Chalmers S. Murray, "Holliday Winters in Peedee," *News and Courier*, Aug. 30, 1931, B11; "Esterville, Transformed Coastal Estate," *News and Courier*, Feb. 15, 1931, B7. See also Alberta Morel Lachicotte, *Georgetown Rice Plantations* (Columbia: The State Printing Company, 1955), 82–84, 88, 155–158.

[12] Chlotilde R. Martin, "Mulberry Castle, Built in Indian Days," *News and Courier*, July 26, 1931, A5. Collectively, these plantations totaled more than 14,000 acres.

[13] Suzanne Cameron Linder, *Historical Atlas of the Rice Plantations of the Ace River Basin – 1860* (Columbia: South Carolina Department of Archives and History Foundation, Ducks Unlimited, and the Nature Conservancy, 1995), 582–589; Robert B. Cuthbert and Stephen G. Hoffius, eds., *Northern Money, Southern Land: The Lowcountry Plantation Sketches of Chlotilde R. Martin* (Columbia: University of South Carolina Press, 2009), 128.

bought Spring Hill Plantation, a 3,500-acre tract lying between the Coosawhatchie River and Bees Creek.[14]

These buyers approached the development of their estates in distinctive fashion. Not all moved quickly to create well-developed settings for leisure and recreation. In fact, some did relatively little beside tend to basic necessities. Compared with people who had acquired land solely for hunting, however, they showed greater concern for aesthetics and sought to conserve material remains from earlier periods. That is, these northerners showed reverence for extant buildings and landscape features. Attention to surviving elements formed an important part of their efforts. In addition, these buyers showed concern for style and spatial organization. They developed domestic spaces with modest levels of refinement and strong attention to practical needs. Some owners also erected new buildings with obvious regard for stylistic precedent. In such cases, they erected buildings that emulated those from earlier periods or exhibited plain styling that neither clashed nor contrasted with surviving structures.

Three distinct approaches developed. One sought to make as few changes as possible. Charles Tuttle took this course at Mansfield. When he acquired the plantation in 1912, virtually all of the buildings from the heyday of rice production remained. The main dwelling, a tripartite structure with a mid-eighteenth-century core, had changed little since about 1850. Nearby stood a building complex that included a rice mill, a winnowing house, and a stable. Most of the slave street also survived and a long avenue of live oaks led to the main house. Save for weathering and plant growth, the buildings and grounds remained largely unaltered.[15]

Tuttle deliberately avoided making changes. From the beginning, he aimed to leave signs of age and decay undisturbed. He removed detritus from the grounds, kept them tidy and well maintained, and made essential repairs to the house and outbuildings. Otherwise, he did little. He purposefully left building exteriors weathered and worn, intent on "preserving" what he called the "old atmosphere." A visitor who saw Mansfield during Tuttle's ownership credited him with "keeping the establishment intact." Tuttle, he observed, "kept the entire place in excellent condition. The lawn is carefully tended, the paths and roadways kept as clear as a race track, while the fences and outhouses are spotless

---

[14] Chlotilde R. Martin, "Spring Hill Plantation Pays," *News and Courier*, Dec. 29, 1930, 3.
[15] "Mansfield, Ancestral Parker Home." See also Mansfield Plantation, National Register of Historic Places Nomination Form, South Carolina Department of Archives and History, Columbia, SC.

in their coat of whitewash." A visit to Mansfield, the visitor concluded, "carries one back to the days when southerners loved their plantation homes with a passionate devotion."[16]

Tuttle was not unique. Hitchcock, his neighbor, showed similar sensibilities. After acquiring Beneventum Plantation in 1916, Hitchcock kept the planter's residence and the surrounding grounds more or less as he found them. At Springfield, Dudley Pickman employed the same strategy. "Wishing to maintain the old atmosphere of the place," he purposefully avoided alterations. He maintained gardens developed by a previous owner and left the exterior of the main house, an "old-fashioned Colonial structure," unchanged. He did, however, remodel the interior. Even though Pickman prized the "old atmosphere," he saw no reason to forgo modern amenities.[17]

A second approach centered on restoration of early houses. Clarence and Adelaide Chapman supplied the pioneering example at Mulberry. When they acquired this plantation in 1915, it stood in disrepair. The main house had been unoccupied for a decade or more, and the grounds, outbuildings, and tidal fields lay in ruin. The Chapmans commissioned Charles Brendon, a New York architect, to rehabilitate and restore the main dwelling. Brendon carried out a careful restoration that removed Victorian-era additions, refurbished the structure, and added a handful of decorative elements. His efforts supplied the Chapmans with an elegant residence. At the same time, laborers rehabilitated the surrounding grounds and a handful of outbuildings. Although the Chapmans made no immediate efforts to develop gardens or formal landscaping, by keeping the lawn in front of the dwelling and an adjoining row of trees well groomed, they created a more refined setting for the main house than had historically existed.[18]

The Chapmans' treatment of Mulberry marked the first self-conscious restoration of a lowcountry plantation dwelling. It set a precedent that would not be rivaled, or even emulated, for roughly a decade. Not until the late 1920s would other sportsmen and sportswomen show similar care and reverence in rehabilitating early buildings. That the Chapmans did not immediately inspire other buyers to act in similar fashion in no way detracts from the significance of their accomplishment. Their actions

---

[16] "Mansfield, Ancestral Parker Home."   [17] Murray, "Springfield on the Peedee."
[18] The Chapmans' remaking of Mulberry Plantation is the subject of Chapter 4. It describes the restoration of the main house and the other changes the Chapmans made in detail.

revealed new sensibilities about the lowcountry and its plantations. By purposefully seeking to return the house at Mulberry to the appearance of an earlier time, the Chapmans identified it as historically valuable. They regarded the house as architecturally and historically important and therefore deserving of specialized treatment. This marked a significant change. For decades, visitors to the lowcountry and lowcountry people had referred to small numbers of plantation houses as "relics" and "ancient." These terms denoted age and anachronism but fell short of explicitly assigning value. The Chapmans did. They showed that at least some people viewed plantations as belonging to history – as tangible remains of the past – and, in turn, possessing cultural significance.[19]

The third approach northerners used proved the most common. It combined rehabilitation with elaboration and embellishment. As an open-ended means of adapting old plantations for new purposes, this approach proved useful where little remained from earlier periods, where northerners found extant remains lacking, or where owners simply saw substantial alterations as desirable. Although ostensibly beset with few constraints, it tended to proceed within predictable aesthetic boundaries. Northerners avoided radical shifts in style. They generally seized upon stylistic precedents and adopted them as guides for further efforts. Northerners referenced extant remains by styling new buildings and landscape features in similar fashion. Material changes thus took on the appearance of following an established course – of continuing within a chosen aesthetic idiom.

Several examples illustrate the basic contours of this approach while highlighting its flexibility. Emerson's rehabilitation of Prospect Hill stands among the earliest and most elaborate. By renovating the main house without making significant exterior changes, building additions with matching styling, and creating extensive gardens, Emerson enlarged, embellished, and adapted the domestic space of his estate without substantially modifying its aesthetic character. His "improvements" echoed the Federal architecture of the house and gardens. Although Emerson dramatically enlarged the building – he erected a ballroom addition on one side and a gymnasium on the other – and poured effort into developing the grounds, all of the improvements followed more or less within the same stylistic mode. The gardens grew larger but maintained consistency in terms of layout, plantings, and organization. Emerson treated surviving

---

[19] On perceptions of the past and gradations therein, see especially David Lowenthal, *The Past Is a Foreign Country* (Cambridge: Cambridge University Press, 1985), 238–243.

outbuildings in similar fashion. By rehabilitating and adapting existing structures and adding new buildings with similar styling, he developed the complex in a manner that maintained aesthetic congruence. Rather than making dramatic alterations, Emerson elaborated and embellished, fusing old and new together in harmonious fashion.[20]

At Annandale Plantation, Dr. J. B. Reeves and his wife, Susan, proceeded along similar lines. After buying the plantation from a hunting club that had owned it since the 1880s, the couple remodeled the main house and developed large gardens. The renovation removed wings added in the late nineteenth century, rearranged the layout of the interior, and installed modern bathrooms. Although not as carefully executed as Brendon's work at Mulberry, the project reclaimed the structure's mid-nineteenth-century appearance. For the surrounding grounds, the Reeveses put in boxwood hedges, planted a variety of shrubs, and laid out brick-lined walkways. Overall, the new owners created a handsome, well-appointed estate. They showed respect and care for select remains while enhancing the setting of the main house. When Charleston *News and Courier* reporter C. S. Murray visited in 1931, he judged Annandale to speak "eloquently of the days when the gentry of Georgetown county lived in baronial splendor."[21]

At Estherville Plantation, John A. Miller began with a different set of circumstances but achieved similar results. When Miller purchased the plantation in 1918, a bungalow built fifteen years earlier served as the main residence. Miller built a larger, more elaborate building around the structure, effectively making it the core of a new residence. The resulting dwelling stood two stories tall beneath a shallow roof. A pair of tall columns below a full entablature framed the main door. In the rear, sunrooms and porches offered opportunities for enjoying the natural surroundings. Palmettos and small shrubs stood close to the building. Miller further developed the landscape by creating extensive gardens and a fifty-acre "deer park." This latter feature, an open expanse of lawn and tall grasses with trees and shrubs set at irregular intervals, attracted large numbers of deer and other wildlife. Inspired by examples at English

[20] Chalmers S. Murray, "Arcadia, Where Lafayette Stopped," *News and Courier*, June 21, 1931, B5; Lachicotte, *Georgetown Rice Plantations*, 18–25; Suzanne Cameron Linder and Marta Leslie Thacker, *Historical Atlas of the Rice Plantations of Georgetown County and the Santee River* (Columbia: South Carolina Department of Archives and History for the Historic Ricefields Association, 2001), 75–78.

[21] Chalmers S. Murray, "Annandale, Secluded in Georgetown," *News and Courier*, May 31, 1931, B3.

country houses, it offered Miller and his guests an inviting setting for walks, picnics, and other activities.[22]

As sportsmen and sportswomen developed other estates, variations appeared. At Nightingale Hall, Jacquelin and Florence Holliday erected a modest dwelling in the form of an upcountry farmhouse. The unassuming building supplied comfortable and commodious lodgings but lacked aesthetic pretention. At the same time, they developed extensive landscaping. Like Miller, they created a deer park, and they also laid out a lawn that stretched fully around the house and down to a river landing. The Hollidays thus reversed the emergent model by developing more refined landscaping than architecture.[23] At Dirleton Plantation, three brothers from Norfolk, Virginia, established a truck farm and entertained friends with duck and quail shoots during the winter months. A "fine old dwelling" overlooking the Pee Dee River supplied comfortable accommodations, and 500 acres of rice fields attracted large numbers of waterfowl. Direlton thus combined commercial and recreational functions on a roughly equal basis.[24] At Spring Hill Plantation in Jasper County, Julian B. Clark did the same. An original member of the Okeetee Club, Clark erected a large house "of Colonial type architecture" and developed a model farm. He spent the better part of the year in the lowcountry with his family. With a crew of African American laborers, Clark planted about 3,500 acres. Revenue from cotton and other crops made Spring Hill a "self-supporting venture."[25]

Each of the approaches that northerners employed differed in orientation and execution. Concern for the "old atmosphere" at some plantations represented one form of reverence for the past. "Restoration" of colonial and antebellum houses supplied another. The notion that surviving remains could be rehabilitated, elaborated, and embellished followed a different logic. It expressed a desire for renewal but showed less reverence for surviving features. By taking a less-strident approach toward the care of aged and decayed structures, it created settings that blended old and new, usually in ways that favored new elements.

---

[22] "Esterville, Transformed Coastal Estate"; Lachicotte, *Georgetown Rice Plantations*, 155–158.

[23] Murray, "Holliday Winters in Peedee."

[24] C. S. Murray, "Dirleton, Pleasant and Profitable," newspaper clipping cited as *News and Courier*, May 3, 1931, in Dirleton Plantation File, South Carolina Vertical File Collection, Charleston County Public Library, Charleston, SC (hereafter SCVFC).

[25] Martin, "Spring Hill Plantation Pays."

Each approach shared a great deal with its counterparts. Each showed reverence for the past while presuming the necessity and legitimacy of adapting extant remains. Each accepted material change as inevitable and desirable. Each saw reverent care as part of a larger strategy for remaking old plantations. Because northerners consistently made use of surviving elements, even the most transformative efforts appeared less than severe. Aesthetic continuity consistently downplayed the scale and extent of the changes that occurred.

At the same time, differing approaches produced different results. The respect that restoration showed for aged buildings and its emphasis on material renewal differed markedly from purposeful retention of age and decay. Whereas restoration created newly refurbished buildings, the latter emphasized picturesque anachronisms. One celebrated an authentic past by making it anew; the other demonstrated authenticity by leaving material remains untouched. By contrast, elaboration and embellishment generally introduced far-reaching changes. Although exceptions existed, it tended to transform surviving remains, usually through a combination of refurbishment and new construction.

In the end, the results varied widely. The weathered, decaying appearance of Mansfield contrasted sharply with the freshly rebuilt look of the house at Mulberry. Similar contrasts could be seen among estates such as Arcadia, Annandale, and Estherville. Invariably, northerners' efforts produced unevenness. New buildings stood alongside old structures and long-deteriorating landscapes abutted newly developed gardens and lawns. The evolutionary process taking place left some areas distressed. So long as the process remained incomplete, it produced conspicuous variations. Each estate thus took on qualities that reflected the pace and extent of the efforts under way. Moreover, the nature of those efforts introduced further distinctions. Whereas Tuttle opted for something akin to preservation in situ at Mansfield, the Chapmans restored an old house to the grandeur they believed it had once possessed and began developing the surrounding landscape. The results not only differed aesthetically but conveyed different messages about the past.

Although the three approaches yielded different outcomes, they also evinced new sensibilities about the treatment of old plantations. Each saw possibilities rooted in acceptance of new uses, new purposes, and new priorities. Each viewed old plantations as suited to redevelopment as rural estates. Cultural barriers imposed few constraints. Although new owners grappled with a host of practical considerations, none seems to have been troubled by the moral implications of remaking sites historically defined by the union of commercial agriculture and slave labor. If any owners harbored apprehensions, their concerns went unrecorded.

Northerners' efforts distinguished themselves in part through a dramatic sense of material transformation. Even where sportsmen and sportswomen sought to make few changes, some alterations proved necessary. Pickman, after all, opted to renovate the house at Springfield, and Tuttle made needed repairs. Purposefully maintaining an old plantation in weathered and worn condition represented a striking change in its own right. Rather than allowing further decay, it imposed stasis for aesthetic purposes. It sought to maintain qualities produced by stagnation for the sake of character and quaintness. Consequently, stasis became a specialized treatment that owners selected rather than a product of social and economic conditions.

Of the three approaches, restoration represented the most calculated and, in some ways, the most complicated. The combination of artifactual treatment and active use invariably produced tensions. Northerners insisted on occupying restored structures and installing modern amenities and conveniences. Use itself militated against the care that restoration showed for material fabric, and installation of new appliances and finishes compounded the problem. In practice, however, the two impulses proved relatively easy to combine. Northerners encountered little difficulty in negotiating the imperatives that each imposed. Adding modern kitchens, bathrooms, and closets rarely posed significant problems, and material renewal not only showed reverence for a valued era but encouraged active use. New owners did make concessions – Tuttle's meals came from "an old fashioned kitchen," for example – but they proved modest in comparison to the changes that occurred. As accommodating gestures rather than self-imposed depravations, such concessions said more about owners' views of behaviors appropriate to seasonal residences than the difficultly of making desired alterations and improvements.[26]

Development of grounds, gardens, and recreational lands showed greater consistency. Although northerners employed elements of each of the same three approaches in treating the landscapes surrounding their dwellings, they also displayed greater flexibility. The major trend lay in rehabilitation and improvement. Northerners arrived in the lowcountry with expectations shaped by country houses in the North. Rather than seeing plantations as spaces of labor and production, they envisioned

---

[26] Martin, "Mulberry Castle, Built in Indian Days"; "Mansfield, Ancestral Parker Home."

settings tailored to their needs. In working to create such settings, owners rehabilitated select features, added new elements, and enlarged grounds and gardens. In turn, they gave their estates more ambitious, more refined appearances than historical plantations had generally possessed.

To the extent that northerners' efforts had a unifying theme, it lay in a general concern for aesthetic continuity. Whether based on solid precedent, idealized notions of "early American" design, or some combination thereof, northerners developed old plantations in ways that appeared consistent with well-established traditions. In effect, northerners revived classicism as a principal mode of architectural expression. Design traditions rooted in Greco-Roman classicism had dominated American architecture for the better part of the eighteenth and early nineteenth centuries. Not until the arrival of the Gothic Revival and Italianate styles in the 1830s did European romanticism exert significant influence. Lowcountry plantation architecture had always displayed considerable variety in terms of style and form. Huguenot émigrés to early Carolina had erected buildings that incorporated French and Dutch traditions and Barbadian settlers had introduced West Indian influences. Lowcountry architecture thus became an amalgam of forms and styles, most of them English in derivation, but not all. Although romantic styles became popular in Charleston in the 1840s and 1850s, they never made significant inroads in the rural lowcountry.[27]

The chaos and destruction of the Civil War brought building to a halt. Afterward, social and economic conditions ensured that little new construction took place. The dominant trend lay in minimal upkeep and abandonment. As long as the prospects for commercial agriculture remained uncertain, new investments in plantation buildings seemed imprudent. Building for reasons other than absolute necessity appeared foolish, given the circumstances.[28]

---

[27] W. Barksdale Maynard, *Architecture in the United States, 1800–1850* (New Haven: Yale University Press, 2002); Mills Lane, *Architecture of the Old South: South Carolina* (Savannah: Beehive Press, 1984); Kenneth Severens, *Charleston Antebellum Architecture and Civic Destiny* (Knoxville: University of Tennessee Press, 1988), chaps. 9–11.

[28] On colonial- and antebellum-era lowcountry architecture in general, see especially Lane, *Architecture of the Old South: South Carolina*; Severens, *Charleston Antebellum Architecture and Civic Destiny*; Jonathan H. Poston, *The Buildings of Charleston: A Guide to the City's Architecture* (Columbia: University of South Carolina Press, 1997); Maurie D. McInnis, *The Politics of Taste in Antebellum Charleston* (Chapel Hill: University of North Carolina Press, 2005); Samuel Gaillard Stoney et al., *Plantations of the Carolina Low Country* (Charleston: Carolina Art Association, 1939); Suzanne Cameron Linder Hurley, *Anglican Churches in Colonial South Carolina: Their History and Architecture*

As the end of the century neared, a few planters erected new dwellings. About 1890, for example, Robert J. Donaldson built a large Queen Anne–style residence for his family at Friendfield Plantation. This house became the dwelling that Bernard Baruch and his family occupied at Hobcaw Barony after about 1905.[29] About 1903, Frank E. Johnson built a modest bungalow at Estherville Plantation – the building that John A. Miller later made the core of his new residence.[30] These efforts grew out of necessity. Rather than creating buildings that projected authority and pretention, they aimed to satisfy modest needs. Moreover, they did little to change the general appearance of the countryside. A handful of new dwellings did nothing to alter the ruined appearance of the region. Material decay supplied the dominant aesthetic; reversing it would require more than a few examples of new construction.

With northerners' arrival, rehabilitation of old plantations began. Whether or not any of the sportsmen who arrived early on saw the plantations they purchased as historically valuable is an open question. Emerson may have seen the house at Prospect Hill in such terms, but the evidence on this point is inconclusive. Regardless, over time, northerners showed clear inclination to rehabilitate old buildings and add to them in a sympathetic manner. None willfully destroyed existing structures. None cleared land and began erecting new buildings. Significantly, none opted to erect houses inspired by contemporary trends in domestic design. The oversized bungalow the McCurdys built at Tomotley is the closest example of a structure that would have fit comfortably in an upscale suburb, but even it displayed features that contrasted with prevailing norms. In the main, northerners rehabilitated buildings from the eighteenth and first half of the nineteenth centuries and renovated others in ways that resulted in somewhat similar styling. Overall, they showed strong preferences for classicism and vernacular building traditions.

What this meant in any sort of immediate sense is less important than the circumstances it created. The degree of intentionality behind northerners' efforts is difficult to discern. None left detailed explanations for the choices they made. Individual tastes and inclinations clearly played a

---

(Charleston: Wyrick and Co., 2000); Shelley Elizabeth Smith, "The Plantations of Colonial South Carolina: Transmission and Transformation in a Provincial Culture" (PhD diss., Columbia University, 1999).

[29] Lee Brockington, *Plantation between the Waters: A Brief History of Hobcaw Barony* (Charleston: History Press, 2006), 39.

[30] Lachicotte, *Georgetown Rice Plantations*, 155–158.

role, as did practical considerations. Reusing existing buildings proved more economical and easier than erecting new structures, and that alone provided an incentive to salvage at least some. Moreover, surviving buildings appear to have played an important role in suggesting possibilities. None of the estates northerners developed in the 1910s and early twenties had an entirely new ensemble of architecture and landscaping. All grew out of selective adaptation, reshaping, and elaboration of existing remains. All used extant remains as a starting point. No matter what northerners chose to do, they began with old plantations in varying states of decay and ruin.

Considered as a whole, northerners' choices closed off what many Americans regarded as an aberrant period in architecture. By the turn of the twentieth century, the eclectic stylings of the Victorian era had fallen from favor. Linked to the excesses of industrialism and social conflict, professional architects and the public alike saw Victorian design as the visual equivalent of chaos. Popular fashion had shifted to the boldly expressive classicism introduced at the 1893 Columbian World's Exhibition and the more restrained variety rooted in new interest in early American architecture. Americans favored Beaux-Arts classicism for public and institutional buildings and the more subdued stylings of the Colonial Revival for domestic architecture and some commercial and institutional buildings. These styles expressed a new and vigorous nationalism. In the wake of the Spanish-American War and decades of political strife, middle- and upper-class Americans turned to architecture that articulated a nascent vision of national identity. Paying homage to America's English roots while proclaiming bold ambitions, varying forms of classicism expressed the aims of a nation on the rise.[31]

Thus, when northerners began rehabilitating and adapting old plantations, their actions took place in a region free of examples of romantic and exotic styles. They started with a built environment virtually bereft of Victorian-era architecture. The last sustained phase of building had

---

[31] On ideological shifts behind the resurgence of classically derived architecture during this period, see especially Richard Guy Wilson, Dianne H. Pilgrim, and Richard N. Murray, *The American Renaissance, 1876–1917* (New York: Brooklyn Museum, 1979); Richard W. Longstreth, "Academic Eclecticism in American Architecture," *Winterthur Portfolio* 17, no. 1 (spring 1982): 55–82. On the relationship between Neoclassicism and broader currents in American society and culture, see Alan Trachtenberg, *The Incorporation of America: Culture and Society in the Gilded Age* (New York: Hill and Wang, 1982), chap. 7.

occurred in the years before the Civil War, when planters continued to build in keeping with established traditions. Half a century later, northerners began a new phase that exhibited strong continuity with the past. Although shrouded in ambiguity for the moment, sportsmen's and sportswomen's actions lent themselves to a number of possible interpretations. Aesthetic and stylistic parallels with surviving buildings meant that northerners could easily be read as continuing past practices, renewing them, or, as some observers ultimately would claim, effecting a "renaissance." Certainly, northerners avoided the appearance of strong disjunctures or differences that might have suggested a radical break with the past.

Northerners' actions thus created the appearance of linkages with the antebellum era. Whether anyone recognized this at the time is unclear, and whether northerners themselves took note of the apparent parallels is no more certain. But for the understandings that would ultimately develop, this pattern had important implications. Stylistic and aesthetic parallels did not ensure people would see northerners as reviving a grand tradition, but neither did they rule that out. In fact, in many ways, northerners encouraged such an interpretation. Cultivated appreciation of buildings and settings from long ago and careful efforts to make sympathetic additions suggested reverence for the past. Steps taken to give new buildings styling similar to their predecessors suggested a desire to revive old traditions. In sum, whether intentional or not, northerners' choices indicated appreciation of surviving remains. By implication, their choices suggested a style of life more in keeping with the past than not. As sportsmen's and sportswomen's efforts proliferated and expanded in new directions, observers would invoke these views time and time again.

CONTINUED DECLINE

As a few sportsmen and sportswomen developed old plantations as winter residences and rural retreats, most of their counterparts continued to see the lowcountry in the same manner as before. Throughout the 1910s, most northerners who traveled to the region came solely to hunt. Their numbers grew at a modest pace and their activities attracted little attention. Recreational use of rural lands remained their priority.

Northerners continued buying land and establishing clubs, retreats, and preserves. In the late 1910s, several friends of John A. Miller, the owner of Estherville Plantation, established the Winyah Gun Club in the Pee Dee section of Georgetown County. It quickly took control of at least

10,000 acres.[32] Clubs founded in earlier years continued to grow and develop. The Kinloch Club, for example, continued adding new members and built a new clubhouse. By about 1923 members and guests relaxed in a comfortable fifteen-room lodge.[33] Meanwhile, sportsmen continued to establish retreats. In Georgetown County, two sportsmen developed a 1,500-acre refuge on the Black River called "Wee-Nee Lodge." The main building featured "wide verandas," a shingled exterior, "rustic porch furniture," and "windows and doors facings framed with twigs." A log house, a garage, a barn, and a superintendent's residence stood nearby.[34] In 1918, Roy A. Rainey, the son of a mining tycoon, bought W. P. Clyde's holdings on Hilton Head Island.[35] The following year, Hunter Grover, a businessman from Ohio, purchased Greenfield Plantation on the Black River.[36] As in earlier years, sportsmen continued buying land at a brisk pace.

The most important developments of the 1910s came not from northern sportsmen's activities but from outmigration of African Americans and further weakening of the regional economy. Blacks began leaving rural areas in large numbers after the devastating hurricane of August 1893. Damage to commercial rice production and the phosphate industry left many with no choice. Berkeley County suffered the greatest losses, with nearly half of its African American residents – more than 23,000 people – departing by 1900. Other counties held steady for a time but then experienced significant decreases. More than 14,000 African Americans left Beaufort County between 1900 and 1920, for example, and

---

[32] "Winter Colonists Now Returning to Their Homes Here," *Georgetown Times*, Dec. 20, 1929, 6.

[33] C. S. Murray, "Kinloch, Exclusive Santee Gun Club," *News and Courier*, July 12, 1931, B7.

[34] Chalmers S. Murray, "Wee-Nee, Black River Lodge," *News and Courier*, Oct. 4, 1931, B11.

[35] Chlotilde R. Martin, "Hilton Head Island Estates Merged," *News and Courier*, Feb. 7, 1932, A3; Robert B. Cuthbert and Stephen G. Hoffius, eds., *Northern Money, Southern Land: The Lowcountry Plantation Sketches of Chlotilde R. Martin* (Columbia: University of South Carolina Press, 2009), 77–78.

[36] Lachicotte, *Georgetown Rice Plantations*, 89; Linder and Thacker, *Historical Atlas of the Rice Plantations of Georgetown County and the Santee River*, 476. Also in the late 1910s, A. Felix DuPont acquired several tracts on the Combahee River, including Hamburg Plantation. These later became the core of an estate he called Long Bow. See William P. Baldwin, *Lowcountry Plantations Today* (Greensboro, NC: Legacy Publications, 2002), 214, 218; Linder, *Historical Atlas of the Rice Plantations of the Ace Basin*, 224; Duncan Clinch Heyward, *Seed from Madagascar* (1937; reprint, Columbia: University of South Carolina Press, 1993), 247–248.

significant numbers also abandoned Colleton and Georgetown counties. In fact, of all the coastal counties, only Charleston and Horry gained population during the 1890–1920 era.[37]

By the early 1910s, only a handful of plantations on the Combahee River continued growing rice on a commercial scale, and most of those ceased operations within a few years. Competition from producers in Louisiana, Arkansas, and Texas left planters with no other choice.[38] Cultivation of Sea Island cotton ended. Output declined sharply after 1910 and the arrival of the boll weevil in 1916 destroyed what remained.[39] Meanwhile, manufacturing also struggled. Throughout the region, a few pockets of activity countered the prevailing trend, but none came close to offsetting the cumulative losses caused by agricultural decline and the demise of the phosphate industry.[40]

By the early 1920s, the rural lowcountry lay in desperate condition. The exodus of African Americans reduced the number of people trying to live off the land, and those who remained showed ingenuity and resolve in sustaining themselves. Subsistence farming, small-scale cotton production, hunting and fishing, and no small number of odd jobs figured in their efforts to put food on the table. Resourcefulness and determination, however, did only so much. Poverty, poor housing, and limited opportunities for wage labor characterized the lives of lowcountry blacks. Few

[37] Calculated from figures in Bureau of the Census, *Fourteenth Census of the United States, Taken in the Year 1920*, vol. I: *Population, 1920, Number and Distribution of Inhabitants* (Washington, DC: Government Printing Office, 1921), 602; Julian J. Petty, *The Growth and Distribution of Population in South Carolina* (Columbia: State Council for Defense, 1943), appendix F, 226–229. On the exodus from lowcountry counties, see George A. Devlin, *South Carolina and Black Migration, 1865–1940* (New York: Garland Publishing, 1989), 357; Rowland and Wise, *Bridging the Sea Islands' Past and Present*, 189–190.

[38] Peter A. Coclanis, *The Shadow of a Dream: Economic Life and Death in the South Carolina Low Country, 1670–1920* (New York: Oxford University Press, 1989), 142; James H. Tuten, *Lowcountry Time and Tide: The Fall of the South Carolina Rice Kingdom* (Columbia: University of South Carolina Press, 2010), chap. 3.

[39] Charles F. Kovacik and Robert E. Mason, "Changes in the South Carolina Sea Island Cotton Industry," *Southeastern Geographer* 25, no. 2 (Nov. 1985): 77–104; Richard Dwight Porcher and Sarah Fick, *The Story of Sea Island Cotton* (Charleston: Wyrick and Co., 2005), 317–318, 323–335; "Passing of Sea Island Cotton," *News and Courier*, March 23, 1925, 4.

[40] Bureau of the Census, *Fourteenth Census of the United States, Taken in the Year 1920*, vol. IX: *Manufacturers, 1919* (Washington, DC: Government Printing Office, 1923), 1383–1384, 1390, 1395; Rogers, *History of Georgetown County*, 498–500; Rowland and Wise, *Bridging the Sea Islands' Past and Present*, 185–189.

Americans of the time lived in conditions as meager. Nearly six decades after emancipation, the region's African American population had freedom of a sort but little more.[41]

The brief upturn of the World War I era bypassed the old plantation districts. As Charleston boomed, the countryside remained stagnant. The main effect of wartime mobilization was increased outmigration. Blacks seized upon opportunities for military service and employment. Military service took young men abroad, opened their eyes to a world of possibilities, and, in many cases, led them away from the lowcountry for good. New job opportunities did the same. Blacks that the US Army judged unfit for service often moved to Charleston or set out for northern cities, eager to take advantage of the wartime labor market. Continued weakness in agriculture and few other options provided strong incentives to leave.[42]

After World War I, then, stagnation and decline advanced to new levels. Although conditions had appeared grim since the 1880s, now they seemed hopeless. Abandonment of tens of thousands of acres left much of the countryside a semi-tropical wilderness. Writing of conditions in these years, Samuel Gaillard Stoney, Jr., described "a region of deserted fields growing up in forest, of ragged dying gardens and grim, cold, pathetic houses, solemnly awaiting their doom by fire or dilapidation" (see Figure 2.1).[43] To this he might have added thousands of African American families struggling to eke by. Although their numbers had declined appreciably, they remained the majority of the population by a large margin. Few people, if any, would have suggested that old plantation districts all along the coast lay on the cusp of becoming seasonal playgrounds for wealthy people from afar. In fact, they did. In many ways, they already were.

[41] On rural life in this era, see Walter B. Edgar, *History of Santee Cooper, 1934–1984* (Columbia: R. L. Bryan Co., 1984), 5, 27; T. J. Woofter, Jr., *Black Yeomanry: Life on St. Helena Island* (New York: Henry Holt and Co., 1930).

[42] Theodore Hemingway, "Prelude to Change: Black Carolinians in the War Years, 1914–1920," *Journal of Negro History* 65, no. 3 (summer 1980): 212–227. Recalling the exodus about a decade later, Samuel G. Stoney and Gertrude Shelby Matthews wrote, "Young men came out from the woods; and the demand for cheap labor kept them from returning." See Samuel Gaillard Stoney and Gertrude Matthews Shelby, *Black Genesis: A Chronicle* (New York: Macmillan Company, 1930), xiii. On conditions in Charleston, see Mamie Garvin Fields, *Lemon Swamp and Other Places: A Carolina Memoir* (New York: Free Press, 1983), 194–196.

[43] Stoney, *Plantations of the Carolina Low Country*, 42.

FIGURE 2.1 Main house at the Wedge Plantation, 1923. By the early twentieth century, plantations throughout the lowcountry lay in ruins. Many observers saw decaying mansions as symbolizing the decline of a once-powerful agricultural empire.
Courtesy of the Charleston Museum, Charleston, SC.

MAKING LEISURE PLANTATIONS

In the mid-1920s, northerners' activities entered a new phase. A sharp spike in real estate transactions heralded its arrival. Northerners suddenly began buying land at a quicker pace and in greater numbers than before. Simultaneously, the hesitancy and tentativeness of earlier years ended. Northerners stopped turning old plantations into hunting retreats and began creating large, well-developed "plantations." Handsome, well-appointed estates quickly became common. In these years, the lowcountry ceased to be a land of ruined plantations. Instead, it became a destination for wealthy, privileged people who passed the fall and winter months at grand sporting estates.

The surge in land acquisition and estate-making grew out of complex circumstances. No single factor sparked the burst of activity. Rather, several combined to encourage investments in land and elegant architecture. First, the strong economy of the postwar era provided capital. Wealthy Americans benefited from the economic expansion that began soon after the war ended and continued virtually unabated into the late 1920s. Sharp advances in the stock market, favorable tax policies, and income from manufacturing all contributed to the growth of personal fortunes.[44] Although wealth alone did not inspire large investments in land and new estates, it nonetheless provided the necessary means.

Second, social conditions also fueled leisure spending. The excesses and indulgences of the 1920s derived in part from conspicuous consumption among the upper classes. As the boom economy and cultural release of the postwar era drove the social antics of the wealthy to new heights, investments in recreation increased. As the lowcountry became a prized destination, pretension grew dramatically.

Third, the continued growth of the lowcountry sporting scene created new opportunities. Northerners increasingly recreated and socialized in heterosocial groups, in circumstances that foregrounded wealth and class. In part, these developments reflected continued evolution of the social milieu of earlier years. As the number of northerners spending the better part of the winter season in the lowcountry grew, so too did opportunities for social intercourse. At the same time, a new emphasis on entertaining – hosting friends, family, and business associates – also played a role. Northerners increasingly saw their "plantations" as venues for socializing with people of comparable social rank and thus made new investments in architecture, landscaping, and infrastructure. New modes of behavior and desires for greater refinement fueled one another, pushing forward the creation of elegant settings and closely aligned spheres of activity.

Northerners' estates quickly took on characteristics similar to country estates in the North. Although they continued to mirror hunting clubs and retreats in some ways, they differed in two respects. First, heterosociality provided the dominant mode of social interaction. Men and women generally recreated and socialized in mixed company, whether for leisure, dining, or a host of casual activities. Second, not only did hunting figure in a broader range of leisure and sporting pursuits, the combination of activities associated with northerners' "plantations" carried symbolic significance. Sport hunting, equestrian outings, and other forms of outdoor recreation evoked allusions to aristocratic traditions and landed

---

[44] George H. Soule, *Prosperity Decade: From War to Depression, 1917–1929* (Armonk, NY: M. E. Sharp, 1947), chaps. 4 and 15.

estates. When enacted in the physical space of a "plantation" – an authentic site of slavery, as northerners and other people of the era understood it – these activities took on additional meaning, replete with connotations of class privilege and hereditary authority.[45]

In the late 1920s, what came to be called "plantation life" reached maturity. Characterized by elaborate symbolism and specific social conditions, plantation life tied select modes of performative display to large, lavishly styled estates. Performance, in this sense, equated to behaviors described by sociologist Thorsten Veblen in his 1899 *Theory of the Leisure Class*. For Veblen, class derived from more than mere wealth. Social elites, he argued, relied on ritualized practices to display status and differentiate themselves from other groups. Plantation life fit squarely within Veblen's reading of upper-class behavior. By combining abstention from productive labor, cultivated appreciation of finery, and conspicuous displays of material wealth, plantation life forged solidarity and asserted class boundaries. It made "plantations" more than mere retreats by assigning them a specialized role in elite culture.[46]

In addition to these developments, the lowcountry's role in an expanding network of leisure destinations further boosted northerners' activities. By the early 1920s, seasonal tourism to a host of destinations along the southeastern seaboard had become well established. Palm Beach, Florida, reigned as the premier destination for status-conscious elites, and other Florida resorts attracted middle- and upper-class northerners in large numbers. "Winter colonies" of polo and golf enthusiasts thrived at Aiken and Camden, South Carolina, and throngs of devoted golfers also flocked to Pinehurst, North Carolina. Meanwhile, Jekyll Island remained an exclusive haven for millionaire industrialists. Whereas the annual flow of travelers to the Southeast had been significantly smaller at the beginning of the twentieth century, by the early 1920s it became a veritable deluge.[47]

---

[45] On country estates and sporting clubs in the North, see especially Clive Aslet, *The American Country House* (New Haven: Yale University Press, 1990); Mark Alan Hewett, *The Architect and the American Country House, 1890–1940* (New Haven: Yale University Press, 1990); Brendan Gill, "Introduction," in *Long Island Country Houses and Their Architects, 1860–1940*, ed. Robert B. MacKay, Anthony K. Barker, and Carol A. Traynor (New York: W. W. Norton and Co., 1997), 11–17.

[46] Thorstein Veblen, *Theory of the Leisure Class* (New York: Macmillan Co., 1899).

[47] Edward N. Akin, *Flagler, Rockefeller Partner and Florida Baron* (Kent, OH: Kent State University Press, 1988), chaps. 8 and 9; George Brown Tindall, *The Emergence of the New South, 1913–1945* (Baton Rouge: Louisiana State University Press, 1967), 103; William Barton McCash and June Hall McCash, *The Jekyll Island Club: Southern Haven*

As winter tourism skyrocketed, the lowcountry sporting scene grew partly because of increased traffic along the Atlantic Coast. From the early 1920s onward, a steady stream of visitors enlivened an already active, still-evolving realm of activity. Some stayed for only a few days at a time; others stayed a week or two or more. No matter what the length of their stays, greater numbers of visitors had immediate effects. During the 1930s, some northerners complained about the unrelenting flow of guests during the early spring, when the winter season at resorts further south drew to a close.[48] The result was a curious irony: although the appeal of the lowcountry lay partly in seclusion and quietude, increased social activity undermined these qualities. At times, northerners' estates became nearly as much a part of an elite social milieu as owners' residences in northern cities.

Meanwhile, the triumphant crescendo of reunion in the 1910s and new perspectives on lowcountry history further advanced the creation of large sporting estates. The fiftieth anniversary of the Civil War and Woodrow Wilson's election to the presidency enshrined white southerners' memory of the conflict in national consciousness and solidified the nation's commitment to white supremacy.[49] The racial violence that erupted during World War I and continued in later years further limited opportunities to recover an emancipationist vision of the Civil War. New scholarship also highlighted the distinctive character of rice slavery and its role in American history. In 1918, Ulrich Bonnell Phillips's *American Negro Slavery*

for America's Millionaires (Athens: University of Georgia Press, 1989); John Hammond Moore, comp. and ed., *South Carolina in the 1880s: A Gazetteer* (Orangeburg, SC: Sandlapper Publishing, 1989), 9–14. On Aiken's "winter colony," see Harry Worcester Smith, *Life and Sport in Aiken and Those Who Made It* (New York: Derrydale Press, 1935). On Pinehurst, see Richard Mandell, *Pinehurst: Home of American Golf* (Pinehurst, NC: T. Eliot Press, 2007).

48 See, for example, Sidney Jennings Legendre, "Diary of Life at Medway Plantation, Mt. Holly, South Carolina," (private collection, copy in author's possession), entries for May 9, 1939 and March 29, 1940; Thomas A. Stone Journal, entries for Feb. 12 and March 17, 1937, in Boone Hall Scrapbook Number 3, South Carolina Historical Society, Charleston, SC. An early mention of plantation life being "almost too gay" appears in Benjamin R. Kittredge to S. Dana Kittredge, n.d. [ca. 1927], folder 4, box 15, Kittredge Family Papers, South Carolina Historical Society, Charleston, SC.

49 David W. Blight, *Race and Reunion: The Civil War in American Memory* (Cambridge: Belknap Press of Harvard University Press, 2001), 338–397; Nina Silber, *The Romance of Reunion: Northerners and the South, 1865–1900* (Chapel Hill: University of North Carolina Press, 1993); Caroline E. Janney, *Remembering the Civil War: Reunion and the Limits of Reconciliation* (Chapel Hill: University of North Carolina Press, 2013), 266–311; Grace E. Hale, *Making Whiteness: The Culture of Segregation in the South, 1890–1940* (New York: Pantheon Books, 1998).

identified the Carolina coast as central to the development of slavery in the southern United States. In chapters on "The Rice Coast" and "Types of Large Plantations," Phillips highlighted the size of low-country plantations, the wealth they produced, and differences between lowcountry slavery and other regions. Grounded in systematic study of plantation records, Phillips's book set a benchmark that would not be rivaled for decades. Although best known for presenting slavery as a benevolent institution, Phillips also stimulated interest in the lowcountry and its past.[50]

Francis Pendleton Gaines's *The Southern Plantation* (1924) did at least as much to focus attention on the lowcountry. Gaines's study charted the origins and development of "the popular conception of the old planta-tion," an "Arcadian scheme" that he saw as illustrating an American "golden age." Gaines identified the Carolina coast as one of a few places where social and material conditions before the Civil War had approxi-mated those commonly depicted in plantation romances. According to Gaines, "unmistakable evidence" showed "an order of life in a few limited localities" that had approximated "in real social charm the trad-itional social charm of the romances" – specifically Tidewater Virginia, "the rice districts of South Carolina," and the lower Mississippi Valley. In these areas, Gaines argued, the "splendor" and "cultured magnificence" commonly associated with the plantation idyll had actually existed. All of the features common to romantic depictions of the antebellum South had grounding in historical fact, he contended. In short, Gaines identified the Carolina coast as one locality where truth matched fiction – where a romantic, gracious South of social harmony and aristocratic grandeur had been real.[51]

Together, Phillips's and Gaines's studies stimulated interest in the low-country and its past. In addition to emphasizing the distinctive character of lowcountry slavery, each book advanced important claims about the wealth and power of the lowcountry planter class, the style and manner in which its members lived, and the conditions of slavery on the rice coast. Read by scholars and popular audiences alike, both books influenced popular views of the lowcountry. Moreover, as journalists and promoters

[50] Ulrich Bonnell Phillips, *American Negro Slavery* (New York: D. Appleton and Co., 1918), 88, 97.

[51] Francis Pendleton Gaines, *The Southern Plantation: A Study in the Development and Accuracy of a Tradition* (New York: Columbia University Press, 1924), first quotation on vii, second and third on 4, fourth on 143, others on 144.

seized upon key points, Phillips's and Gaines's writings informed popular discourse. By the early 1930s, points of emphasis derived from *American Negro Slavery* and *The Southern Plantation* figured at the center of popular representations of the lowcountry and its past.[52]

Meanwhile, deepening material decay and stagnation continued to shape popular views of the region. By World War I, the lowcountry increasingly appeared as land mired in a different time. Especially in rural areas, little seemed to have changed for decades. Popular writers, artists, and amateur historians made images of a grand and gallant past central to depictions of the region.[53] In the mid-1920s, efforts of this kind took on new focus and assumed new prominence. The crucial shift came with the beginning of sustained efforts to promote Charleston as a tourist destination and the outpouring of visual and literary art today recalled as the Charleston Renaissance. Together, these developments inspired a wave of representations that cast the lowcountry as a land apart. Although portrayals of the region varied, all emphasized its unique history and character. All cast the region as unique in the history of the South and the nation. All emphasized the enduring influence of the lowcountry past. Although deeply biased, rife with omissions, and characterized by varying degrees of romanticism, these portrayals played a crucial role in shaping popular views of the lowcountry at a time of growing interest in the nation's heritage.[54]

---

[52] The historical writings of Charleston novelist and historian Herbert Ravenel Sass offer the best example. Sass used Gaines's claims about the romantic splendor of the pre–Civil War lowcountry to legitimate his view of the region's history and seized upon Phillips's characterization of the historical relationship between Charleston and the surrounding countryside. See Herbert Ravenel Sass, "The Low-Country," in Augustine T. Smythe et al., *The Carolina Low-Country* (New York: Macmillan Co., 1931), 4, 6–7. Cf. Gaines, *The Plantation Tradition*, 143, and Phillips, *American Negro Slavery*, 96–97, respectively. More generally, Sass relied on these and other scholarly studies to craft a narrative of regional history that emphasized the exceptionalism of the lowcountry. In addition to "The Low-Country," see also Herbert Ravenel Sass, "The Rice Coast: Its Story and Its Meaning," in Alice Ravenel Huger Smith, *A Carolina Rice Plantation of the Fifties* (New York: William Morrow and Co., 1936), 3–55; Herbert Ravenel Sass, "The Ten Rice Rivers," *Saturday Evening Post*, Dec. 13, 1941, 20–21, 105–108; Herbert Ravenel Sass, "South Carolina Rediscovered: A Native Son Finds Spectacular Changes in the 'Moonlight and Magnolia' State, Scene of Huge H-Bomb Project," *National Geographic* 103, no. 1 (March 1953): 281–321.

[53] See, for example, Stoney, *Black Genesis: A Chronicle*, xii; Julia Wood Peterkin, *Roll, Jordan, Roll* (New York: Robert B. Ballou, 1933).

[54] On these developments, see especially Yuhl, *Golden Haze of Memory*; W. Fitzhugh Brundage, *The Southern Past: A Clash of Race and Memory* (Cambridge: Belknap Press of Harvard University Press, 2005), chap. 5; Martha R. Severens, *The Charleston*

In sum, in the years immediately following World War I, several developments brought new attention to the lowcountry. Continued expansion of northerners' activities created new opportunities for self-representation and fueled the creation of larger, more ambitious estates. Existing estates took on new potential and estate-making increasingly focused on conspicuous display. The social realm associated with northerners' estates became more status-conscious, and scholarly and popular writing emphasized the significance of lowcountry history. Moreover, material decay continued to advance. As it did, the need to make sense of it – to understand what it signified – became acute. In this context, historical writing, visual and narrative representations, and sensory awareness produced new depictions of the lowcountry and its past. Writers and artists portrayed the region as a land of enduring charm and timeless appeal. By setting the history of the lowcountry against the broad sweep of the southern past, popular writers, artists, and performers called attention to its role in American history and its distinctive characteristics.

The differing impulses that came together in these years simultaneously inspired and formed the backdrop for a spectacular burst of activity. In one three-week period in 1925, northerners acquired more than 1,000,000 acres of coastal land. In the first six months of 1927, out-of-state sportsmen and sportswomen purchased more than 60,000 acres in Beaufort, Colleton, and Jasper counties.[55] Even more significant, northerners began amassing large acreages specifically for the purpose of creating large estates. In 1925 alone, northerners acquired four tracts centered on old plantations that soon became country estates. Two more followed in 1926, five the following year, and another in 1928. By the end of the decade, northerners had acquired twenty-six tracts where efforts to develop handsome estates had started or would soon begin.

The land-buying and estate-making spree attracted widespread attention. Lowcountry newspapers began reporting on real estate transactions and the plans of wealthy sporting enthusiasts. Readers of the *News and Courier*, the *Beaufort Gazette*, and the *Georgetown Times* soon grew accustomed to reports of northerners buying large tracts of land and

Renaissance (Spartanburg, SC: Saraland Press, 1998); James M. Hutchisson and Harlan Greene, eds., *Renaissance in Charleston: Art and Life in the Carolina Low Country, 1900–1940* (Athens: University of Georgia Press, 2003).

55 Mary Katherine Davis Cann, "The Morning After: South Carolina in the Jazz Age" (PhD diss., University of South Carolina, 1984), 429.

announcing intentions of building stately mansions at once.[56] Some purchases made the pages of the *New York Times*.[57] In these years, lowcountry people became well acquainted with the activities of "winter colonists," and other Americans also heard about the group.

As real estate deals made headlines, a rush of activity began in the countryside. At dilapidated plantations, laborers began clearing brush, tearing down old barns and sheds, pruning trees, and cleaning out weathered houses. Workers laid foundations for new dwellings and stripped rotted clapboards from buildings deemed worthy of restoration. Laborers trudged into rice fields to begin the difficult and messy work of rebuilding dikes, cleaning drains, and building new trunks. These efforts marked the beginning of a surge of activity that would continue into the mid-1930s.

What began in these years was a veritable estate-making boom: a sustained campaign aimed at creating large, elegant, and sometimes opulent estates. Northerners acted with firm resolve and deliberate intent. In contrast to the uncertain efforts of earlier years, they showed little hesitation in remaking old plantations as they wished. Nor did northerners show restraint in creating estates where little or nothing survived from before – or where plantations of the traditional variety had never existed. One of the most important innovations of the era was the creation of nearly or completely "new" plantations. Where northerners acquired lands without surviving remains, they routinely created stylized ensembles of architecture and landscape. Where they acquired lands that had never been part of plantations, they followed the same course. In earlier years, surviving remains had been crucial. Material vestiges of past eras continued to have strong appeal and remained useful starting points, but they no longer provided the sole basis for creating large estates. Ideas about "plantations" had become sufficiently well established to allow for wholesale fabrication. If plantations lying in almost any stage of decay could

[56] See, for example, "Real Estate Change Hands," *Beaufort Gazette*, April 7, 1927, 1; "Big Land Deal Made Last Week," *Beaufort Gazette*, Aug. 4, 1927, 1; "Big Real Estate Deal Completed," *Beaufort Gazette*, Oct. 27, 1927, 1; "Northerner Buys Beaufort Lands," *Beaufort Gazette*, Nov. 18, 1927, 1; "Another Big Land Deal Made," *Beaufort Gazette*, Dec. 22, 1927, 1; "Buys Large Plantations," *Beaufort Gazette*, Jan. 12, 1928, 3; "Activity in Real Estate," *Beaufort Gazette*, March 29, 1928, 1.

[57] "Buys Hunting Lands," *New York Times*, Oct. 27, 1927, 51; "Northerners Buying Big Southern Estates," *New York Times*, Aug. 18, 1929, E2; "New Yorker Buys Plantation," *New York Times*, March 14, 1934, 36; "New York Woman Buys a 775-Acre Plantation," *New York Times*, Oct. 29, 1935, 37; "Plantation of 1,000 Acres Sold to Du Pont Official," *New York Times*, June 15, 1937, 43.

be rehabilitated and put to new use, new "plantations" could also be created. Northerners employed both approaches. Together, they supplied viable formulas for the creation of dozens of new estates.

Northerners showed ingenuity and confidence in creating their estates. Greater stylistic and aesthetic coherence became apparent by the end of the 1920s. Strong contrasts and marked variations in style and form remained present, but they diminished as estate-making proceeded. The overall trend showed commitment to refinement and conspicuous display. Whereas estates developed in earlier years had relatively modest aims, the majority created after 1925 showed strong ambition.

As in earlier years, Georgetown County remained a center of activity. In the late 1920s, northerners developed at least four plantations that equaled or surpassed those established earlier. At Litchfield Plantation on the Black River, Henry Norris, a physician from Bryn Mawr, Pennsylvania, remodeled and enlarged a dwelling originally built about 1820. He made extensive repairs to the aging structure, added a large wing, and erected a garage and stables. Norris also developed extensive landscaping in the immediate vicinity of the house. A visitor who saw Litchfield in the early 1930s called Norris's residence one of "the most delightful houses on the Waccamaw peninsula."[58] At nearby Ponemah Plantation, Pennsylvanian Willis E. Fertig erected a handsome residence with a monumental portico. A veranda set atop the main block provided superb views of the Black River and surrounding landscape. Sunrooms in subordinate wings allowed Fertig and his family to relax while enjoying similar vistas. Near the house stood a group of outbuildings that included an old rice mill – "a reminder of the days when rice planting was at its height" – four employee's houses, a boat house, a garage, generator and pump houses, several barns, and a kennel.[59]

Farther south, similar estates developed. In 1925, George D. B. Bonbright acquired Pimlico Plantation, a tract of 3,000 acres situated about halfway up the western branch of the Cooper River. The existing house stood in poor condition, so Bonbright had it demolished. In its place he

---

[58] Chalmers S. Murray, "June Finds Norrises at Leitchfield," *News and Courier*, Aug. 16, 1931, A6. Norris strived to retain the styling of the original house and carry it over to new addition. According to Murray, visitors found it difficult to tell that the addition was "not part of the old structure." Norris also developed an inventive landscaping scheme. On the site of the old slave street he created a flower garden, and in the rear he built a reflecting pool with a bed of African lotus and a "garden of gay flowers." Murray opined that "a more peaceful and restful scene can hardly be imagined."

[59] Chalmers S. Murray, "Ponemah, 'Happy Hunting Ground,'" *News and Courier*, April 26, 1931, A6.

erected a large Colonial Revival dwelling. With a broad façade and offset wing, the building possessed an unusual form but projected a handsome appearance. Grounds "made lovely" with flowers and shrubs beautified the setting.[60] At the upper end of the Cooper's western branch, Nicholas G. Roosevelt purchased Gippy Plantation. He rehabilitated a mansion built about 1840, surrounded it with a generous lawn, and developed a dairy farm. Milk and butter from his heard of eighty cows – forty-one Guernseys and thirty-nine "good grade cows" – provided sufficient revenue to offset the cost of maintaining his estate.[61] At the Bluff Plantation, J. Fritz Frank chose to leave decay in place. He and his wife took pride in keeping the main house – "a regular, old-fashioned plantation home" – just as they had found it. They treated thirty-one "large and venerable" live oaks on the grounds the same way. Eventually, however, Frank found it necessary to make minor changes. Tired of eating "eggs served cold or with rain in them," he moved an old detached kitchen to the rear of the house and connected it as an addition in order to give his servants direct access to the room where he and his wife took their meals (see Figure 2.2).[62]

In the lower southeastern corner of the state, in Jasper, Colleton, and Beaufort counties, sportsmen created large estates with stately mansions, elegant landscaping, and spectacular views of coastal waterways. These made the duck- and quail-hunting lands between the St. Helena Sound and the Savannah River an even greater focus of activity. Most of the estates created in this area featured stylish new houses. Northerners often had no choice but to build new. Planters had built fewer houses in these counties than higher up the coast, and Sherman's march through the Carolinas destroyed many of them. Others later fell victim to fire, decay, and storms. In the Georgetown area and closer to Charleston, large numbers of early houses survived. In the southeastern corner of the state, few did. Consequently, sportsmen and sportswomen found it necessary to erect new buildings.[63]

Several of the earliest estates northerners created established a model that would be replicated over and over again. At Mackey Point Plantation, a 5,700-acre tract made of up five old plantations between the

---

[60] Chlotilde R. Martin, "New Yorker Reclaims Pimlico," *News and Courier*, May 24, 1931, B2.

[61] "Gippy and Its Guernsey Herd," *News and Courier*, Jan. 25, 1931, B7.

[62] "Bluff Plantation Overlooks Cooper," *News and Courier*, June 14, 1931, A2.

[63] For contemporary observations about the fate of plantation houses in this area, see Sass, "The Low-Country," 13; John R. Todd and Francis M. Huston, *Prince Williams Parish and Plantations* (Richmond: Garrett and Massie, 1935), 106.

FIGURE 2.2 Main house at the Bluff Plantation, 1928. Photograph by Ben Judah Lubschez.
Courtesy of the Gibbes Museum of Art/Carolina Art Association, Charleston, SC.

Pocotaligo and Tullifinny rivers, George D. Widener built a huge Colonial Revival–style residence. Designed by Philadelphia architect Charles N. Read, it featured a two-story main block with adjoining wings. A hipped roof with tall interior end chimneys defined its roofline. Weidner, a wealthy "sportsman and philanthropist" from Philadelphia, did little to the surrounding grounds. He planted a lawn of Italian ryegrass in the rear, set shrubs close around the house, and created a winding driveway that looped its way through a grove of towering oaks before arriving at the main entrance. In contrast to most of his counterparts, Weidner opted not to create a formal garden or add extensive plantings. Even so, his estate impressed visitors. When Chlotilde R. Martin saw it in 1930, she described it as "one of the show places of the coastal country."[64]

Directly across the river to the west, Bayard Dominick, a financier, built an equally impressive house with a similar design on his Gregorie

---

[64] Chlotilde R. Martin, "Widener Builds at Mackey Point," *News and Courier*, Dec. 25, 1930, B3; Cuthbert and Hoffius, eds., *Northern Money, Southern Land*, 145–147.

FIGURE 2.3 Main house at Gregorie Neck Plantation, circa 1930.
From John R. Todd and Francis M. Hutson, *Prince Williams Parish and Plantations*
(Richmond: Garrett and Massie, 1935), 238.

Neck Plantationn (see Figure 2.3). It also featured a central block-with-wings plan, but, unlike Weidner's residence, the wings swept back at oblique angles. Located in a grove of oaks overlooking the river, the house blended the character of an upscale residence with the form and scale of an institutional building. Martin judged it "somehow suggestive of Mount Vernon." The interior, she noted, showed "simple but exquisite taste." Dominick carried out a more elaborate beautification campaign than his neighbor across the river. He had three acres of Italian ryegrass laid out around the house and planted azaleas, camellias, and magnolias at irregular intervals. Three double-red camellia trees came from a site in Hampton County about twenty-five miles away. Each stood over twenty feet tall and weighed twenty-five tons. According to the *News and Courier*, they were probably the largest trees of their kind ever transplanted in South Carolina.[65]

---

[65] Chlotilde R. Martin, "Dominick Builds among Giant Oaks," *News and Courier*, Feb. 8, 1931; "Japonicas Moved to Gregorie Neck," *News and Courier*, Feb. 7, 1932, A11.

The basic formula evident at Widener's and Dominick's estates – a handsome Colonial Revival–style house set amid a verdant landscape at an opportune riverfront location – would appear over and over again. It became especially conspicuous in Jasper, Beaufort, and Colleton counties, but northerners also developed examples farther up the coast. Ultimately, Colonial Revival–style houses stood on nearly half of the estates that sportsmen and sportswomen created. Although these buildings varied significantly in scale, style, and form, all projected poise and refinement. By evoking visions of a genteel past, they situated the lowcountry within an idealized narrative of American origins, effectively eliding the fierce debate over slavery, the travails of Reconstruction, and the continued subjugation of African Americans.[66]

As the estate-making boom gathered momentum, variations appeared. The general trend toward grandeur and overt display found a strong counterpoint in continued adherence to earlier approaches and the often idiosyncratic preferences of individual owners. Franklyn L. Hutton's Edisto River estate, Prospect Hill Plantation, affords an apt example. Dubbed a "little kingdom" by the *News and Courier*, Prospect Hill encompassed seven antebellum plantations that together totaled over 5,500 acres. Hutton turned the antebellum house on the property into a residence for his superintendent. He built a spacious new mansion, "Curley Hut," for himself and his family. An unusual example of Shingle-style architecture, the building stood near the site of an old rice mill. The main block featured an inset porch with thin columns and a row of bold dormers. Offset wings and a rambling form echoed the great "cottages" of Mount Desert and other New England resorts. Although the clubhouse at Okeetee possessed greater sophistication, Hutton's house nonetheless represented a highly developed example of the style.[67]

Curley Hut marked the transition between two eras. Possibly the last large Shingle-style building erected in the lowcountry, it straddled the line between informality and the new emphasis on order and refinement. The only comparable structure built during the period is the clubhouse at Yeamans Hall, a huge building with a deep porch and massive roof.

---

[66] For examples, see Chlotilde R. Martin, "20-Acre Lawn to Surround Bingham House in Beaufort," *News and Courier*, Dec. 2, 1930, 5; Chlotilde R. Martin, "Clay Hall, Redmond Mansion Lodge," *News and Courier*, Dec. 7, 1930, A11; Chlotilde R. Martin, "The Delta – Home of H. K. Hudson," *News and Courier*, Jan. 20, 1931, 7; Chlotilde R. Martin, "Henry W. Corning's Place," *News and Courier*, Jan. 24, 1932, A11.

[67] J. V. N., Jr., "Do You Know Your Lowcountry? Prospect Hill," *News and Courier*, May 9, 1938, 10.

Described by the *News and Courier* as "typical of New England," the clubhouse stood as an example of the handsome, conservatively styled buildings that had become the norm for country clubs across the North and Midwest.[68] By the time Hutton built Curley Hut, northerners had abandoned rusticity as an appropriate idiom for their lowcountry holdings. Buildings erected in earlier years continued to serve the same purposes as before, but the new emphasis on grandeur underscored the growing importance of social activity. No longer merely a sporting destination, the lowcountry had become a stage for more elaborate forms of recreation and social intercourse.

Where northerners opted for simplicity in these years, they erected large bungalows or plain "cottages" derived from vernacular forms. These buildings provided comfortable accommodations but lacked the styling and sophistication of classical architecture. Some incorporated modest Colonial Revival influences, but none came close to matching the poise and stature of houses with towering porticos and wings. At Laurel Spring Plantation on the Combahee River, financial tycoon Edward F. Hutton built a "low, white rambling bungalow in the shape of a 'Z.'"[69] At an estate near Gray's Hill in Beaufort County, Cleveland manufacturer Henry W. Corning passed his winters in a group of small, simply styled, one-story cottages. The "main building" contained living and dining rooms and a kitchen; sleeping quarters occupied the other two. The plain and unassuming appearance of these structures led one commentator to observe that "there is nothing ornate or pretentious about the Corning place."[70] At Long Bow Plantation, Canadian scientist and inventor Harold Ashton Richardson remodeled and enlarged a modest farmhouse. Although the building appeared small when viewed from the façade, it stretched out in the rear for "some distance," with a total of nineteen rooms inside. Instead of creating ornamental gardens, Richardson and his family opted for practicality by growing a "great variety" of fruits and vegetables.[71]

---

[68] On Yeamans Hall, see "Do You Know Your Charleston? Yeamans Hall Club," *News and Courier*, Nov. 23, 1931, 10.

[69] Chlotilde R. Martin, "The Hutton Estate of 16,000 Acres," *News and Courier*, Jan. 25, 1931, B3. Hutton made an early nineteenth-century house on the Lowndes Plantation the residence of his superintendent, J. A. Gibbes. This structure was reputedly the only house on the Combahee River spared by Union soldiers during Sherman's march. Martin described it as "large and square and white ... [like] most of the old Southern homes."

[70] Martin, "Henry W. Corning's Place."

[71] Chlotilde R. Martin, "Inventor Lives Quietly in Colleton," *News and Courier*, March 29, 1931, B6.

In the early 1930s, northerners developed several plantations that broke new ground in terms of lavishness, extravagance, and grandeur. William R. Coe's Cherokee Plantation offers a prime example. In 1930, Coe, a financier, bought Board House Plantation on the Combahee River. He and his wife, Caroline, immediately began turning it into a showplace. An avid horticulturalist, Coe named the plantation for the Cherokee rose, a species with resplendent white petals that thrives in the southeastern United States. The Coes hired the Olmsted Brothers Company to craft a grounds and garden plan. At the same time, they sent architects to Europe to buy rooms out of old castles. Once these men had obtained several suitably impressive interiors, they proceeded to design a house around them. The mansion at Cherokee is a hulking Neo-Georgian edifice set amid elaborately landscaped grounds. Built of red brick and adorned with a full-height pedimented portico, the house contains thirty-two rooms. Mansard-roofed flankers stand alongside the main block, connected by delicate hyphens. Keystones accent the façade windows and a broken pediment rests above the entry door. A model of symmetry, balance, and rational order, the house at Cherokee stands as an opulent yet formulaic interpretation of a plantation mansion. More a Georgian-era country manor than a lowcountry plantation house, it brought an amalgam of English and Tidewater Virginia–inspired elements to a former rice plantation.[72]

The grounds at Cherokee are as striking as the main house. Visitors approached the Coes' residence by traveling along a causeway buttressed by cypress trees. Rows of tall cedars line a broad lawn in front of the house, creating a setting that exudes formality and ensures a view from a distance. The resulting effect adheres to a long tradition of European landscape design. The open expanse highlights the monumental size and architecture of the house and its integration with the surrounding landscape. Visitors are compelled to consider the entire composition from a distance, thereby accentuating the scale and majesty of the ensemble. In the rear of the house, a meandering lawn bounded by azaleas stretches out toward old rice fields.[73]

The Coes created one of the most elaborate estates in the lowcountry. It is also among the most unusual. Unabashedly extravagant, it displayed pretention in ways not previously seen. Even the most elegant houses

---

[72] Linder, *Historical Atlas of the Rice Plantations of the Ace River Basin*, 72–73; Baldwin, *Lowcountry Plantations Today*, 250–263.
[73] Linder, *Historical Atlas of the Rice Plantations of the Ace River Basin*, 72–73.

built before the Civil War lacked the scale, grandeur, and ornamentation that distinguished the dwelling at Cherokee. The same held true for the grounds, which exceeded the most pretentious antebellum examples. Although only a few northerners created similarly ambitious estates, Cherokee did not stand alone. Increasingly, northerners created truly opulent plantations that showed comparable attention to refinement and material display. Characterized by large mansions, elaborate landscaping, and extensive use of historical motifs, these estates outstripped the comparatively modest efforts of earlier years. They expressed grandeur clearly and conspicuously and showed the lowcountry to have arrived as a seasonal destination for monied, status-conscious elites.

Estates such as Cherokee heralded the crossing of yet another threshold. If the making of entirely new "plantations" represented one innovation of the estate-making boom, the arrival of estates whose form and features ignored regional traditions supplied another. Cherokee did not incorporate remains from earlier periods or draw inspiration from historical varieties of lowcountry architecture. Instead, it brought a "manorial" vision to the lowcountry, evoking the country houses of rural England, the Georgian plantation mansions of Tidewater Virginia, and the great estates of Long Island. Rather than growing out of indigenous traditions, estates such as Cherokee showed the increasing influence of designers who served an elite clientele. Architects and landscape architects hired by the wealthiest sportsmen and sportswomen relied on a semi-standardized architectural vocabulary and took a plainly formulaic approach to their work. Eager to please well-heeled clients, they drew upon a wide variety of historical styles in crafting large, lavishly styled houses and landscapes. The results satisfied owners while showing the growing influence of broad trends in design and new standards for conspicuous display.

Even as estates such as Cherokee took shape, most northerners exercised greater restraint. Many preferred to rehabilitate old plantations, which provided a greater range of options and avoided the "newness" that came with new construction. The approaches established in earlier years continued to guide northerners' efforts but the lines between them blurred. Whereas in the 1910s the differences among them had been clearly drawn, continued evolution created strong overlap. Restoration, retention of decay, and elaboration and embellishment remained part of northerners' repertoire but became differing points of emphasis more than independent approaches. Increasingly, northerners blended all three.

Restoring old houses remained central to sportsmen's and sports-women's efforts. At the Wedge Plantation on the Santee River, E. G. Chadwick rehabilitated an elegant residence built in 1826. When he acquired the plantation in the spring of 1929, the main house stood in an advanced state of decay. Workers carefully restored the building to its antebellum appearance and built a new addition in the rear. Nearby they erected a group of matching outbuildings. "Velvety green" lawns and camellias, magnolias, and small shrubs occupied more than ten acres. Viewing the estate after about two years of intensive work, *News and Courier* reporter Chalmers S. Murray credited Chadwick with capturing "the atmosphere of Colonial America." He viewed Chadwick's estate as belonging to "a different age."[74]

At neighboring Harrietta Plantation, Horatio Shonnard, a stock-broker, undertook the single most conscientious restoration of a planta-tion dwelling in the Georgetown area. Shonnard and his wife, Sophie, acquired Harrietta in August 1929. Eager to begin wintering on the Carolina coast, they immediately launched an ambitious restoration and grounds rehabilitation campaign. Workers carried out a painstaking renovation of the main house, a two-story central-hall dwelling built in the late eighteenth century. W. C. Bujac, the Shonnards' superintendent, beautified the grounds with 6,000 azaleas (twenty-six different varieties in all), 300 wisteria vines, 56 sweet olives, and 150 dogwood trees. These plantings supplemented a large number of existing camellias, all "perfect specimens" at least 100 years old. When completed, Harrietta ranked among the most elegant estates on the entire coast. The Shonnards' efforts won accolades from locals and inspired a feature article in *House Beautiful* magazine.[75]

Farther south, Donald Dodge purchased the "old William Seabrook Plantation" on Edisto Island. Revered locally for its association with Lafayette's visit to America in 1824, the plantation stood in severe distress. The Dodges rehabilitated the main dwelling, redecorated the interior, and installed modern amenities. They also developed the grounds with lawns of ryegrass, brick and crushed-shell walkways, and

---

[74] Chalmers S. Murray, "The Wedge on the South Santee," *News and Courier*, Sept. 20, 1931, A7. See also J. M. L., Jr., "Do You Know Your Lowcountry? The Wedge," unidentified newspaper clipping in SCVFC.

[75] Chalmers S. Murray, "Harrietta on Santee Blooms Again," *News and Courier*, Sept. 13, 1931, A7; Sara Furman, "Harrietta: An Old Plantation House on the Santee River," *House Beautiful* 70, no. 6 (Dec. 1931): 475–480.

ornamental plantings. Through their efforts, the Seabrook estate again became what it had been in its heyday: a showplace.[76]

The impulse to leave decay in place remained influential but ceased to be used as a wholesale approach. Instead, it became a specialized treatment applied to discrete buildings and settings. The Franks may have been the last to employ it in an all-encompassing manner. From the mid-1920s onward, northerners singled out buildings they considered especially quaint and left them untouched. By situating such structures as picturesque ruins, northerners highlighted their appearance, decayed state, and contrasts with newly built and "restored" buildings.

At Waverly Plantation on the Waccamaw, for example, Reverend J. D. Paxton rehabilitated a surviving domestic complex and the surrounding grounds but deliberately left an old slave chapel untouched. He repaired the roof of the chapel but took pains to "proceed carefully" in order to leave the "original effect" undisturbed.[77] At Clay Hall Plantation in Beaufort County, G. L. Redmond incorporated four "ancient" slave cabins into his estate. Redmond built a Colonial Revival–style mansion, a stable, kennels, and a caretaker's residence. At the same time, he left the slave cabins exactly as he found them. Built of English brick and reputedly erected before the Revolution, they stood in an "excellent state of preservation." Commenting on Redmond's estate, *News and Courier* reporter Chlotilde R. Martin wrote, "the new has been delightfully blended with the old, the comforts of the one and the charm of the other uniting to create an atmosphere that is altogether charming."[78]

Rehabilitation coupled with elaboration and embellishment remained a reliable method of turning aged plantations into well-developed estates. As in earlier years, northerners made use of surviving remains, refashioned them to meet their needs, and added new elements as they saw fit. At Friendfield Plantation on the Sampit River, Radcliffe Cheston, Jr., erected a handsome new house as a replacement for an antebellum dwelling that had burned in 1926. Taking pains to "recreate as nearly as

---

[76] Chalmers S. Murray, "Lafayette House on Edisto Restored," *News and Courier*, July 19, 1931, B3. Murray wrote that because of the Dodges' efforts, "the old house stands today in a perfect state of preservation." For a similar example, see Linder, *Historical Atlas of the Rice Plantations of the Ace River Basin*, 201–217; and Isabella G. Leland, "Do You Know Your South Carolina? Edisto's Grove Plantation Was Built after Revolution," *News and Courier*, June 2, 1958, B1.

[77] Chalmers S. Murray, "Waverly, Gem of the Waccamaw," *News and Courier*, July 26, 1931, A6.

[78] Martin, "Clay Hall, Redmond Plantation Lodge."

possible the atmosphere of the former mansion," he erected a "spacious frame house" fashioned "along colonial lines." With twenty-two rooms, an elegant entrance, and a grand portico, the house closely resembled the original and afforded Cheston and his family enviable accommodations for their stays on the Carolina coast.[79] At Hope Plantation in Colleton County, Z. Marshall Crane had the Charleston architectural firm of Simons and Lapham turn a sturdy but undistinguished house into a columnar mansion. Simons and Lapham incorporated the existing structure into the new dwelling. Whether any of the former remained visible afterward is unclear. The house they created stood two stories tall beneath a hipped roof. Built of frame construction with a monumental portico, it supplied Crane and his family with a stately residence for their winter sojourns. A long avenue of mature oaks and twelve-acre lawn added formality to the site and accentuated the building's statuesque appearance.[80]

In the mid-1930s, the pace of land acquisitions slowed and the final phase of estate-making began. It brought further innovation in the form of elaborate, well-appointed estates and new architectures. In general, estates created during this period occupied two categories. One continued the "manorial" theme with formulaic ensembles of architecture and landscape. These brought opulent but stereotyped designs for large houses, outbuildings, and gardens to the rural lowcountry. They showed that northerners increasingly conceived of the lowcountry less as a world unto itself and more as one destination within an expanding network of seasonal retreats. The second variety of estates introduced modern architecture and a few exotic styles to the region. These employed imaginative building programs that broke decisively with regional traditions. Some introduced architectural styles never before considered appropriate to the lowcountry – or, for that matter, plantations anywhere – while others showed inspired attempts to accommodate northerners' functional and aesthetic demands.

Sportsmen and sportswomen created estates modeled on the manorial idiom throughout the lowcountry. In the wake of the Coes' efforts, northerners created at least a dozen estates with similar characteristics. In Georgetown County, Walker P. Inman, a half-brother of tobacco heiress Doris Duke, purchased Greenfield Plantation on the Black River.

---

[79] Chalmers S. Murray, "Cheston Builds on Old Plantation," *News and Courier*, Aug. 2, 1931, B2.

[80] Chlotilde R. Martin, "Crane House Reflects Old South," *News and Courier*, March 1, 1931, A6.

He immediately erected a "stately" Neoclassical mansion composed of a large, two-story central bock with projecting wings. The Palm Beach–based firm of Wyeth and King, which catered to society clients, supplied the design. According to one account, the architects studied "an old Georgian homestead" to determine the appropriate scale and profile for window and door openings. The New York firm of Innocenti and Webel crafted a grounds plan, and Inman spared neither effort nor expense in putting its recommendations into place. When completed, a well-developed sequence of ornamental gardens, groomed lawns, and artificial lakes surrounded the residence. The most impressive expanse lay in the rear of the building, where a formal garden filled with camellias descended to a large lake lined with "hundreds of azaleas."[81]

Less than a year later, Robert Goelet created a similar estate at Wedgefield Plantation. Heir to a Manhattan real estate fortune, Goelet continually made headlines for his brazen driving habits and playboy lifestyle. He and his wife, Ann, bought Wedgefield from William Winchester Keith, a Baltimore man who had owned it since 1932. During his use of the property, Keith had rehabilitated the grounds and main house. The dwelling, built along "quaint colonial lines," dated to the early nineteenth century. A visitor who saw the building in the fall of 1931 judged it "one of the most attractive structures of its kind in the country" and noted that it never failed "to interest the lovers of early American architecture." The Goelets promptly had it demolished. They commissioned William L. Bottomley, a prominent New York architect, to design a domestic complex suited to their needs. Bottomley crafted a one-story dwelling with wings and an elaborate entrance. Regency-style accents gave the structure an elegant appearance. As at Greenfield, Innocenti and Webel designed the grounds plan, which included a spacious forecourt and extensive gardens. In the rear of the house, flowering shrubs wrapped around a terrace that the Goelets used for outdoor entertaining. Broad lawns stretched away into the distance, descending toward a cloverleaf lake and continuing to rice fields at the edge of the Black River.[82]

---

[81] Lachicotte, *Georgetown Rice Plantations*, 89–93; "Carolina Classic," *House and Garden* 73, no. 2 (Feb. 1938): 34–35.

[82] Chalmers S. Murray, "Wedgefield Planted for Game," *News and Courier*, Sept. 27, 1931, A9; "Goelet Plantation Mansion Started Near Georgetown," *News and Courier*, Nov. 10, 1935, C2; "Wedgefield Plantation," *Country Life*, Dec. 1938, 53–57, 101–102. See also Linder and Thacker, *Historical Atlas of the Rice Plantations of Georgetown County and the Santee River*, 441–444; Lachicotte, *Georgetown Rice Plantations*, 71.

Other estates struck a similar chord. At Poco Sabo Plantation on the Ashepoo River, Silas Wilder Howland erected a spacious neo-Palladian mansion. Designed by Simons and Lapham, it stood two stories tall with a center hall plan, delicate classical details, and gleaming white elevations. Howland turned to a local firm for landscaping, which resulted in a plainer grounds scheme than found at many estates.[83] At Dixie Plantation on the Stono River, Vincent Friermonte and his wife, the former Mrs. John Jacob Astor, hired the New York firm of Polhemus and Coffin to turn an existing dwelling into a handsome mansion. The remodeled house featured a soaring Corinthian portico and sumptuous interior decor.[84] In Jasper and Colleton counties, architect Willis Irvin of Augusta designed at least five houses for northern owners. Irvin billed his residential designs as "characteristic of the Old South." He excelled at producing plans for handsome mansions in what had become popularly known as the "Southern Colonial" style and found strong demand for his services among northern sportsmen. A skilled purveyor of houses that featured bold entablatures, denticulated cornices, and sumptuous interiors, Irvin catered to the conservative tastes of an established elite.[85]

At the opposite end of the stylistic spectrum, two plantations with modernist building complexes supplied different solutions to sportsmen's and sportswomen's needs. Stylistically, neither showed discernible respect for tradition. Like all modernist designs, they purposefully avoided historical references. Nonetheless, both estates showed thoughtful consideration of the practical and symbolic demands associated with plantations used for leisure. The relative success of each depended largely on personal taste. Still, by seeking to break free of historical constraints and confronting familiar problems head-on, these plantations made innovative bids to move beyond established formulas. In fact, each showed greater clarity of purpose than most of their counterparts.

In 1936, Clare Boothe Luce and Henry R. Luce acquired Mepkin Plantation on the Cooper River. Once the property of the slave-trading patriot Henry Laurens, Mepkin had been owned since about 1912 by J. W. Johnson, one of three brothers who founded the Johnson &

[83] Alexander Moore, *Poco Sabo Plantation: A Place in Time* (N.p.: H. Anthony Ittleson, 2005), 50–54.

[84] "Fiermonte Buys 600-Acre Place," *News and Courier*, Dec. 1, 1935, B1; "The Fiermontes at Dixie Plantation," *News and Courier*, Jan. 7, 1935, 14; J. V. N., Jr., "Do You Know Your Charleston? Dixie Plantation," *News and Courier*, March 1, 1937, 10.

[85] Irvin designed houses for the following plantations: Clarendon, Gregorie Neck, Davant, Castle Hill, and Bonnie Doone. See Willis Irvin, *Selections from the Work of Willis Irvin – Architect* (N.p.: n.p., 1937).

FIGURE 2.4 Main house at Mepkin Plantation, 1936. Photograph by Samuel Gottscho, Gottscho-Schleisner, Inc.
Courtesy of the City Museum of New York, New York, NY.

Johnson Company, manufacturers of surgical bandages. When the Luces acquired the plantation, they opted to demolish the house that Johnson had built, a two-story dwelling with a shingled exterior. They commissioned Edward Durrell Stone, then a rising star among architects, to design a residential complex. Stone drafted plans for a group of buildings centered on a large green and framed by a series of garden walls. His composition drew upon what he conceived of as the traditional hierarchy of plantation architecture. The Luces' residence stood as the largest and most prominent of the buildings in the group; Stone considered it the "big house." Nearby he placed three guest cottages, which Stone likened to slave cabins. Garden walls modeled on designs typically seen in Charleston tied the group together, enclosing the green and creating visual unity. Nearby stood a group of outbuildings that included a stable, kennels, and a row of workers' houses (see Figure 2.4).[86]

---

[86] Edward Durell Stone, *The Evolution of an Architect* (New York: Horizon Press, 1962), 48–51; "Mepkin Plantation, Moncks Corner, S.C.," *Architectural Forum* 66, no. 6 (June 1937): 515–522; "Modern in South Carolina, Winter Home of Mr. and Mrs. Henry R. Luce, Moncks Corners, South Carolina," *House and Garden* 72 (Aug. 1937): 36–39.

Stone's complex articulated a restrained version of domestic modernism. Less austere than his other residential designs of the era, the buildings at Mepkin combined avant-garde aesthetics with a spatial order inspired by popular views of the plantation South. The main buildings, built of brick and painted white, had angular features and block-like profiles. Casement windows further accentuated the industrial aesthetics. Most observers found the complex bewildering. One journalist characterized it as having "severely simple lines" and considered the overall effect "debatable." He did, however, add that the buildings were "not unpleasing to the eye" and appeared comfortable.[87] In a derisive reference to contemporary gas station designs, locals reputedly referred to the complex as the "Luce filling station."[88] Although Stone's complex satisfied the Luces' needs, it also made Mepkin one of the most unusual of all northern-owned estates. By rejecting some historical traditions and embracing others, Stone's complex offered a novel solution to the challenge of a lowcountry sporting retreat.

Auldbrass, the other estate with a modernist complex, lay far to the south in Beaufort County. In the mid-1930s, Charles Leigh Stevens, a New York business consultant, acquired a 4,253-acre tract on the Combahee River east of Yemassee. From the beginning, Stevens envisioned a refuge where he could escape the frenetic pace of life in New York City. He intended to hunt, relax, and run a model farm for pleasure and profit. Stevens commissioned Frank Lloyd Wright, then in his early seventies, to design a building complex for the estate. Working closely with his client, Wright devised plans for a group of low, linear buildings. Whereas Stone's complex at Mepkin stood apart from its surroundings, Wright embraced the tonal palette and visual textures of the lowcountry. He fashioned the buildings out of cypress and took cues from the natural environment in developing their general form and most prominent features. The exterior walls, for example, lean out at angles of eighty-one degrees. Wright saw this as a nod to the live oaks at the site – which, he observed, never grew straight up. Decorative elements show similar influences. The copper downspouts of the main house form stylized interpretations of Spanish moss, and the cypress panels that frame the clerestory

[87] Thomas R. Waring, Jr., "Do You Know Your Charleston? Mepkin Plantation," *News and Courier*, Jan. 24, 1938, 10.
[88] Sylvia Jukes Morris, *Rage for Fame: The Ascent of Clare Boothe Luce* (New York: Random House, 1997), 297.

windows are fashioned to look like feathered arrows, in reference to the Yemassee Indians native to the area. Wright thought deeply about the design of the complex and its relationship to the surroundings. He created a building group that integrated itself more fully, and more successfully, with the lowcountry landscape than any other.[89]

Recognizing leisure as the primary purpose of Stevens's estate, Wright did away with any sense of hierarchy in order to create a thoroughly informal setting. In this sense his design differed from Stone's, which saw no conflict between traditional patterns of plantation organization and the Luces' needs. The main house at Auldbrass stretches out in a long, slender shape that resembles a ship more than a domestic building. Wright designed it around a gridiron of hexagons four feet across, deliberately without a discernible center. He favored a "general-plan form" to establish intimacy with the surrounding forest glade. Large window bays serve the same purpose. Aesthetically, the main house differs little from the other large buildings at the site, which include a mule and cow barn, a chicken coop, a kennel, and a caretaker's cottage. The building group is therefore decidedly egalitarian. Set amid a grove of live oaks, it forms an irregular sequence of angular, oddly shaped buildings that seem to rise gently out of the earth to shallow peaks a short distance above.[90]

Auldbrass is one of a handful of estates that became as much folly as active retreat. Stevens's business commitments and unstable personal life hampered construction of the complex, and Wright's idiosyncrasies and heavy commission schedule introduced further delays. Rising costs and a sharp downturn in Stevens's income brought construction to a halt in December 1942. The interior of the main house remained unfinished and work on several of the farm buildings had not begun. Construction did not resume until the late 1940s. By then the lowcountry had lost much of its cachet as an upper-class destination. Although some sportsmen and sportswomen continued wintering in the region, their numbers had declined and the lowcountry sporting scene had lost much of its vigor.[91]

Although Auldbrass never fulfilled its promise, it still spoke plainly of the bold ambitions northerners brought to the lowcountry at the peak of the estate-making boom. As new sporting estates proliferated, ideas about plantations became increasingly diffuse. Some owners pushed aesthetic

[89] David G. De Long, *Auldbrass: Frank Lloyd Wright's Southern Plantation* (New York: Rizzoli, 2003), chaps. 1–5.
[90] Ibid., 76–78, 86–89, and passim.   [91] Ibid., chap. 4.

and practical boundaries in new directions. Estates as such as Auldbrass and Mepkin figured among the results. Although neither qualified as a complete success, each explored important terrain. As examples of what open-minded owners and two exceptionally talented designers created, they showed how far northerners' efforts had progressed and how carefully some winter colonists thought about the design problems associated with "plantations" devoted to leisure.

Beyond these examples, a few northerners created estates that made equally decisive breaks with tradition. Atalaya, a Moorish Revival "castle" built by Archer M. Huntington and his wife, Anna Hyatt Huntington, supplies a telling example. In January 1930, the Huntingtons purchased four plantations on the upper end of the Waccamaw Neck. This transaction assembled an estate of about 6,600 acres on an exceptionally scenic stretch of coastline. The Huntingtons subsequently expanded their holdings to more than 9,100 acres. Unlike virtually all of their peers, they had no interest in hunting. Archer preferred to watch wildlife "under natural conditions" and Anna devoted her time to artistic pursuits. An accomplished sculptor, she created figurines, animals, and other subjects in bronze and other metals. She also supported the work of other artists. The Huntingtons named their estate "Brookgreen" after one of the plantations they purchased.[92]

Although an old mansion survived at Brookgreen, the Huntingtons decided to demolish it. Amid the surviving grounds and landscaping they developed a sculpture garden that gave Anna a forum for displaying her artwork and sculptures by other artists. As the garden took shape, the Huntingtons began construction of a new residence. At a beachfront site they erected a Moorish Revival–style building. Built of brick and reinforced concrete with a stuccoed exterior, the structure has a square footprint measuring two hundred feet on each side. More than thirty rooms situated along three sides of the interior perimeter supply the principal living spaces. The beachfront side enclosed the Huntingtons' dining and breakfast rooms, a library, a secretarial office, and, at the far southeastern corner, the master bedroom. At the rear of the south side, two large studios gave Anna ample working space. A large courtyard occupies the center of the structure. This space is bisected by a covered

---

[92] Chalmers S. Murray, "Huntington House Reminds One of Africa," *News and Courier*, June 7, 1931, A8; Atalaya and Brookgreen Gardens, National Register of Historic Places Nomination, South Carolina Department of Archives and History, Columbia, SC. Atalaya is Spanish for watchtower.

walkway lined with open arches and brick screenwalls. A tower nearly forty feet tall rises from the midpoint. A tank mounted inside supplies water to the complex. In the rear of the building, a small courtyard created an open-air workspace in the center of a service area. Nearby stood stables, kennels, woodsheds, and supply rooms.[93]

Work on Atalaya began in January 1931 with Archer personally supervising construction. The project represented a boon to the local area, for it supplied paying jobs where few had existed for some time. The Huntingtons brought skilled craftsman in from the Newport News shipyard to train local workers in bricklaying and other skills. By March 1932, the mammoth structure neared completion. *News and Courier* reporter Chalmers S. Murray aptly observed that the house had "no rival in this part of the country." He described it as "built along modified Spanish lines," similar to "the general designs used in ... North Africa," and reminiscent of "scenes in Tunis and Algiers."[94]

Dramatically spare in its appearance, Atalaya is arguably the most unusual of all the buildings northerners erected. Its design expressed appreciation for Iberian history and culture, one of Archer Huntington's major interests. It also introduced an entirely foreign architecture to the lowcountry. To many observers, Atalaya appeared odd and virtually inexplicable. Other northerners, however, also introduced new architectures to the region. Collectively, these buildings challenged prevailing assumptions about plantations and raised questions about the aesthetic parameters of the type. As a group, they suggested how malleable the idea of a "plantation" had become.

At Richmond Plantation on the eastern branch of the Cooper River, George Ellis, Jr., built a Tudor-style residence that would have been more at home in the English countryside than in the rural lowcountry. His "lodge," as one account characterized it, took a long, linear form. Built of salvaged eighteenth-century bricks, it had squat walls and a large, steeply pitched roof of purple-gray slate. Uneven surfacing accentuated the mottled appearance. Window frames and porch posts fashioned out of salvaged wood complemented the aged look of the brickwork, and several old oak doors imported from England enhanced the overall effect. Essentially an attempt to create a late-medieval residence from scratch, Ellis's house pushed aesthetic boundaries in a new and unusual direction.

---

[93] Atalaya and Brookgreen Gardens, National Register of Historic Places Nomination, sec. 7, pp. 2–4, sec. 9, pp. 3–4.
[94] Murray, "Huntington House Reminds One of Africa."

By symbolically linking Richmond to the sporting traditions of the English gentry, the house showed how freely some northerners associated the lowcountry with particular varieties of upper-class life.[95]

At Hasty Point Plantation, Jesse Metcalf built a similar dwelling. Metcalf, a wealthy playboy with a healthy appetite for sporting pursuits, bought five plantations along the Pee Dee River in the late 1920s. By 1930 he and his wife, Deborah, owned about 15,000 acres. Their residence combined the rambling form of an English cottage with a scale characteristic of the classically influenced mansions standing on many nearby estates. Set amid a grove of live oaks, the house had a bowed façade that looked out upon the river. Brick masonry and cypress-timbered construction gave the building a weighty, solid appearance. Inside, "wrought iron hinges and latches, massive cross beams, brick and heart pine floors, [and] generously proportioned fireplaces" carried out an "old English" theme, similar to Ellis's estate. Throughout the house, the Metcalfs displayed hunting trophies, rifles, and "other paraphernalia dear to the heart of the sportsmen."[96]

By the late 1930s, the estate-making boom had largely run its course. Northern "winter colonists" had become an important part of the lowcountry scene. From mid-November through the late spring of each year, their activities enlivened usually quiet stretches of territory. Hunting parties tromping through the woods and paddling across shimmering marshes symbolized the lowcountry's resurgence. In the same moment as Charleston became a popular tourist destination and Americans consumed tales of a romantic southern past with new enthusiasm, the lowcountry gained recognition for grand and elegant "plantations." Northerners' estates stood at the forefront of popular views of the region and its past. As symbols of a land with a storied history, they revealed the lowcountry's emergent roles within the South and the nation.

Northerners' estates represented the single most dramatic development to occur in the rural lowcountry for decades. Not since the advent of large-scale tidal rice production during the second half of the eighteenth century had such a profound transformation taken place. Although the destruction of the Civil War and the decline that set in afterward had introduced extensive changes, the overall effects paled by comparison.

---

[95] "Shooting Lodge in Carolina: 'Richmond,' the Estate of George A. Ellis, Jr., Esq., Near Moncks Corner, South Carolina," *Country Life* 65, no. 2 (Dec. 1933): 60–61.

[96] Chalmers S. Murray, "Metcalf Builds on Peedee Bluff," *News and Courier*, March 22, 1931, B14.

Northerners' estates surpassed everything in recent history. Just as the rise of tidal rice culture had reorganized land and space, so too did the making of plantations devoted to leisure. Northerners did more than simply rehabilitate old plantations; they reshaped surviving remains, erected new buildings, treated old houses with reverence and respect, created park-like settings, and strived for visual and stylistic coherence. They managed large acreages for recreation, rehabilitated tidal fields, and allowed forest and swamplands to revert back to a seemingly "natural" state. They created estates that accommodated upper-class protocols while displaying wealth and power. In short, the estate-making campaign had profound consequences. Although contemporaries marveled the process, its full significance proved illusive. Only when viewed critically, with the benefit of historical distance, would the overall effects become clear.

Northerners' activities transformed large stretches of the lowcountry. As late as World War I, the region had been a near-endless expanse of ruined plantations and overgrown fields. By the mid-1930s, large, lavishly styled estates could be found all along the coast. Although large stretches of countryside still lay idle and many plantations lay in ruins, the former sense of isolation and stasis had passed. Signs of decay had diminished appreciably. As northerners brought new vitality to the region, material renewal became ubiquitous. Manicured lawns, well-kept domestic complexes, and tidal fields maintained as waterfowling domains signaled the arrival of recreation as a major social and economic force.

Evolution of plantations beyond their existing form and meaning figured as the most significant consequence of the estate-making boom. Northerners' estates transformed plantations in multiple ways. Although the process of material change captured the most attention, the underlying assumptions proved equally significant. During the late 1920s, northerners' estates crossed a threshold. For the first time in history, it became possible to think of plantations principally in terms of physical appearance. For centuries, the union of staple agriculture and unfree labor had anchored ideas about plantations and their physical expression. The transition from slave to free labor in societies throughout the Americas had introduced profound changes, but the survival of intensive agriculture and coerced labor left the fundamental precepts of the concept in place. Northerners' estates moved beyond the old idiom. By making material form and spatial order principal characteristics, they moved a select group of plantations onto new terrain. Moreover, they showed how little Americans recalled about plantation slavery. For Americans to accept, let alone celebrate, the remaking of former sites of racialized exploitation offers an object lesson in humans' capacity to rationalize injustice. By World War I, most Americans accepted the fiction of an aristocratic slave society as the true history of the pre–Civil War South. Few Americans, north or south, saw Reconstruction as anything but a horrific mistake. Once these beliefs became widely accepted, the rehabilitation of former sites of slave labor for new purposes lay only a short step beyond.[97]

The estate-making boom marked a watershed in lowcountry history. Although plantations had dominated the lowcountry landscape since the early eighteenth century, northerners' estates differed from their historical predecessors. Although many of the names and locations remained the same and much of the material fabric did also, otherwise, the new estates differed profoundly. In use, material form, and meaning, northerners' plantations occupied a unique realm. For most of lowcountry history, plantations had produced agricultural staples for distant markets. Through systematic exploitation of black labor, planters had achieved levels of production and efficiency that would otherwise have been impossible. Northerners' estates supplied people of wealth and privilege with seasonal retreats and venues for social intercourse. Between the two types

---

[97] Attempts to establish all-encompassing definitions of plantations are notoriously problematic. In emphasizing the union of unfree labor and staple agriculture, I stress social and cultural factors fundamental to the majority of plantations during the era of the trans-Atlantic slave trade and the aftermath of the emancipations of the nineteenth century. For relevant definitions, see Philip D. Curtin, *The Rise and Fall of the Plantation Complex: Essays in Atlantic History*, 2nd ed. (New York: Cambridge University Press, 1990); Robin Blackburn, *The Making of New World Slavery: From the Baroque to the Modern, 1492–1800* (New York: Verso, 1997), chap. 8; Ira Berlin, *Many Thousands Gone: The First Two Centuries of Slavery in North America* (Cambridge: Belknap Press of Harvard University Press, 1998), 95–108; Charles S. Aiken, *The Cotton Plantation South since the Civil War* (Baltimore: Johns Hopkins University Press, 1998); Jack Temple Kirby, *Rural Worlds Lost: The American South, 1920–1960* (Baton Rouge: Louisiana State University Press, 1987), chap. 1; Stuart B. Schwartz, *Sugar Plantations in the Formation of Brazilian Society: Bahia, 1550–1835* (New York: Cambridge University Press, 1985), chap. 1; Bureau of the Census, *Plantation Farming in the United States* (Washington, DC: Government Printing Office, 1916), 7–14; Bureau of the Census, *Special Study: Plantations* (Washington, DC: Government Printing Office, 1940), 5; Edgar T. Thompson, *The Plantation* (1932; reprint, Columbia: University of South Carolina Press, 2010), 3–22. Recent scholarship has emphasized the instrumentality of plantations. See S. Max Edelson, *Plantation Enterprise in Colonial South Carolina* (Cambridge: Harvard University Press, 2006), 2–6; Joyce E. Chaplin, "Tidal Rice Cultivation and the Problem of Slavery in South Carolina and Georgia, 1760–1815," *William and Mary Quarterly* 49, no. 1 (Jan. 1992): 50. On remembrance of slavery, see especially Blight, *Race and Reunion*; Silber, *Romance of Reunion*; Janney, *Remembering the Civil War*.

lay a gulf of ideas, experience, and belief. Although people continued to speak of plantations as though no dramatic changes had occurred, in fact, many had. The differences between old plantations and new estates could hardly have been greater. Throughout the region, plantations afforded venues for distinctive varieties of leisure and recreation. In them, new views of the lowcountry and its past coalesced.

3

New Lowcountry, New Plantations

Contemporary views of northerners' plantations masked complicated realities. Although most observers saw the new estates as old plantations rehabilitated and adapted for new purposes, they differed from earlier plantations and those that remained agricultural enterprises in several ways. Larger, more sumptuous, and characterized by different patterns of spatial organization, northerners' "plantations" broke new ground. Moreover, although beliefs about plantations of past eras and the presence or absence of surviving remains influenced the material form of northerners' estates, other factors also played a role. The single most important source of inspiration came from contemporary country estates, a type well known to estate owners. Men and women who created lowcountry estates had close ties to estates on Long Island, on the outskirts of Philadelphia, and in Delaware's Brandywine Valley. A few owned "cottages" at Newport, Rhode Island. In short, although northerners referred to their lowcountry properties as "plantations," just like other people of the era, the ideas expressed in their creation drew heavily upon contemporary idioms.[1]

---

[1] Information about owners' backgrounds is derived from a sample of 104 for whom reliable biographical information is available. A general overview is Daniel Vivian, "Who Were the 'Rich Yankees?': Owners of Sporting Plantations in the South Carolina Lowcountry, 1900–1950," paper delivered at the annual meeting of the Agricultural History Society, Briarcliff Manor, NY, June 24, 2016. See also George C. Rogers, Jr., *The History of Georgetown County, South Carolina* (Columbia: University of South Carolina Press, 1970), 489–496; John H. Tibbetts, "The Bird Chase," *Coastal Heritage* 15, no. 4 (spring 2001): 3–13; Virginia C. Beach, *Medway* (Charleston: Wyrick and Co., 1999), 38–39.

Considering northerners' estates as a group brings several unifying characteristics into view. Collectively, these demonstrate the "newness" of northerners' "plantations." Nomenclature proved deeply misleading. Although the term "plantation" remained widely used, its meaning morphed as northerners created their estates. Not only did the new "plantations" differ from historical examples in purpose, they occupied more land, displayed greater refinement, and organized themselves according to owners' priorities. Although owners' social status and relations between owners and laborers echoed antebellum precedents, otherwise, northerners' estates occupied their own realm. As products of upper-class social practices, common beliefs about historical plantations, and efforts to combine valued remains with new features, northerners' plantations assumed unique form.

PLANTATIONS OLD AND NEW

During the late antebellum era, plantations in areas where northerners later developed large estates ranged in size from several hundred to several thousand acres. On the barrier islands, Sea Island cotton plantations averaged about 750 acres. A few exceeded 1,000 acres but most occupied less territory. Of the 359 recorded by the 1860 census, 52 percent occupied between 500 and 999 acres.[2] Tidal rice plantations exhibited greater variation. Plantations near Georgetown averaged between 800 and 930 acres, with a median of 730 acres. A few encompassed between 400 and 600 acres, for rice fields at favorable locations could be highly productive despite limited acreage. Along the Ashepoo, Combahee, and Edisto rivers, rice plantations proved somewhat larger, with an average size of 1,376 acres. Throughout the lowcountry, plantations of several thousand acres could be found, but numbers alone do not tell the full story. Especially large plantations, virtually to a rule, encompassed forest and other unimproved lands that planters used mainly as foraging grounds and for timber. Acreages devoted to rice did not expand significantly as overall size did.[3]

[2] Charles F. Kovacik and Robert E. Mason, "Changes in the Sea Island Cotton Industry," *Southeastern Geographer* 25, no. 2 (Nov. 1985): 90.
[3] Statistics derived from figures presented in Suzanne Cameron Linder, *Historical Atlas of the Rice Plantations of the ACE River Basin – 1860* (Columbia: South Carolina Department of Archives and History for the Archives and History Foundation, Ducks Unlimited, and the Nature Conservancy, 1995); Suzanne Cameron Linder and Marta Leslie Thacker, *Historical Atlas of the Rice Plantations of Georgetown County and the Santee River* (Columbia: South Carolina Department of Archives and History for the Historic Ricefields

Northerners' estates generally occupied larger land areas. The sixty-six for which reliable figures are available averaged 4,315 acres. Considered on a county-by-county basis, averages ranged from 1,841 acres in Williamsburg County to 5,969 in Colleton. Eighteen estates exceeded 5,000 acres, and eight possessed more than 10,000. Although eighteen (27 percent) occupied 1,400 or fewer acres, most encompassed considerably more. Twenty-eight (42 percent) had between 1,500 and 5,000 acres, and another nine (13 percent) occupied between 5,001 and 10,000 acres. Put simply, most northern-owned estates exceeded the size of typical antebellum plantations by a significant margin. Although slightly more than a quarter mirrored antebellum norms, the remainder occupied more land, some dramatically so.[4]

The basic reason for the difference in size lies with northerners' commitment to recreation. Obtaining sufficient acreage for high-quality hunting involved large-scale land ownership. Needs varied depending on priorities. Northerners who favored duck hunting privileged tidal fields and marshes in locations that made attractive feeding grounds. Upland terrain and forests offered habitat for quail, wild turkey, deer, fox, and doves. Landholdings that offered opportunities for multiple varieties of hunting, as many northerners favored, encompassed huge acreages. The sprawling estates owned by figures such as Bernard M. Baruch, Isaac Emerson, Edward F. Hutton, Jesse Metcalf, and Marshall Field III, all of which encompassed 14,000 acres or more, offer apt illustrations.[5] Smaller estates proved adequate for many northerners' needs, however. The twenty-seven that ranged between 3,000 and 10,000 acres, for example, offered sizable hunting domains, even if they fell short of their larger counterparts. Moreover, the variety of activities

---

Association, 2001); Alberta Morel Lachicotte, *Georgetown Rice Plantations* (Columbia: The State Printing Co., 1955); John R. Todd and Francis M. Hutson, *Prince Williams Parish and Plantations* (Richmond: Garrett and Massie, 1935); Henry A. M. Smith, *The Historical Writings of Henry A. M. Smith* (Spartanburg, SC: Reprint Co., 1988), vol. 1.

[4] Statistics derived from analysis of acreages obtained from the following sources: Linder, *Historical Atlas of the Rice Plantations of the ACE River Basin*; Linder and Thacker, *Historical Atlas of the Rice Plantations of Georgetown County and the Santee River*; Lachicotte, *Georgetown Rice Plantations*; *News and Courier* articles on individual estates cited in chaps. 2 and 6.

[5] Linder and Thacker, *Historical Atlas of the Rice Plantations of Georgetown County and the Santee River*, 53–54, 77–78, 370; Lachicotte, *Georgetown Rice Plantations*, 9–15, 18–25, 192–131; William P. Baldwin, *Lowcountry Plantations Today* (Greensboro, NC: Legacy Publications, 2002), 280; Grace Fox Perry, *Moving Finger of Jasper* ([Ridgeland, SC?]: n.p., 1962), 141–142.

that eventually became the norm for many owners also benefited from sizable landholdings. Horseback and carriage rides became more pleasurable with large acreages, and access to varied terrain resulted in improved hunting and fishing opportunities.[6]

Although discourses surrounding the new estates generally downplayed differences across time and space, the historical record reveals strong contrasts between northerners' estates and antebellum plantations. In most cases, northerners did not buy individual plantations but instead aggregated land associated with multiple historical plantations. Franklyn Hutton, for example, "wielded together seven plantations totaling 5,500 acres." C. W. Kress's Buckfield Plantation encompassed four "old" plantations, Savannah, Retreat, Recess, and Stokes. The Pratt Family's Cheeha-Combahee Plantation took in "a number of old plantations," including "Magwood, Wickman, Woodburn, Oak Hill, Middleton, Minot, Whaley, Townsend, Tarr Bluff, Riverside and Fields Point." In Georgetown County, Jacquelin Holliday's Nightingale Hall encompassed its namesake and two other plantations. Throughout the lowcountry, consolidation figured in the creation of a majority of northern-owned estates. Although some northerners purchased plantations whose boundaries had not changed since the antebellum era and others added only small tracts, most took a different course.[7]

Naming practices obscured the significance of consolidation. When northerners aggregated old plantations, they usually retained the historical name of the plantation where they maintained their residence and applied it to the resulting whole. Thus, under Hutton's ownership, Prospect Hill, a plantation of 2,440 acres before the Civil War, became an estate of 5,500 acres. H. K. Hudson turned Delta Plantation, originally a

---

[6] For discussions suggestive of the relationship between landholdings and recreation, see Neal Cox, *Neal Cox of Arcadia Plantation: Memoirs of a Renaissance Man* (Georgetown, SC: Alice Cox Harrison, 2003), 35–39, 91–92; Frances Cheston Train, *A Carolina Plantation Remembered: In Those Days* (Charleston: History Press, 2008), 54, 82–84, 92–97, 102; Sidney J. Legendre, "Diary of Life at Medway Plantation, Mt. Holly, South Carolina," private collection (copy in author's possession), May 17, 21, and 28 and Nov. 22, 1937, and Feb. 8, 1940.

[7] J. V. N., Jr., "Do You Know Your Lowcountry? Prospect Hill," *News and Courier*, May 9, 1938, 10; Linder, *Historical Atlas of the Rice Plantations of the ACE River Basin*, 477; Chlotilde R. Martin, "Buckfield, Kress Narcissus Farm," *News and Courier*, Apr. 12, 1931, B14; Chlotilde R. Martin, "Chee-ha – Combahee Plantation," *News and Courier*, Feb. 15, 1931, B10; Chalmers S. Murray, "Holliday Winters in Peedee," *News and Courier*, Aug. 30, 1931, B11. See also Linder, *Historical Atlas of the Rice Plantations of the ACE River Basin*, 69–75.

2,700-acre tract, into a 5,300-acre estate by combining it with three other plantations. Mepkin Plantation, historically a 3,100-acre tract, became a 7,200-acre holding under the ownership of Henry and Clare Boothe Luce. Rhetorical practices obscured the significance of these changes. Although some newspaper articles noted instances of consolidation, others did not, and all referred to new estates by the owners' preferred name. Representation thus implicitly presented northerners' estates as similar to their antebellum namesakes, despite fundamental differences in size and form.[8]

In other instances, northerners gave entirely new names to their estates, a practice that also proved misleading. Bernard Baruch took the title "Hobcaw Barony" from a colonial-era land grant that had encompassed slightly less than 14,000 acres at the bottom of the Waccamaw Neck. When Baruch began buying land on the peninsula in 1905, he made it his goal to own all the land contained in the original grant, a feat he achieved by about 1907. Baruch and his guests referred to his estate as "Hobcaw Barony" or simply "Hobcaw," but others added "plantation" to the title. References to "Hobcaw Barony Plantation" thus became common, even though no such plantation had existed historically. Similarly, Isaac Emerson named his estate "Arcadia," a term he took from Greek mythology that had no relation to any of the historical plantations it encompassed. Newspaper reports quickly turned "Arcadia" into "Arcadia Plantation."[9]

Organizational patterns also distinguished historical plantations from new estates. During the antebellum era, tidal rice and Sea Island cotton plantations generally possessed a dispersed layout, with buildings, work-yards, and productive activities spread over a large area. Although variations existed, the most common arrangements used a large portion of the total acreage. On Sea Island cotton plantations, planters built dwellings near the edge of marshes to take maximum advantage of sea breezes. Slave houses and outbuildings usually stood nearby. Planters initially

[8] J. V. N., Jr., "Do You Know Your Lowcountry? Prospect Hill"; Linder, *Historical Atlas of the Rice Plantations of the ACE River Basin*, 477, 479; Chlotilde R. Martin, "The Delta – Home of H. K. Hudson," *News and Courier*, Jan. 20, 1931, 7; Richard C. Rhett, Map of Mepkin Plantation, 1930, Plat Book F, p. 67, Berkeley County Register of Deeds, Moncks Corner, SC; "Mepkin Is Bought by N.Y. Publisher," *News and Courier*, Apr. 12, 1936, B5.

[9] Chalmers S. Murray, "Baruch Rebuilds at Hobcaw Barony," *News and Courier*, May 17, 1931, A8; Chalmers S. Murray, "Arcadia, Where LaFayette Stopped," *News and Courier*, June 21, 1931, B5; Chalmers S. Murray, "Attracted by Climate Northerners Become Land Barons of Carolina Coast," *News and Courier*, Dec. 15, 1929, A3; Chlotilde R. Martin, "Low-Country Plantations Stir as Air Presages Coming of New Season," *News and Courier*, Oct. 16, 1932, A6.

cultivated Sea Island cotton on high ground but later shifted portions of their crops to reclaimed salt marsh after recognizing that the latter yielded higher output. Elsewhere, planters grew provision crops such as corn and sweet potatoes.[10]

Tidal rice plantations exhibited similar intensity of land use with different organizational patterns. Tidal fields lay clustered along rivers and streams, stretched out in geometric grids bounded by embankments and canals. On many plantations, freshwater reserves and older inland fields lay farther inland. Slave houses stood nearby sites of labor, typically about 500 yards from the planter's residence. Especially large plantations and those that spanned rivers and streams often had multiple slave streets or "settlements," as contemporaries referred to complexes of slave housing and associated yards and garden plots. Barns, storehouses, and rice mills stood nearby, generally situated near sites of labor and, in the case of mills, near the river for the sake of water power. Planters cultivated food crops in upland fields and maintained pasture land for livestock. Forest provided foraging lands for hogs and supplied firewood and timber for fencing and building materials. Docks provided points of departure for clean rice and other products and sites for receiving supplies and some foodstuffs.[11]

Variations existed throughout the lowcountry, and it is important to recognize that any general portrait obscures important differences. Topography and hydrology shaped the layout and form of tidal rice plantations in ways that historians have yet to fully understand. Landforms that

[10] Kovacik and Mason, "Changes in the Sea Island Cotton Industry," 83–90; Richard D. Porcher and Sarah Fick, *The Story of Sea Island Cotton* (Charleston: Wyrick and Co., 2005), 138–144, 148–154, 359–380.

[11] Morgan, *Slave Counterpoint*, 104–107, 110–123; Edelson, *Plantation Enterprise in Colonial South Carolina*, chap. 3; Joyce E. Chaplin, *An Anxious Pursuit: Agricultural Innovation and Modernity in the Lower South, 1730–1815* (Chapel Hill: Institute of Early American History and Culture by the University of North Carolina Press, 1993), chap. 7; James L. Michie, *Richmond Hill Plantation, 1810–1868: The Discovery of Antebellum Life on a Waccamaw Rice Plantation* (Spartanburg, SC: Reprint Co., 1990), chap. 4; Charles Joyner, *Down by the Riverside: A South Carolina Slave Community* (Urbana: University of Illinois Press, 1984), 117–126, 139–140; Mart A. Stewart, *"What Nature Suffers to Groe": Life, Labor, and Landscape on the Georgia Coast, 1860–1920* (Athens: University of Georgia Press, 1996), 87–94, 96–116, 146–149, 151–160, 168–174. On inland rice production and associated technologies, see Hayden R. Smith, "Reserving Water: Environmental and Technological Relationships with Colonial South Carolina Inland Rice Plantations," in *Rice: Global Networks and New Histories*, ed. Francesca Bray et al. (New York: Cambridge University Press, 2015): 189–211.

facilitated control of water offered a starting point for cultivation and thus influenced organizational patterns over time. As important, the "pitch of the tides" proved crucial to the development of plantations throughout the region. Tidal cultivation depended on estuaries where saltwater dammed up part of the freshwater flow and raised river levels. Planters tapped into the freshwater layer to inundate fields. Thus, natural conditions limited the geographic range of tidal fields and influenced productivity. Not all rivers and creeks afforded comparable opportunity. The volume and flow of freshwater proved crucial, as planters recognized. The enormous output of plantations along the Waccamaw, the Pee Dee, and the Santee rivers, for example, stemmed partly from favorable hydrologic conditions.[12]

Several other caveats apply. Although popular imagery consistently depicted stately mansions as ubiquitous, a significant number of lowcountry plantations never had planter's dwellings of any kind. S. Max Edelson has found that a limited number of planters built substantial plantation houses by the end of the eighteenth century. The majority lay within a fifty-mile radius of Charleston, a "core zone" of settlement where social conditions led planters to erect handsome dwellings and develop formal landscaping. The "civility and refinement" of core zone plantations contrasted with those farther afield, which concentrated on intensive production of staple crops. Although planters later erected well-built houses on plantations near Georgetown and Beaufort, a considerable number never became more than clusters of fields amid wilderness, some with a slave settlement and a few makeshift buildings, others without. Other plantations had only modest planter's houses, not buildings of significant stature.[13]

Accounting for the full range of plantation products adds further complications. Although Sea Island cotton and rice made up the majority of lowcountry plantations' output, some turned out other goods. Between the 1740s and the Civil War, for example, brickmaking flourished in the

---

[12] Edelson, *Plantation Enterprise in Colonial South Carolina*, 103–117; Stewart, "*What Nature Suffers to Groe*," 99; Chaplin, *An Anxious Pursuit*, 231–232, 239–243. The importance of topography and related factors such as soil type are underscored in Smith, "Reserving Water," 193–211.

[13] Edelson, *Plantation Enterprise in Colonial South Carolina*, chap. 4 (first quotation on 130, second on 136); Lachicotte, *Georgetown Rice Plantations*; Linder, *Historical Atlas of the Rice Plantations of the ACE River Basin*; Linder and Thacker, *Historical Atlas of the Rice Plantations of Georgetown County and the Santee River*; Todd and Hutson, *Prince Williams Parish and Plantations*.

Wando River basin north of Charleston. In some areas, planters derived a third or more of their annual income from brick sales. Meanwhile, some planters pursued diversified production over staple crops. During the second quarter of the eighteenth century, Henry Laurens geared his Mepkin Plantation to producing corn, beans, firewood, and lumber for his three other plantations and for markets in Charleston. Although the number of plantations operated in this fashion remained limited, their significance should not be overlooked, for they highlight responses to changing market conditions and strategies that may have become more common as planters coordinated activities across multiple plantations. Different patterns of production necessarily produced different built environments. Put simply, although production of agricultural staples dominated the processes that shaped lowcountry plantation landscapes, other activities also proved significant.[14]

Overall, three factors distinguished Sea Island and tidal rice plantations of the late antebellum era. First, the geographic constraints associated with both crops and the relative stability of plantation agriculture on the South Atlantic coast created well-developed built environments that generally possessed more numerous and more substantial buildings than commonly found on upland plantations. The landscape transformations that rice and Sea Island cotton demanded added depth and complexity. Although historians have devoted far greater attention to rice culture, Sea Island cotton planters also devised impressive systems for controlling water. The early twentieth-century agricultural historian Lewis Cecil Gray judged ditching on Sea Island cotton plantations "almost as elaborate as in the rice industry." Although more recent analyses suggest that Gray may exaggerate somewhat, the basic point remains salient: both of the lowcountry's major staples dramatically refigured the natural landscape.[15]

Second, cultivation of food crops and staples and judicious use of pastures and forest produced patterns of intensive land use. Although some especially large plantations included terrain that saw limited activity, planters generally made use of nearly all available acreage. Patterns of

---

[14] Lucy B. Wayne, "'Burning Brick and Making a Large Fortune at It Too': Landscape Archaeology and Lowcountry Brickmaking," in *Carolina's Historical Landscapes: Archaeological Perspectives*, ed. Linda F. Stine et al. (Knoxville: University of Tennessee Press, 1997): 97–111; Edelson, *Plantation Enterprise in Colonial South Carolina*, chap. 6.

[15] Kovacik and Mason, "Changes in the South Carolina Sea Island Cotton Industry," 85, 87–90; Lewis Cecil Gray, *History of Agriculture in the Southern United States to 1860* (Washington, DC: Carnegie Institution, 1933), II: 734 (quotation).

labor and land use demonstrated the scale of plantation operations, the effort required for production of staple crops in large quantities, and the resources involved.

Finally, as countless observers noted, lowcountry plantations exhibited impressive order and authority. Carefully arranged fields, rigorous maintenance, and machine-like efficiency created an atmosphere of disciplined operation that underscored planters' control of labor and constant vigilance against environmental threats. Planters fretted endlessly about the challenges they faced, slaves' indolence, and countless risks, but the broad contours of lowcountry history reveal the success of the plantation regime. As high-stakes, capital-intensive operations, rice and Sea Island cotton plantations sought to leave as little to chance as possible. The results manifest themselves in conditions that evinced the scale of enterprise present, ties to international markets, and the intricacy of daily operations.[16]

Northerners' estates reflected different priorities. With leisure, domestic comfort, and status display as principal goals, owners developed environments suited to their needs. In general, three qualities distinguished northerners' estates as a group. First, well-developed building complexes with substantial material refinement proved common. Even the least pretentious of the new estates displayed greater refinement than the majority of antebellum plantations, and the most ambitious far surpassed any of their predecessors. The majestic styling of the domestic complexes at estates such as Cherokee, Windsor, Arcadia, and Dixie Plantation echoed country estates in the North and those of European nobles.[17] More modest

---

[16] T. Addison Richards, "The Rice Lands of the South," *Harper's Monthly Magazine* 114, no. 19 (Nov. 1859): 730–731; Edward King, *The Great South* (Hartford, CT: American Publishing Co., 1875), 429–437. For a more nuanced perspective that recognizes slaves' agency, see Stewart, "*What Nature Suffers to Groe*," chap. 3; Mart A. Stewart, "Rice, Water, and Power: Landscapes of Domination and Resistance in the Lowcountry, 1790–1880," *Environmental History Review* 15, no. 3 (autumn 1991): 47–64.

[17] Linder, *Historical Atlas of the Rice Plantations of the ACE River Basin*, 72–23; Linder and Thacker, *Historical Atlas of the Rice Plantations of Georgetown County and the Santee River*, 77–78, 437; J. V. N., Jr., "Do You Know Your Charleston? Dixie Plantation," *News and Courier*, March 1, 1937, 10. For examples of similarly sumptuous complexes, see Chlotilde R. Martin, "Widener Builds at Mackey Point," *News and Courier*, Dec. 25, 1930, B3; C. B. Williams, "Ashepoo River Plantation Restored by the Alfred H. Casparys, of New York," *News and Courier*, March 4, 1934, B3; "Doubleday House, $30,000 Job Done," *News and Courier*, Jan. 5, 1936, C3; Beulah Glover, "New Yorkers Built Big Plantation House on Bellinger Barony in Colleton," *News and Courier*, Feb. 17, 1935, C2; Alexander Moore, *Poco Sabo Plantation* (N.p.: H. Anthony Ittleson, 2005), 54–55.

estates possessed main dwellings and outbuildings with a quality of construction and stylistic sophistication that stood well above most antebellum plantations. Second, northerners devoted greater attention to formal landscaping than the vast majority of their antebellum predecessors. Larger, better developed grounds, gardens, and allées typified the new estates, revealing owners' commitment to leisure and social ambitions.[18]

Third, clear divisions between domestic and recreational spaces resulted in distinctive patterns of land use. Owners' residences and associated outbuildings marked the center of spaces where intensive activity took place. Maintenance of houses and surrounding grounds, labor required for domestic operations and recreation, and the activities of owners and their guests all took place near main dwellings. Beyond lay domains that owners and guests saw mainly as recreational spaces. To be sure, more than recreation took place on such lands. Laborers planted feed crops to attract wildlife, cultivated food crops, repaired and maintained tidal fields, cut timber, maintained roads and trails, and carried out sundry other activates. Moreover, tenants lived on these lands. Small farmsteads lay scattered at irregular intervals, typically with cotton, rice, and row crops planted nearby. To see the majority of northerners' acreage as simply serving recreational purposes, then, is erroneous. Nonetheless, use of such lands fell below antebellum norms. Patterns of activity left most of northerners' holdings less intensively used as compared with earlier periods.[19]

Northerners' estates thus assumed a two-part organization with clear divisions between domestic spaces and other domains. Although variations existed, the overall pattern showed remarkable consistency. The specialized purposes for which northerners created and used their estates manifest themselves in patterns that became apparent early on and grew more pronounced over time. Especially during the estate-making boom of the late 1920s, as domestic spaces became larger, more elaborate, and more refined, the contrast with other lands became clear and distinct.

---

[18] See, for example, Chlotilde R. Martin, "The Delta – Home of H. K. Hudson," *News and Courier*, Jan. 20, 1931, 7; Chlotilde R. Martin, "Clay Hall, Redmond Mansion Lodge," *News and Courier*, Dec. 7, 1930, A11; Chlotilde R. Martin, "New Yorker Reclaims Pimlico," *News and Courier*, May 24, 1931, B2; "Frelinghuysen Builds at Rice Hope," *News and Courier*, Aug. 9, 1931, B12; Chalmers S. Murray, "Cheston Builds on Old Plantation," *News and Courier*, Aug. 2, 1931, B2.

[19] "Diary of Life at Medway Plantation"; Murray, "Attracted by Climate"; Martin, "Low-Country Plantations Stir."

Surveying differences between northerners' "plantations" and those of earlier periods raises vital questions about the choices northerners made in creating their estates. Although northerners saw their estates as plantations and coveted buildings and landscape features from earlier periods, they also came to the lowcountry with extensive knowledge about recreational settings elsewhere. Contemporaries showed remarkably little interest in the knowledge northerners brought to the region and how it might have informed their activities. Avoidance of the subject apparently revealed less about the incuriosity of journalists and other observers than presumptions about the inherent significance of plantations. That northerners might do something besides rehabilitate the revered estates of yesteryear seemed beyond comprehension. When viewed critically, however, northerners' actions demonstrate the influence of the American country house. At a time when handsome estates in bucolic settings anchored a vibrant sphere of upper-class life, northerners' appropriation of design influences from such estates is hardly surprising. In creating their "plantations," northerners merged material remains of plantation slavery with formulas derived from contemporary country estates. The results yielded material environments suited for upper-class leisure that most observers also saw as embodying an aristocratic past.

COUNTRY HOUSES

Country houses developed with the spectacular growth of industrial fortunes after the Civil War. As new strength in manufacturing and finance produced a class of exceptionally wealthy Americans, its members broke with long-standing traditions of privacy and restraint. Elegant townhouses, fancy carriages, lavish dinners, and elaborate balls became common. Newly wealthy Americans displayed their wealth openly and deliberately, eager to tout their success. Wealthy Americans also developed institutions and behaviors to differentiate themselves from other classes. By the 1880s, exclusivity became an overriding concern. As the years passed, the wealthy developed elaborate mechanisms for policing social boundaries. Refinement, manners, and family reputation became essential for differentiating members of "society" from parvenus.[20]

---

[20] Eric Homberger, *Ms. Astor's New York: Money and Social Power in a Gilded Age* (New Haven: Yale University Press, 2002); Sven Beckert, *The Monied Metropolis: New York and the Consolidation of the American Bourgeois, 1850–1896* (New York: Cambridge University Press, 2001), chaps. 5–8; Frederic Cople Jaher, "Style and Status: High Society

Sport became a vital forum for displaying wealth and status. Wealthy Americans gravitated to pastimes traditionally favored by European nobles such as horse racing and breeding, hunting, and yachting. Newer pursuits also became popular. Polo, first played in the United States in 1867, quickly won a large following. Elites also took up tennis and golf, two other new arrivals. Tennis, first played in America in 1874, quickly became a mainstay of upper-class recreation. Golf, introduced to the United States in 1888, attracted participants at a slower pace but nonetheless became popular.[21]

Sport served the interests of status-conscious elites because of its capacity to demonstrate status and foster solidarity. The ability to abstain from productive labor reified economic independence. Not having to earn income to provide for life's necessities made wealth clear. The time required to develop the skills and specialized knowledge necessary for basic competency of recreational activities, let alone mastery, did the same. Like sport hunting, all of the pastimes that became popular among the upper classes required practice, discipline, and consistency. None proved especially accommodating to novices, and proficiency demanded time, effort, and persistence. Demonstrating prowess before peers, then, revealed significant investments of time and effort. Making leisure a priority required resources of a kind that few people possessed, and newly wealthy Americans did so with enthusiasm, eager to indulge in activities that distinguished them from other groups.[22]

in Late Nineteenth-Century New York," in *The Rich, the Well Born, and the Powerful: Elites and Upper Classes in History*, ed. Frederic Cople Jaher (Urbana: University of Illinois Press, 1973): 258–284; Frederic Cople Jaher, "The Gilded Elite: American Multimillionaires, 1865 to the Present," in *Wealth and the Wealthy in the Modern World*, ed. W. D. Rubinstein (New York: St. Martin's Press, 1980), 189–276.

[21] Robert B. MacKay, "Introduction," in *Long Island Country Houses and Their Architects*, ed. Robert B. MacKay, Anthony K. Baker, and Carol A. Traynor (New York: Society for the Preservation of Long Island Antiquities in association with W. W. Norton and Co., 1997), 22–26; Clive Aslet, *The American Country House* (New Haven: Yale University Press, 1990), 24–25, 71–77; Mark A. Hewitt, *The Architect and the American Country House* (New Haven: Yale University Press, 1990), 11–14, 127; Dixon Wecter, *The Saga of American Society: A Record of Social Aspiration, 1607–1937* (New York: Charles Scribner's Sons, 1937), chap. 11.

[22] Wecter, *The Saga of American Society*, chap. 11; Thorstein Veblen, *The Theory of the Leisure Class* (New York: Macmillan Co., 1899), 41–52, 101–106, 169–187; Mark Rothery, "The Shooting Party: The Associational Cultures of Rural and Urban Elites in the Late Nineteenth and Early Twentieth Centuries," in *Our Hunting Fathers: Field Sports in England after 1850*, ed. R. W. Hoyle (Lancaster, UK: Carnegie Publishing, 2007): 96–118; Emma Griffin, *Blood Sport: Hunting in Britain since 1066* (New Haven: Yale University Press, 2007), 55–56, 79–81.

The finery associated with each pursuit offered additional opportunities to display wealth and specialized knowledge. Just as elite hunters donned fancy attire and trekked into the field with pedigreed horses and dogs, other pursuits offered opportunities for status display. Yachting, for example, required seagoing vessels, and devotees quickly became connoisseurs of fine boats, specialized equipment, and nautical regalia. Polo especially demonstrated wealth because of the resources required. Competitive matches generally required participants to have six ponies at their disposal, each representing a substantial capital investment. Pastimes such as tennis and golf, although modest by comparison, nonetheless required specialized equipment and attire.[23]

Upper-class Americans' enthusiasm for sport developed alongside growing concern about the chaos and crowding of large cities and deepening convictions about the wholesomeness of rural life. Confidence in the revitalizing influence of nature grew in direct proportion to apprehensions about pollution, noise, congestion, and social problems. Beginning in the 1880s, wealthy Americans seized opportunities to create estates dedicated to leisure and sport on the periphery of major cities. Commonly referred to as "country houses," properties of this type quickly became markers of status and symbols of a rising elite. Typically located beyond the suburbs, at a distance from other houses and farms, they afforded opportunities for recreation, relaxation, and outdoor activity in settings characterized by natural beauty, quiet, and seclusion.[24]

In 1904, architectural writer Barr Ferree captured upper-class Americans' zeal for the countryside in surveying the results of what he termed a "great new building energy." In "the very brief space of ten years," he wrote, the wealthy had created "a new type of American country house," "a sumptuous house, built at large expense, often palatial in its dimensions, furnished in the richest manner, and placed on an estate, perhaps large enough to admit of independent farming operations." In most cases, Ferree added, a garden formed "an integral part of the architectural scheme." Ferree attributed the rise of such houses to a "new conception of country life" and a "new appreciation and realization of its manifold joys and pleasures." The "old idea of country life as synonymous with the farm no longer prevails," he argued. "The rich," Ferree wrote, had

---

[23] Wecter, *The Saga of American Society*, chap. 11; Veblen, *Theory of the Leisure Class*, 41–52, 101–106, 169–187; Aslet, *American Country House*, 70–78.

[24] Aslet, *American Country House*; Hewitt, *The Architect and the American Country House*, 12–14, 153–193.

discovered "the simpler and more natural life of the country." At handsome rural estates, they enjoyed "varied sports and open-air activities."[25]

Wealthy Americans built estates of the kind Ferree described on Long Island, along Philadelphia's Main Line, in the Brandywine Valley of Delaware, on Boston's North Shore, in Cleveland's Chagrin Valley, and in Chicago's Lake Forest and Lake Geneva.[26] Long Island quickly became the "epicenter" of what one historian has called "well-heeled America's new experiment in country living." Between the end of the Civil War and World War II, leading figures in business and industry built 975 estates between the city limits of New York and the town of Montauk, at the far eastern tip of the island. "County homes," the *New York Herald* reported in 1902, "are continuous for a hundred miles." Some featured palatial dwellings, lavish gardens, large acreages, and dozens of servants and groundskeepers. Others assumed more modest form but followed the same basic organizational scheme. Variations notwithstanding, all accommodated the upper-class enthusiasm for vigorous recreation out of doors. As part of wealthy Americans' quest to style themselves as an aristocratic class, country houses afforded venues for leisure, recreation, and performative socializing.[27]

Country houses varied in size and appearance but shared several unifying characteristics. The guise of self-sufficiency proved most important. Although the ideal of a self-supporting estate inspired many owners, in practice it proved nearly impossible to attain. As an organizational formula, however, it offered expansive possibilities. Wealthy Americans saw acreage sufficient to suggest landed wealth as symbolizing elite privilege. Clusters of outbuildings where barns, granaries, and stables stood alongside sheds and storehouses suggested operations characteristic of a working estate. The combination of land and agricultural buildings placed estate owners at the helm of sizable undertakings and demonstrated authority that reached beyond wealth alone. Moreover, many owners engaged in gardening or farmed for domestic consumption as part of their embrace of rural life. In sum, American country houses invited comparisons to the estates of European nobles. Although most commentators saw English country estates as the principal inspiration, other traditions of elite residency also proved important.[28]

[25] Ferree, *American Estates and Gardens*, 1–2.   [26] Aslet, *American Country House.*

[27] MacKay, Baker, and Traynor, eds., *Long Island Country Houses and Their Architects* (quotations on 19).

[28] Aslet, *American Country House.*

Houses styled to project wealth and power also proved crucial. In 1903, Harry Desmond and Herbert Croly described the dwellings on country estates as "stately homes." "These great modern dwellings," they observed, served as residences for their owners and their families, echoed European antecedents, yet expressed distinctively American "social and aesthetic ambition[s]." Historical styles predominated, especially those associated with landed elites. Tudor Revival designs, for example, implicitly linked newly wealthy Americans to the gentry of fifteenth-century England, and imitation French chateaux, Renaissance villas, and Neoclassical mansions also proved popular. Stylistic variations notwithstanding, country estates strived to project an aura of confidence and self-assurance. Desmond and Crowly saw an expression of great wealth and "the freedom of American life."[29] Others likened country houses to the character of the archetypical English gentleman, particularly the qualities of dignified reserve, decorum, and studied detachment.[30] Whatever the case, country houses strived for seemingly irrefutable cultural authority, usually in opposition to the relatively new fortunes of their owners.

Coordinated ensembles of architecture and landscaping also typified American estates. Ferree's observation that gardens usually formed part of the "architectural scheme" hints at the effort owners and designers put into creating outdoor pleasure grounds. Gardens, lawns, and allées lay close beside main dwellings, often surrounding them entirely and stretching outward toward scenic views or forested tracts. Especially opulent estates featured large, well-maintained gardens with extensive plantings, manicured beds and shrubs, stepped terraces, pergolas, and balustraded walkways. More modest estates opted for smaller, less elaborate assemblages, albeit on a similar pattern. Regardless of scale, the overall effect fulfilled several aims. Stylistic coordination gave domestic spaces a monumentality suggestive of landed wealth. Scale, complexity, and historical styles suggested age, evolutionary development, and stability. Moreover, gardens and lawns accentuated the architecture of main dwellings while simultaneously attracting interest on their own. Stylistic integration invited close study of coordinated features and transitions between natural and domestic spaces. It also demonstrated refinement that seemed, at least on the surface, too intricate to have been created hastily. In addition,

[29] Harry W. Desmond and Herbert Croly, *Stately Homes in America* (New York: D. Appleton and Co., 1903) (quotations on 12). See also Hewitt, *The Architect and the American Country House*, chap. 4.
[30] Aslet, *American Country House*, chap. 5.

landscaping served important recreational and social roles. The sequence of outdoor spaces allowed owners to display power to guests subtly, without making overt claims, and lawns and gardens provided outdoor venues for recreation and socializing.[31]

Beyond aesthetic considerations, American country houses served vital social functions. Upper-class enthusiasm for rural life produced a vibrant sphere of activity that bound elites together through sport, leisure, and copious socializing. House parties, country weekends, hunts, polo matches, regattas, and no small number of other events made country houses central to marking and maintaining class boundaries. As venues where elites gathered for activities among peers, country houses signified privilege, exclusivity, and ritualized performance. Related institutions such as country clubs and sporting clubs reinforced and extended social networks. Like fancy balls, society dinners, and private clubs, country houses played a crucial role in upper-class life.[32]

Discerning parallels between country houses in the North and lowcountry sporting "plantations" requires little effort. Although significant differences existed, the overall pattern is clear. As northerners created their estates, they relied on design principles from outside the region. Especially during the late 1920s, country houses became a major influence.

Greater refinement and stylistic sophistication offered conspicuous evidence of the trend. Estates featuring classical revival houses multiplied during the estate-building boom. William and Caroline Coe's Cherokee, Robert and Ann Goelet's Wedgefield, Walker P. Inman's Greenfield Plantation, Vincent and Madeline Fiermonte's Dixie Plantation, and Marshall and Ruth Field's Chelsea Plantation offer telling examples. At these and other estates, northerners erected houses with sumptuous finishes, elegant styling, and elaborate interior and exterior ornamentation. All surpassed the grandest lowcountry dwellings of the antebellum era and showed the dominance of Beaux Arts design, particularly its emphasis on formality and rational order. They also underscored northerners' enthusiasm for styling characteristic of upper-class residences and institutional buildings throughout the United States. As new sporting "plantations" grew more

---

[31] Ibid.; Hewitt, *The Architect and the American Country House*; MacKay, Baker, and Traynor, eds., *Long Island Country Houses and Their Architects*.

[32] Aslet, *American Country House*, esp. chap. 9; Hewitt, *The Architect and the American Country House*, 10–14.

pretentious, owners increasingly expected architecture and interiors like those in other elite havens.[33]

Professional designers played a crucial role in delivering increased sophistication and grandeur. The number of prominent architects and landscape designers who worked on behalf of northern sporting enthusiasts grew dramatically during the estate-making boom. The Charleston firm of Simons and Lapham designed at least four houses at northern-owned estates and remodeled or restored at least five others. Albert Simons, a Charleston native who received his architectural training at the University of Pennsylvania, excelled at Neoclassical interiors with delicate door and window surrounds, paneled rooms, and arched passageways. Samuel Lapham, also a native Charlestonian, had studied architecture at the Massachusetts Institute of Technology. He proved equally adept with the colonial-inspired classical vocabulary. The pair formed a partnership in 1920 and later developed a strong clientele among northern millionaires. Although Simons complained about "Wall Street Planters" and the annual "swarm of Yankees," he also recognized that he and Lapham would have struggled if not for monied northerners. Especially during the Depression, other clients proved scarce.[34]

Other architects and architectural firms that designed houses for northerners included Charles N. Read and Mellor and Meigs of Philadelphia; Polhemus and Coffin, William L. Bottomley, and Russell and Clinton of New York; Olaf Otto of Savannah; and Wyeth and King of Palm Beach.[35] Landscape architects also contributed. Loutrel Briggs, a New York native with a degree in landscape architecture from Cornell University, visited Charleston for the first time in 1927. Two years later, he

---

[33] Linder, *Historical Atlas of the Rice Plantations of the ACE River Basin*, 72–23; "Wedgefield Plantation: The Shooting Lodge of Mr. Robert Goelet," *Country Life* 75, no. 2 (Dec. 1938): 53–57, 101–102; "Carolina Classic," *House and Garden* 73, no. 2 (Feb. 1938); "Do You Know Your Charleston? Dixie Plantation"; Baldwin, *Lowcountry Plantations Today*, 280; Perry, *Moving Finger of Jasper*, 141–142.

[34] Stephanie E. Yuhl, *A Golden Haze of Memory: The Making of Historic Charleston* (Chapel Hill: University of North Carolina Press, 2005), 36–39.

[35] Chlotilde R. Martin, "Widener Builds at Mackey Point," *News and Courier*, Dec. 25, 1930, 33; Frances Cheston Train, *A Carolina Plantation Remembered: In Those Days* (Charleston: History Press, 2008), 34–37; Susan Hume Frazer, *The Architecture of William Lawrence Bottomley* (New York: Acanthus Press, 2007), 282–289; "Do You Know Your Charleston? Dixie Plantation"; "Shooting Lodge in Carolina: 'Richmond,' the Estate of George A. Ellis, Jr., Esq., near Moncks Corner, South Carolina," *Country Life* 65, no. 2 (Dec. 1933): 60–61; Martin, "The Delta – Home of H. K. Hudson"; "Carolina Classic," *House and Garden* 73, no. 2 (Feb. 1938).

opened a seasonal office in the city. Briggs generally resided in Charleston from November to April and returned to New York for the rest of the year. Although his Charleston practice concentrated on gardens at urban townhouses, he also developed landscape designs for several northern-owned plantations. By the late 1930s, gardens and landscapes of Briggs's design lay at Mepkin, Mulberry, and Rice Hope plantations.[36]

Innocenti and Webel proved more influential. Umberto Innocenti, an Italian who came to America on a scholarship from the University of Florence, and Richard K. Webel, a Harvard-trained landscape architect, met while working for the New York firm of Vitale and Geiffert. In 1931 the pair decided to form a partnership. Early commissions included several estates on Long Island, a municipal park in Pittsburgh, and pavilions for the Netherlands and Italy at the 1939 World's Fair. Innocenti and Webel began working in the lowcountry at the behest of Landon K. Thorne, an early client. In the early 1930s, Thorne, a financier, had commissioned Innocenti and Webel to design gardens at his estate at Bayshore, Long Island. After Thorne and his brother-in-law, Alfred L. Loomis, purchased about 15,000 acres on Hilton Head Island, he asked the firm to beautify the grounds of two hunting lodges. Webel managed the firm's main office on Long Island, while Innocenti supervised field crews on Hilton Head. Soon, commissions for other northern-owned properties followed. By the end of the decade, Innocenti and Webel had created gardens and landscaping at three estates near Beaufort and three near Georgetown.[37]

Briggs and Innocenti and Webel took contrasting approaches to commissions at northern-owned "plantations." Briggs believed a garden should function as an outdoor room and be visible from the house in order to establish a close interior-exterior relationship. He employed walls, fences, and evergreens to define edges and added pools, fountains, and small statues for visual interest. Limited use of ornamental plants created unity and an intimate sense of scale. As a group, Brigg's plantation designs made outdoor spaces extensions of main dwellings and

---

[36] James R. Cothran, *Charleston Gardens and the Landscape Legacy of Loutrel Briggs* (Columbia: University of South Carolina Press, 2010), 6–23, 53–59; Folder "Briggs, Loutrel W., 1929–1930," box 1, Joseph S. Frelinghuysen Papers, Archibald S. Alexander Library, Rutgers University, New Brunswick, NJ.

[37] Gary R. Hilderbrand, ed., *Making a Landscape of Continuity: The Practice of Innocenti and Webel* (Cambridge: Harvard University Graduate School of Design, 1997); Chlotilde R. Martin, "Hilton Head Island Estates Merged," *News and Courier*, Feb. 7, 1932, A3.

sought to create enveloping settings where hosts could entertain guests amid lush color.[38]

Innocenti and Webel's commissions exhibited greater majesty. Innocenti bore primary responsibility for the firm's lowcountry designs. He used rows of trees to define space and large, open lawns to create unencumbered vistas that allowed visitors to admire dwellings and owners and guests to take in the surrounding landscape. He also proved adept at creating landscapes that appeared mature in a relatively short period of time. By transplanting large trees with masses of soil around the roots, for example, he developed features that appeared decades old.[39]

By the late 1930s, the combination of majestic architecture and extensive landscaping had become so common that northerners who made different choices stood out as unusual. Soon after Harry F. Guggenheim purchased property on the Wando River in 1936, for example, he received a letter from Norman Armstrong, a Charleston arborist. Armstrong explained the importance of giving trees regular care and discussed his services. He concluded with a request to inspect Guggenheim's estate and prepare a treatment plan. Guggenheim, a scion of the Guggenheim family of industrialists, responded with a letter that sought to correct Armstrong's misapprehensions. "I think you must have a wrong impression about my plantation," Guggenheim wrote. "It is not the typical winter-summer home of the northerner," he explained. "I have a simple house with old oak trees and a few shrubs," he noted. Guggenheim closed by thanking Armstrong for his interest but made it clear that he would not need his services.[40]

Further evidence of country houses' influence is provided by the shifting composition of estate owners as a group. Although exceptionally wealthy men and women had a part in northerners' activities from the beginning, their numbers grew dramatically during the estate-making boom. Moreover, socially prominent couples became more common. Bona fide members of society increased with the development of large, opulently styled estates. Robert and Ann Goelet, Marshall and Ruth Field, Gertrude and Sidney Legendre, George and Jessie Widener, Nelson

---

[38] Cothran, *Charleston Gardens and the Landscape Legacy of Loutrel Briggs*, chap. 6; Loutrel W. Briggs, *Charleston Gardens* (Columbia: University of South Carolina Press, 1951), 136–141, 146–147, 151–153.

[39] Hilderbrand, ed., *Making a Landscape of Continuity*, 11–16.

[40] Norman Armstrong to Harry F. Guggenheim, Feb. 22, 1938; Harry F. Guggenheim to Norman Armstrong, March 2, 1938, both in folder "'A': Miscellany, 1936–49," box 133, Harry Frank Guggenheim Papers, Library of Congress, Washington, DC.

and Ellen Doubleday, and William and Caroline Coe all fit into this category. These men and women and others like them belonged to the highest echelons of the American upper class. As members of a wealthy, privileged elite, they organized their lives according to conventions regarding seasonal residency, exclusivity, and prestige-based associational networks.[41]

All of the socially prominent men and women who became part of the lowcountry sporting scene had ties to estates on Long Island or similar properties elsewhere. The Coes, for example, owned Planting Fields, one of the grandest of all Long Island estates. Marshall and Ruth Field summered at "Caumsett," a spectacular 2,000-acre estate at Lloyd's Point. Gertrude Legendre's parents rented estates on Long Island in some years; in others they summered in Europe. Nelson Doubleday owned "Effendi Hill," a renowned retreat at Mill Neck with a Georgian colonial-inspired residence. Frederick Pratt and his family, owners of Cheeha-Combahee Plantation, spent their summers at Poplar Hill, one of several Pratt estates on Long Island. J. Cornelius and Nancy Rathbourne, owners of Beneventum Plantation, also owned Pelican Farm, an estate near Old Westbury, New York. Meanwhile, the Goelets owned Ochre Court at Newport, Rhode Island, and Robert and Charlotte Montgomery resided at Ardrossan, the largest estate on Philadelphia's Main Line. George Widener, the owner of Mackey Point Plantation in Jasper County, had grown up at Lynnewood Hall, another Main Line estate. Put simply, by the mid-1930s, finding northern "plantation" owners who did not also own or have close connections to a country estate in the North proved difficult.[42]

Recognizing estate owners' ties to elite havens in the North raises crucial questions about their "plantations." How did owners view the estate-making process? Did they see themselves as rehabilitating old

[41] For background on these figures, see *National Cyclopedia of American Biography*, 63 vols. (New York: James T. White and Co., 1892–1984), XXXV: 417 and LVII: 264–265; "N. Doubleday, 59, Publisher, Dead," *New York Times*, Jan. 12, 1949, 27; Steven D. Becker, *Marshall Field III: A Biography* (New York: Simon and Schuster, 1964); Alex M. Robb, *The Sanfords of Amsterdam: The Biography of a Family in America* (New York: William-Frederick Press, 1969).

[42] MacKay, Baker, and Traynor, eds., *Long Island Country Houses and Their Architects*, 59, 79, 238–239, 265, 326–328, 425–427; Michael C. Kathrens, *Newport Villas: The Revival Styles, 1885–1935* (New York: W. W. Norton, 2009), 58–67; William Morrison, *The Main Line: Country Houses of Philadelphia's Storied Suburb, 1870–1930* (New York: Acanthus Press, 2002), 134–137; Gertrude S. Legendre, *The Time of My Life* (Charleston: Wyrick and Co., 1987), 4–11, 16.

plantations, creating estates reminiscent of those on Long Island, or some combination of the two? None of the men and women who created lowcountry estates left clear statements about these matters, so direct answers are illusive. The overall pattern strongly suggests northerners saw themselves as doing both. Northerners saw plantations a regional type and sought to retain key attributes. Careful treatment of edifices from the colonial and antebellum periods, cultivation of aged aesthetics, and maintenance of existing organizational patterns showed commitment to fundamental qualities. To be sure, northerners saw plantations principally through the lens of built form. They prioritized architecture, landscapes, and spatial organization and paid little attention to other qualities. In their view, the historical nexus of staple agriculture and unfree labor figured simply as part of the past, not as a defining characteristic. Material form and acreage mattered most; other qualities figured as incidental.

At the same time, northerners relied heavily on influences drawn from northern country houses. Expectations regarding architectural style, domestic amenities, landscaping, gardens, upkeep, and recreation all derived from contemporary practices in the North. The refined styling and sumptuousness that characterized many estates did also, as did the ornate Neoclassicism evident at plantations such as Cherokee, Wedgefield, and Chelsea. Put simply, northerners could not have created their "plantations" without the Gilded Age enthusiasm for "country life." The great estates of the North provided the basic template for northerners' lowcountry estates.

Northerners saw no contradiction in viewing their lowcountry properties simultaneously as "plantations" and estates. Their sensibilities not only reflected romanticization of the Old South and historical amnesia but perspectives on historical patterns of elite residency. Architects, architectural writers, and journalists cast American country houses as the latest iteration of an established type with deep roots. According to the conventional wisdom, American country houses owed their origins to the fortunes produced by the surging advance of industrial capitalism after the Civil War. The strongest historical influence came from the great estates of rural England, and most observers also saw European chateaux, hunting lodges, and castles as related. Some even saw select types from the ancient world as belonging to a broader genealogy. Moreover, although informed commentators consistently emphasized the recent origins of American country houses, they also identified antebellum southern plantations as an exception. Some classified plantations as landed estates like those of rural England; others left their exact status unspecified.

Regardless, virtually all saw them as somehow related to the country estates of America's new elite.[43]

For the purposes of understanding northerners' actions, the haziness of such views is less important than what they encouraged. Americans' tendency to recall antebellum plantations as quasi-feudal estates placed them in the same general category as English country houses, their American counterparts, and a host of other elite properties. Whether many sportsmen or sportswomen recognized this is uncertain, and whether any of the designers who worked on their behalf did is also unclear. Regardless, contemporary perspectives saw enough of a relationship between American country houses and historical plantations to encourage rehabilitation of the latter. Certainly, nothing stood in the way. Viewing historical plantations as aristocratic seats effectively encouraged rehabilitation of surviving examples. Northerners' "plantations" demonstrated the results. By merging romanticized views of the southern past with the form, purpose, and styling of contemporary country estates, they created a new hybrid with potent associations.

CONCLUSION

Viewing northerners' "plantations" in a critical perspective helps to explain how people of the 1920s and 1930s understood them and their relationship to historical plantations and country estates in the North. Although contemporary discourses cast the new "plantations" as unique to the lowcountry, their roots reached far and wide. No matter how thoroughly some sportsmen and sportswomen may have romanticized

---

[43] Donn Barber, "Introduction," in *American Country Homes and Their Gardens*, ed. John Cordis Baker (Philadelphia: House and Garden, 1906), 9–10; Ferree, *American Estates and Gardens*, 1–2; William Herbert, *Houses for Town or Country* (New York: Duffield and Co., 1907), 63–88; Fiske Kimball, "The American Country House," *Architectural Record* 46, no. 4 (Oct. 1919): 308–322. For contemporary views of the history of country houses, see the following essays by Antoinette Perrett: "History of the Country Estate – I. Egypt," *Country Life* 68, no. 5 (Sept. 1935): 27–29, 74; "History of the Country Estate – II. Babylonia," *Country Life* 68, no. 6 (Oct. 1935): 34–36, 72, 75; "History of the Country Estate – III. The Roman Farm," *Country Life* 69, no. 1 (Nov. 1935): 37–39, 72–76; "History of the Country Estate – IV. The Roman Pleasure Villa," *Country Life* 69, no. 2 (Dec. 1935): 45–47, 76–77; "History of the Country Estate – V. Moorish Empire in Spain," *Country Life* 69, no. 5 (March 1936): 49–50, 70–71; "History of the Country Estate – XIII. The Southern Estates," *Country Life* 71, no. 1 (Dec. 1936): 45–46, 115–118. For historians' perspectives on the origins of American country houses, see Aslet, *American Country House*, chap. 2; Hewitt, *The Architect and the American Country House*, 10–14.

the Old South, even the most elaborate imaginings provided an inadequate template for northerners' actions. The estates of Long Island, the Berkshires, the Brandywine Valley, and the Main Line supplied models that northerners emulated, adapted, and expounded upon. Yet if the broad contours of northerners' actions are clear enough, how the estate-making process worked at its most basic level is less so. How sportsmen and sportswomen reshaped individual plantations requires close scrutiny to understand.

The next two chapters examine the remaking of two plantations in detail. They reveal an uneven, incremental process steered partly by owners and partly by various other agents. Neither uniform nor concerned solely with aesthetics, it took different forms at different locations and yielded different outcomes across estates. Material change figured as its most conspicuous manifestation, but historical narration followed close behind. Northerners' activities led journalists and interested amateurs to write new histories of plantations undergoing wholesale transformation. The results bound together newly remade plantations with nostalgic accounts of a glorious southern past.

4

Creating Mulberry Plantation, 1915–1935

The Colonial Revival as an Estate-Making Idiom

> One of the earliest and most architecturally interesting houses on the
> Cooper River is Mulberry.
>
> – Loutrel Briggs, 1951[1]

When northerners turned old plantations into new estates, they did more than simply rehabilitate and renew. They did more than refurbish, repair, and build. They also bonded narratives to place. Northerners' activities drew attention to long-dormant sites. As material change proceeded, journalists, avocational historians, and interested onlookers chronicled the history of plantations undergoing wholesale transformation. Most such narratives focused on elite actors and associated edifices. They foregrounded planter families, big houses, and nearby spaces. At the same time, such histories paid little attention to slaves, labor, or commerce. As devotees of lowcountry history crafted celebratory accounts of early plantations, idealized histories became closely tied to new estates with grander, more sumptuous material environments than had historically existed.

The remaking of Mulberry Plantation exemplifies this process. One of the most revered plantations on the South Atlantic coast, Mulberry became acclaimed for its role in regional history long before northerners' arrival. Its stature rose dramatically, however, as Clarence and Adelaide Chapman turned it into a seasonal residence and hunting retreat. The Chapmans purchased Mulberry in November 1915. They immediately

---

[1] Loutrel W. Briggs, *Charleston Gardens* (Columbia: University of South Carolina Press, 1951), 136.

began restoring the main dwelling to its early nineteenth-century appearance, developing the surrounding grounds, and refurbishing several surviving outbuildings. In later years they continued making improvements.
By the early 1930s they created a showplace. Alongside their efforts,
commentators celebrated Mulberry as an important example of South
Carolina's colonial heritage. Extolled for the architecture of the main
house, a storied history, and the quality of the Chapmans' efforts, Mulberry became renowned as one of the most important plantations in the
coastal region. By the late 1920s it exemplified the newly established
category of "historic" plantations as clearly as any.

Mulberry casts problems associated with reverent treatment of aging
remains in sharp relief. Restoration of the main dwelling made a revered
example of colonial architecture the focal point of an elegant estate. The
stylistic idiom of the Colonial Revival inspired, provided guidance for
subsequent actions, and conveyed symbolic power. New buildings, new
gardens, and new landscaping followed in the same stylistic vein. New
histories of Mulberry focused on surviving features and the Chapmans'
activities. In effect, onlookers celebrated the Chapmans' efforts while
ignoring Mulberry's historical extent and the full range of activities that
had historically taken place on its lands. In this fashion, narrative and
material form became closely intertwined. In the same moment as the
Chapmans remade Mulberry, its history became a story of elegant architecture, planter families, and perseverance. Meanwhile, slaves, labor, and
commercial activity fell from view. Material absences mirrored narrative
omissions. By the 1930s, Mulberry became an elegant country estate with
a history that matched its material form and appearance.

BEGINNINGS

Mulberry may well be the best known of all early rice plantations on the
Carolina coast. During the antebellum era, it became a landmark for
people traveling on the Cooper River. By 1900, architectural historians
cited the house as an important example of Georgian design. It has since
remained a staple of state and regional surveys of historic architecture.[2]

---

[2] See, for example, William Roach Ware, ed., *The Georgian Period: A Collection of Papers
Dealing with "Colonial" or XVIII-Century Architecture in the United States*, 6 vols. (New
York: U.P.C. Book Co., 1923), III: 112; Harold Donaldson Eberlin, *The Architecture of
Colonial America* (1915; reprint, New York: Johnson Reprint Corp., 1968), 97–98; Fiske
Kimball, *Domestic Architecture of the American Colonies and of the Early Republic*
(New York: Charles Scribners' Sons, 1927), 69, 72, 285; Henry Chandlee Forman,

FIGURE 4.1 Thomas Corum, Mulberry Plantation, circa 1800.
Courtesy of the Gibbes Museum of Art/Carolina Art Association, Charleston, SC.

Recent interest in a circa 1800 painting by itinerant engraver Thomas Corum has brought Mulberry new attention (see Figure 4.1). Between about 1795 and 1800, Corum produced small paintings of plantations along the Cooper River with intentions of producing a book of engraved views. He never completed the volume, but his paintings constitute an early record of lowcountry agricultural landscapes. Corum's painting of Mulberry shows an early slave street and the main dwelling. The latter stands in the distance but is plainly visible. Slaves stand at the center of the painting and in a yard at right. A pair of figures in the distance appear to depict a slave and an overseer, or possibly the plantation owner, in conversation. The slave cabins appear to have clay walls and are topped by thatched roofs. The juxtaposition of the slave street and the main dwelling aptly illustrates the social and racial hierarchies of early Carolina.[3]

Coram's painting of Mulberry has been widely reproduced. It appears on the dust jacket of Ira Berlin's *Many Thousands Gone*, in the South Carolina volume of Mills Lane's *Architecture of the Old South* series, and

---

The Architecture of the Old South: The Medieval Style, 1585–1850 (Cambridge: Harvard University Press, 1948), 182; Thomas T. Waterman, *The Dwellings of Colonial America* (Chapel Hill: University of North Carolina Press, 1950), 31–38.

[3] John Michael Vlach, *The Planter's Prospect: Privilege and Slavery in Plantation Paintings* (Chapel Hill: University of North Carolina Press, 2002), 8–9.

in John M. Vlach's study of plantation paintings, *The Planter's Prospect*.[4]
Consequently, the painting is one of the most familiar images of an early
rice plantation known to scholarly and popular audiences. In recent years,
the image has taken on added significance because of discoveries about
the slave cabins it depicts. Archaeological investigations indicate that the
clay walls and basic form reflect African building practices. In conjunction with attention to black agency in the origins and development of the
lowcountry, scholars have used Corum's painting to illustrate the inter-
section of African and Anglo cultures. In this sense, the painting symbol-
izes the influence of African peoples on lowcountry society and culture.[5]

Familiarity and sustained interest, however, has done little to make
Mulberry an object of serious investigation. Even as symbolic power of
the plantation has grown, its history has remained obscure. Since the late
nineteenth century, narratives emphasizing select owners and a handful of
well-remembered events have dominated. Interest in the main dwelling is
a hallmark of many, and others highlight Mulberry's early origins and
longevity. Some accounts mention that the dwelling served as a garrison
during the Yemassee War of 1715–1717, and some tout the Chapmans'
restoration. Otherwise, Mulberry's story remains untold. That it func-
tioned mainly as a site of staple agriculture and changed dramatically over
time is rarely mentioned. Although scholarship published since about
1970 has implicitly linked Mulberry to perspectives emphasizing the
importance of African peoples and other non-Anglo actors, popular
accounts continue to tell its history in the same way as became common
a century ago: as the story of a plantation without slaves, labor, profit
motives, or hardships of any kind.[6]

Mulberry lies on the western branch of the Cooper River, just past the
"T," the point where the western and eastern branches meet. It began as a
diversified agricultural enterprise during the early stages of settlement in
St. John's Berkeley Parish and grew to become a large and productive rice
plantation. At its peak during the first half of the nineteenth century,

---

[4] Ira Berlin, *Many Thousands Gone: The First Two Centuries of Slavery in North America*
(Cambridge: Belknap Press of Harvard University Press, 1998); Mills Lane, *Architecture
of the Old South: South Carolina* (Savannah: Beehive Press, 1984), 21–23; Vlach, *The
Planter's Prospect*, 8–9.

[5] Leland G. Ferguson, *Uncommon Ground: Archaeology and Early African America,
1650–1800* (Washington, DC: Smithsonian Institution Press, 1992), 63–79.

[6] For recent historical sketches, see William P. Baldwin, *Lowcountry Plantations Today*
(Greensboro, NC: Legacy Publications, 2002), 34–45; J. Russell Cross, *Historic Ramblin's
through Berkeley* (Columbia: R. L. Bryan Co., 1985), 69–73.

Mulberry encompassed 945 acres, with 120 devoted to rice. More than fifty-five slaves worked its lands, making it one of the largest plantations in the area. The main dwelling, a hybrid of Georgian and Jacobean architecture with four corner turrets, is among the most unusual domestic buildings on the South Atlantic coast. Characterized by a squat profile and heavy construction, it reflects the conditions that greeted early settlers to the Carolina frontier.[7]

The Broughton family figured among early Carolina's political and economic elite. Thomas Broughton emigrated from the West Indies in the early 1680s. About 1683 he married Anne Johnson, the daughter of Sir Nathaniel Johnson, who later served as governor of the colony. Broughton initially pursued the Indian trade, a lucrative form of commerce on the colonial frontier. His political connections aided his commercial ambitions and he soon became one of a small number of men who dominated the trade. In the early 1710s, Broughton turned his attention to planting and began developing Mulberry. The site that Broughton chose for his residence lies atop a bluff overlooking the river. Construction of the dwelling began about 1710 and reached completion by about 1714. One of the earliest brick buildings erected outside Charleston, it came to be called "Mulberry Castle" after sheltering settlers during the Yemassee War and because of its turreted form.[8]

---

[7] R. M. McKelvey, "Plat of Mulberry Plantation," 1808, South Carolina Historical Society, Charleston, SC (hereafter SCHS); Seventh Census of the United States, Slave Schedule, St. John's Berkeley Parish, Charleston District, 4–5, microfilm (M432, roll 862), National Archives and Records Administration, Washington, DC (hereafter NARA).

[8] For use of the term "Mulberry Castle," see John Beaufain Irving, *A Day on Cooper River* (Charleston: A. E. Miller, 1842), 12; Constance Woolson, "Up the Ashley and Cooper," *Harper's New Monthly Magazine*, Dec. 1875, 22; H. R. Dwight, *Some Historic Spots in Berkeley* (Moncks Corner, SC: Monck's Corner Drug Co., 1921), 14. A variation is "Some Estates on the Ashley and Cooper Rivers," which refers to the house as Mulberry Towers. See Ware, ed., *Georgian Period*, III: 112. One account that interprets the corner pavilions as intended for "military defence" is Daniel Elliot Huger Smith, "The Mulberry," 4, Alice Ravenel Huger Smith Papers, SCHS. Other early brick buildings include the second house at Medway Plantation and the main dwelling at Exeter Plantation. Brick dwellings appeared in Charleston in the seventeenth century; see Peter A. Coclanis, *The Shadow of a Dream: Economic Life and Death in the South Carolina Low Country, 1670–1920* (New York: Oxford University Press, 1989), 178, no. 176. On the origins of the house, see Carter L. Hudgins et al., *The Vernacular Architecture of Charleston and the Lowcountry, 1670–1990: A Field Guide* (Charleston: Historic Charleston Foundation, 1994). On Mulberry's role in the Yemassee War, see "A Statistical Account of St. Stephens' District," in David Ramsay, *Ramsay's History of South Carolina* (Newberry, SC: W. J. Duffie, 1858), II: 292.

Thomas Broughton intended Mulberry to become his family's seat in the New World. When he died in 1737, Mulberry passed to his eldest son, Nathaniel, who focused his energies on planting. He owned 128 slaves and, in addition to Mulberry, an adjoining tract of 711 acres and another 2,530 acres in Wassamsaw Swamp. He died late in 1754 or during the first few days of 1755.[9] The plantation then passed to Nathaniel Broughton II (Nathaniel's first son, Thomas, predeceased him). Nathaniel II held Mulberry for barely a decade before he died. By the terms of his will, Mulberry became the property of his son, Thomas. The fourth Broughton to own the plantation, Thomas oversaw its operations for half a century. When he died around 1808, the first significant division of the property occurred. "North Mulberry," a 482-acre tract that included the 1714 house and its associated outbuildings, became the property of his eldest son, Thomas IV. The adjoining tract, "South Mulberry," passed to his son Phillip Porcher.[10] Although legally separate, the two parcels remained linked by family relations and returned to joint ownership in the 1870s. Ultimately, Mulberry remained in the hands of the Broughton family for 195 years.[11]

During the late antebellum era, Mulberry operated with an efficiency typical of plantations on the Cooper. In 1849 the plantation turned out 159,000 bushels of clean rice, a figure that put it in the top 15 percent of plantations in St. John's Berkeley Parish. Rice grew on 154 acres. Upland fields produced corn, oats, peas, and other foodstuffs. Large stands of timber supplied firewood and building materials. The Broughtons' practice of willing slaves to descendants led to the formation of a close-knit

---

[9] Ruth Holmes Whitehead, *Broughton Family Sourcebook* (N.p.: n.d., [1998?]), 11–21. Nathaniel's other holdings included Ruinville Plantation, a 566-acre property on Sandy Island in All Saints Waccamaw Parish. Nearby was a 764-acre tract owned by his brother Andrew. See Suzanne Cameron Linder and Marta Leslie Thacker, *Historical Atlas of the Rice Plantations of Georgetown County and the Santee River* (Columbia: South Carolina Department of Archives and History for the Historic Ricefields Association, 2001), 223. For a detailed description of Mulberry's output in the 1750s, see George D. Terry, "'Champaign Country': A Social History of an Eighteenth Century Lowcountry Parish in South Carolina, St. Johns Berkeley County" (PhD diss., University of South Carolina, 1981), 264–267.

[10] Conveyance of South Mulberry from Mary Broughton to Samuel Barker, Feb. 6, 1836, Book Q10, 502–506; Conveyance of Mulberry Plantation from Alexander Mazyck, executor, to Thomas Milliken, July 27, 1835, Book K10, 309–311; both in Charleston County Register of Mesne Conveyance, Charleston, SC (hereafter Charleston RMC).

[11] Theodore G. Barker to Pine Timber Corporation, May 4, 1909, Deed Book C-14, 73–78, Berkeley County Register of Deeds, Moncks Corner, SC (hereafter BCRD).

group of slave families. In 1860, for example, the sixty-seven slaves at South Mulberry all belonged to families the Broughtons had owned for generations.[12]

Mulberry's material form and spatial organization adhered to patterns common throughout the lowcountry. The complex shown in Corum's painting and the core activities associated with rice production became more dispersed over time. As of 1808, the main dwelling stood at the center of a building group that included barns, sheds, and eight slave cabins. Immediately adjacent to the house lay two workyards, and in the rear stood an orchard and a vegetable garden. Tidal fields hugged the river's edge, below the house and outbuildings. In addition to the 120 acres then planted in rice, slaves had cleared another 135 acres of river swamp and 63 more lay uncleared.[13]

By the 1850s, the slave street occupied a location about a quarter mile downslope from the main dwelling, closer to the river and several tidal fields. Nearby stood a rice mill and several barns. Development of these facilities moved productive activities away from the main house and separated the domestic space of the slaves from that of the master and his family.[14]

The Civil War and emancipation threw Mulberry into chaos. Black freedom made production of staple crops more difficult, less remunerative, and prone to conflict. Rice cultivation resumed at Mulberry early, in 1866, albeit on a greatly reduced scale. On January 1, 1866, eighteen free men and women entered into a labor contract with John B. Milliken, the son of then-owner Thomas Milliken. The group consented to produce rice and other crops in return for a one-third share.[15] This arrangement,

[12] Federal Censuses of Agriculture, 1850 and 1860, St. John's Berkeley Parish, Charleston District, MS, South Carolina Department of Archives and History, Columbia, SC; Sanford William Barker, "List of Negroes Belonging to South Mulberry, 1860," Agricultural Experiments and South Mulberry Plantation Journal, SCHS.

[13] McKelvey, "Plat of Mulberry Plantation"; Will of Thomas Broughton, recorded Aug. 19, 1738, in Whitehead, *Broughton Family Sourcebook*, 5. On typical patterns of spatial organization, see Morgan, *Slave Counterpoint*, 110–123.

[14] Smith, "The Mulberry," 7.

[15] S.C. Labor Contracts, Jan.–May 1866, 23, vol. 237, Records of the Bureau of Refugees, Freedmen, and Abandoned Lands (hereafter RBRFAL), RG 105 (M1910, roll 89), NARA. For insight into the severity of conditions at Mulberry, see John B. Milliken to Capt. F. W. Liedtke, Oct. 20, 1867; John B. Milliken to Capt. F. W. Liedtke, Nov. 4, 1867; J. B. Milliken to Freedmen's Bureau Agent, Jan. 3, 1868, all in Letters Received by Subassistant Commander at Monck's Corner, RFRFAL (M1910, roll 88), NARA; John B. Milliken to Capt. F. W. Liedtke, Jan. 14, 1867, Letters Relating to Rations; and Register of Destitutes, vol. I, 56, in RBRFAL, roll 89.

typical of the immediate postbellum era, initiated a tenuous resumption of operations that endured for several years. In the early 1870s, Thomas G. Barker, a Broughton family member, purchased Mulberry and launched efforts to revive it as a commercial enterprise. He managed to increase production but never came close to matching the output of the antebellum era. Conditions at Mulberry reflected the general state of agriculture in the region. With freedpeople committed to producing for themselves and eager for autonomy, planters struggled to maintain their fields and get crops to market.[16]

Soon after purchasing Mulberry, Barker made extensive renovations to the main dwelling. He added a tin-shingled roof to the main block and installed new windows on the ground floor. The windows had two-over-two sashes, consistent with the Victorian styling of the day. Barker also added small rectangular windows in the gable ends of the attic story and may have replaced the entry doors.[17] These changes altered the appearance of the house, protected it from weather, and made it more comfortable. They also indicated that Barker expected the plantation to be remunerative for some time. Barker apparently believed the worst conflicts of the postwar era had passed and that rice planting again held promise.

Although increased productivity affirmed Barker's optimism for a time, conditions at Mulberry eventually collapsed. The turning point came in the 1890s. The hurricanes of 1893, 1894, 1896, and 1898 damaged the rice fields, felled trees, and wreaked havoc on the barns and sheds. By the beginning of the twentieth century, Mulberry stood in disrepair. Barker shuttered the main house; the rice fields lay flooded and overgrown.

---

[16] Lease, T. Alexander Broughton to Theodore G. Barker, Feb. 5, 1874, Book P16, 47–49; Lease, John B. Milliken to Theodore G. Barker, March 6, 1874, Book P16, 50, all at Charleston RMC. Barker was the grandson of Thomas Milliken, an Irish immigrant who married a Broughton daughter. See Henry Edmund Ravenel, *Ravenel Records* (Atlanta: Franklin Printing and Publishing Co., 1898), 193; *Cyclopedia of Eminent and Representative Men of the Carolinas of the Nineteenth Century*, 2 vols. (Madison, WI: Brant and Fuller, 1892), I: 133. On labor at Mulberry during the era, see John B. Milliken, Mulberry Journal, 1863–1889, SCHS.

On the resumption of production at Cooper River plantations, see Sanford William Barker, "Rice Land Cultivated on Cooper River in 1866 and 1867," Agricultural Experiments and South Mulberry Plantation Journal, 15.

[17] *Cyclopedia of Eminent and Representative Men*, I: 132–137; "Observations by C. N. Bayless, A.I.A.," Mulberry Plantation vertical file (30-15-30), SCHS (hereafter MPVF); Chapman Family Photo Collection, SCHS.

Tall grass and weeds covered the grounds. Mulberry, once a well-maintained site of commercial enterprise, had become an emblem of wrenching decline.[18]

Rice production at Mulberry probably ended by 1909, the year that Barker sold the plantation to a timber company.[19] Other plantations in the area ceased operations after the catastrophic hurricane of 1911, which pushed saltwater far up the Cooper and rendered most fields sterile.[20] By then, Mulberry had long since become an object of retrospection. In the 1870s, journalist Constance Woolson had identified it one of many "old houses and gardens" on the Cooper that called to mind "colonial memories and Revolutionary legends." Her nostalgic portrait cast it as one of about a dozen sites along the river that recalled the nation's beginnings.[21] Two decades later, another account characterized Mulberry as the "old home of the Broughtons." It highlighted the architecture of the main house and the variety of plantings found nearby.[22] These and other writings revealed the outlines of an emergent model. Recollections centered on architecture, aesthetics, and a handful of owners soon became the hallmark of histories about a place with a more elaborate and complex past. Stories of a plantation without slaves, labor, or commercial activity illustrated the narrowness of retrospective views of slavery and the absence of African American voices in public discourse. As the Civil War and emancipation receded in time, so did memory of the purposes plantations had historically served and the conflicts slavery had produced (see Figure 4.2).

[18] "Mulberry Plantation, Pawley Plantation, Salt Point Plantation," TS [ca. 1935], MPVF; Chapman Photo Collection, SCHS.

[19] Theodore G. Barker to Pine Timber Corporation, May 4, 1909, Deed Book C-14, 73–78, BCRD.

[20] The claim that the 1911 hurricane ended rice production on the Cooper River comes from Dorothy P. Legge Interview (TS, Aug. 21, 1987), Historic Charleston Foundation Archives, Charleston, SC. On the damage caused by the 1911 hurricane, see "Rice Crop Wiped Out," *News and Courier* (Charleston, SC), Aug. 31, 1911, 10. The demise of rice production is vividly shown in federal census records. In Berkeley County, the number of acres planted in rice declined from 9,210 in 1899 to 1,408 a decade later and only 319 in 1919. See *Twelfth Census of the United States, Taken in the Year 1900: Agriculture*, Pt. II, *Crops and Irrigation* (Washington, DC: U.S. Census Office, 1902), 193; Bureau of the Census, *Thirteenth Census of the United States, Taken in the Year 1910*, vol. VII: *Agriculture, 1909 and 1910* (Washington, DC: Government Printing Office, 1913), 516.

[21] Woolson, "Up the Ashley and Cooper," 22–23.

[22] Ellen Porcher, TS of "Old Home of the Broughtons," *The Meteor* (Charleston, SC), Apr. 6, 1895, MPVF.

FIGURE 4.2 During the late nineteenth century, old-line families began visiting ancestral homesteads on fall weekends and holidays. Shown here is a gathering at Mulberry circa 1910.
Courtesy Historic Charleston Foundation, Charleston, SC.

MAKING MULBERRY ANEW

When the Chapmans purchased Mulberry, they joined a small but growing group of northerners who owned estates on the Cooper River. The Kittredges had wintered at Dean Hall for almost a decade, and J. W. Johnson had acquired Mepkin Plantation in 1912. In rehabilitating Mulberry, the Chapmans became the first northerners to turn a plantation on the Cooper into an elegant estate. In later years, others would follow suit, but at the time they stood alone.[23]

In terms of background, the Chapmans fit the emerging profile for owners of sporting estates. Wealthy, enthusiastic about recreational

---

[23] On the Kittredges' ownership of Dean Hall, see Chlotilde R. Martin, "The Cypress Gardens, in Berkeley," *News and Courier*, undated newspaper clipping, Dean Hall Plantation Vertical File, SCHS. On Mepkin Plantation while owned by the Johnsons, see Chlotilde R. Martin, "Mepkin, Henry Laurens Plantation," *News and Courier*, May 17, 1931, B14.

hunting, and charmed by the coastal region, they reveled in their annual visits to Mulberry. Clarence Chapman had spent his career in finance. He became a member of the New York Stock Exchange in 1900 and continued trading securities until 1936, when he retired from business. The Chapmans resided on a 350-acre estate in Bergen County, New Jersey.[24] They used Mulberry as a winter home, a private hunting retreat, and a venue for entertaining. Like other northerners, their aims centered on leisure, recreation, and escape from cold and snow.[25]

The Chapmans began their efforts by restoring the main house to its circa 1800 appearance (see Figure 4.3). For this project they hired Charles Brendon, an English-born architect practicing in New York. Brendon may be the least acclaimed of all the designers who worked for northerners who created lowcountry estates. Although little is known about his career, the work he carried out at Mulberry shows him to have been a conscientious student of early American architecture. Brendon returned the house to its late eighteenth-century appearance, made needed repairs, and added a handful of decorative elements. In the process, he removed features added during the last quarter of the nineteenth century while retaining as much eighteenth-century fabric as possible. These measures suggest a great deal about his intentions. Brendon and the Chapmans saw themselves as treating an important example of early American architecture with care and respect, literally as a historical artifact. They showed concern for its early form and appearance by salvaging original materials, replacing deteriorated elements in kind, and highlighting eighteenth-century decorative features. At the same time, they showed no hesitation in making alterations for the sake of comfort and convenience. Paradoxically, they strived for authenticity while making substantial changes. Most likely they saw themselves as striking an appropriate balance – restoring the house to something close to its original appearance while modifying it for domestic use. The project embodied tensions seen in countless

[24] "Stock Exchange News," *New York Times* (New York, NY), Jan. 14, 1917, E4; "Exchange Firm Retires," *New York Times*, Dec. 11, 1936, 43; "Clarence Chapman, Once a Stockbroker," *New York Times*, Apr. 14, 1947, 27; Phil Samenuk to author, June 5, 2007, in author's possession. On the Chapmans' home in New Jersey, see Leo Bugg, "Northern Jersey Big Home Center," *New York Times*, Apr. 23, 1911, sec. 8, p. 12; "Developers Take Chapman Estate," *New York Times*, Aug. 20, 1950, sec. 8, p. 1.

[25] The Chapmans' version of plantation life is shown in a compilation of home movies shot at Mulberry circa 1930. Historic Charleston Foundation has transferred these to a DVD. I viewed the disc in its entirety at the Historic Charleston Foundation Archives on January 22, 2008. See "Mulberry Castle" (ca. 1930), DVD recording, HCFA.

FIGURE 4.3 Main house shown immediately before the Chapmans' purchase,
circa 1915. Note standing-seam metal roof and ancillary buildings in rear.
Courtesy Historic Charleston Foundation, Charleston, SC.

restorations: a desire to reclaim the past while making it part of the
present, a zeal for authenticity amid liberal alterations, and installation
of some modern amenities alongside deliberate avoidance of others.

The dwelling had long since ceased to be the house Thomas Broughton
built. In addition to Barker's renovation, a remodeling carried out circa
1800 had installed new interior finishes, a new roof, a new stair, and a
handful of other elements, including ornate mantels, denticulated cornices, and paneled walls in the parlor and dining room. Workers had
also installed new joists and flooring in the parlor and mantles in the
library and upstairs rooms. The new finishes added Federal styling to the
principal entertaining rooms, thus bringing them into step with contemporary tastes. The new stair replaced a narrower, steeper original and
used a gabled landing to ease the transition between the two floors. The
new roof stood several feet taller than the original and thus added
headroom to the second story. The portico and stair porch were also

rebuilt and possibly modified. In all, the renovation thoroughly reworked the original design.[26]

Brendon began by removing almost all of the features added by Barker's remodeling. These included the tin roof and two-over-two windows. Brendon left the small attic windows in the gable ends untouched, apparently because he considered them functional and unobtrusive. Meanwhile, he left earlier surviving elements in place, including the Federal mantles and cornices, the staircase, and mantles in the library and upstairs rooms. Brendon had these refinished but left them unchanged.

Brendon made relatively few changes to the exterior. He focused on restoring eighteenth-century features that had been lost or obscured and replacing deteriorated elements in kind. Workers installed a new cedar shake roof that replicated the appearance of the original, repointed all of the masonry, and replaced cracked and spalled bricks. They also installed new windows throughout: twelve-over-twelve sashes on the ground floor and nine-over-nines above, as appropriate for the colonial period. Louvered wooden shutters of uncertain origin were present on all of the exterior windows. Brendon replaced these with paneled (i.e., solid) shutters, which lacked functionality (they did not allow for air passage) but had an appearance that fit contemporary expectations for "colonial" design. In addition, some of the dormers had suffered water damage and needed repairs. Photographs taken during the restoration indicate that workers rebuilt most of them front and back. The portico and entrance stair received similar treatment. Carpenters fashioned a new roof and columns, and bricklayers rebuilt the stairs and covered them with cement to stave off decay.

The most extensive alterations took place inside. Brendon directed a thorough refinishing of the entire house. Workers patched plaster; replaced deteriorated woodwork; stripped and sanded moldings, door-jambs, and paneling; and installed new floors in the pavilions. All rooms received fresh paint, and the floors in the principal rooms were refinished. Brendon left the layout of the first floor unchanged but remodeled the

---

[26] Unless otherwise indicated, the analysis of changes made to the house over time and the renovation that Brendon oversaw is based on the following sources: Architects' Emergency Committee, *Great Georgian Houses of America*, 2 vols. (1937; reprint, New York: Dover Publications,1970), II: 49–54; Lawrence A. Walker, "Observations of Paul Buchanan and Charles Phillips," TS, 1987, in Mulberry Plantation History File, HCFA; "Observations by C. N. Bayless, A.I.A.," MPVF; "Old Home of the Broughtons"; firsthand inspection by author, Jan. 23, 2008.

second to accommodate modern bathrooms and closets. The second story
had previously possessed two large bedrooms in the rear and two rooms
of unequal size in the front. Brendon modified this arrangement by creating a center-hall plan. This made space for bathrooms and larger closets
and may have seemed characteristically "colonial," given the prevalence
of central-hall plan dwellings in early Virginia. The stairhall reached to
the front wall of the house and changed the size and orientation the
bedrooms. The new rear rooms measured about 13 by 18 feet; the two
in front were slightly larger at 14 feet 9 inches by 18 feet. All ran longest
from the front to the rear of the house. Between each pair of adjoining
bedrooms, Brendon installed modern bathrooms.

For all practical purposes, Brendon redesigned the second story of the
house. He salvaged doors, paneling, and fireplace surrounds in order to
maintain the eighteenth-century aesthetics but otherwise used new materials. These changes qualified as the most extensive of the entire project.
Outfitting the Chapmans' private spaces required substantial alterations,
and Brendon showed no hesitation in carrying them out.

An impulse to add new decorative elements manifest itself throughout
the restoration. In the pediment of the portico, for example, Brendon
added a carved medallion composed of two fruited sprigs of mulberry
set in a horseshoe. This feature paid homage to the name of the plantation and referenced its agricultural origins. Between the columns below,
Brendon added a kneewall – a subtle alteration that gave that gave the
entrance a more domestic character. The paneled shutters on the exterior
windows represented the single most noticeable attempt to enhance the
appearance of the house. These presumed to replicate whatever shutters
had existed historically, almost certainly on the basis of conjecture. In the
dining room, Brendon installed an elegant sculpted shell cupboard, thus
adding further refinement to what was already the most formal room in
the house.

Even as the Chapmans endorsed the addition of these elements
and installed indoor plumbing, they decided to forgo electricity in the
name of authenticity. As the *News and Courier* reported, the Chapmans
used "lamps and candles" for lighting, just as earlier owners had done.
"Mrs. Chapman," the newspaper noted, had "a unique collection of
old lamps."[27]

---

[27] Chlotilde R. Martin, "Mulberry Castle, Built in Indian Days," *News and Courier*, July
26, 1931, A5.

FIGURE 4.4 Main house shown soon after restoration, probably early 1920s. This image shows the house in its restored form but before the development of the grounds and gardens. Photograph by Harriette Kershaw Leiding.
Courtesy of the Charleston Museum, Charleston, SC.

The restoration amounted to a sizable undertaking. It took two years to complete, in part because of the scale of the project, but also because Brendon hired "northern men" as craftsmen and laborers. As a group, these workers refused to "work down South in the summer" and thus slowed progress. The project also proved costly. Materials costs totaled $45,000. Brendon's fees added another $10,000. The superintendent, a "Mr. Poppenheim," earned $1,200 annually, and the laborers – 80 hands in all – each earned $1.00 per day for 100 days of work. Ultimately, the renovation cost the Chapmans slightly more than $65,000.[28]

The project altered the appearance of the house in two ways (see Figure 4.4). Although most observers assumed the building had regained its eighteenth-century appearance, close examination reveals a different story. Because it had stood in dilapidated condition, extensive repairs had to be made inside and out. Some portions of the house had to be

[28] "Mulberry Plantation, Pawley Plantation, Salt Point Plantation."

completely rebuilt, others nearly so. Written and photographic sources reveal substantial reconstruction. The work that Brendon oversaw did not rebuild the house from the ground up, but in some ways it came close.

By replacing elements they considered inappropriate, Brendon and the Chapmans created a colonial aesthetic that echoed but did not replicate the dwelling's circa 1800 appearance. Brendon worked from a combination of physical evidence and knowledge of historical design principles. No written documentation for the historical form of the house is known to exist, and it is unlikely that any ever did. In determining how to reconfigure the dwelling, he likely relied on study of existing elements and informed judgment. Most of the features Brendon installed probably amounted to educated guesses at what had historically been present. How close the restoration came to replicating the circa 1800 appearance of the house is an open question. In the end, the overall form remained unchanged, but some of the most prominent stylistic details may have been new.

Material refurbishment proved more important. The thoroughness of the restoration left no part of the house untouched. Masonry repairs, reconstruction of the portico and dormers, installation of new windows and shutters, and fresh paint inside and out – these created a level of fit and finish that few eighteenth-century buildings had possessed. In addition to addressing practical problems, they erased all signs of time's passage. By refurbishing a ruined building, they radically altered its appearance. In turn, the house assumed an entirely new condition. Although its form and features identified it as the product of a different time, the overall appearance situated it squarely in the present. Reverent treatment combined with preparation for ongoing use created a distinctive look. No longer a decaying vestige of a bygone era, the house possessed a gloss and shine that belied its actual age.

Contemporary observers recognized the change. One commentator wrote, "You would not suspect its age from its outward appearance."[29] This observation spoke volumes about the effects of the restoration. It simultaneously denied the age of the house while giving it new emphasis. For decades, decay had dominated. A byproduct of economic decline, decay revealed Mulberry's fate as a commercial enterprise. By contrast, Brendon's restoration gave the house a crisp and polished "newness." Visually, it wrenched the house out of a distant era and pulled it squarely

---

[29] J. O. Mosely, "Mulberry on Cooper River Noted for Its Great Beauty," unidentified newspaper clipping dated Feb. 11, 1923, MPVF.

into the present. At the same time, highlighting eighteenth-century features identified the house as belonging to a revered era. Removal of weathering and decay made style the sole temporal marker. The former had aggregated slowly over time, in tandem with the decline of rice production; the latter appeared virtually overnight, as a product of the Chapmans' efforts. One resulted from social and economic upheavals; the other developed with purposeful treatment of a valued artifact.

Ultimately, the restoration produced all of the contradictions and tensions that characterize any effort to return a revered edifice to a particular era. Rather than recapturing a select piece of the past, the project remade the building for contemporary needs. Brendon conserved much of the original fabric and most of the features installed circa 1800. He left the original design intact and strived to make alterations sensitively, so as not to introduce discordant styling. The aim lay in returning the structure to what it had been a century or more before. Yet by virtue of extensive reconstruction and refurbishment, the project created a substantially new building with a "historic" appearance. The change stemmed in part from the removal of additions and alterations that had accrued over time, but it also reflected the addition of new features and complete refurbishment. The Chapmans' dwelling belonged simultaneously to a colonial past and a present that valued select remains. No longer representative of its full history, the house assumed a dualism characteristic of revered edifices remade for contemporary needs.[30]

THE LANDSCAPE BEYOND

The Chapmans reshaped the landscape at Mulberry as deliberately and extensively as they did the main house. Their initial efforts focused on rehabilitating the surrounding grounds and a few other spaces. Over time, they made additional improvements. Their overall goal lay in turning Mulberry into a site of seasonal residency and a venue for leisure, entertaining, and sport.

The Chapmans first turned their attention to the grounds while work on the main dwelling proceeded. They began by creating automobile access to the house. Workers cut and graded two miles of road through forested land to reach US Highway 52. Previously, the principal access

---

[30] The cultural consequences of restoration have received inadequate attention from scholars. The best treatment is David Lowenthal, *The Past Is a Foreign Country* (Cambridge: Cambridge University Press, 1985), 263–282, 325–362.

FIGURE 4.5 Black labor made northerners' estates possible. Here, workers clean a drain in a rice field at Mulberry circa 1917.
Courtesy Historic Charleston Foundation, Charleston, SC.

had been from the river. At least some portions of the road followed a path or road that dated to the early nineteenth century. When finished, the new road meandered through the forest on the west side of the plantation and swept gracefully past huge oaks as it approached the house. J. O. Mosely visited Mulberry soon after its completion. He described the road as a "lane forty feet wide, with graceful curves every hundred yards or so." Each curve offered a "different panoramic view." Live oaks dramatized the approach to the house by keeping it out of sight until the last minute. As Mosely explained, "as you near the mound you behold the stately evergreen oak trees with moss-covered limbs." A small rise "lands you in front of the magnificent mansion."[31]

Rehabilitation of the rice fields proved a far larger undertaking. Workers began by cleaning out drainage ditches and rebuilding collapsed sections (see Figure 4.5). Then they made repairs to about four miles of dikes. To ensure control of water levels, crews built thirty-one new trunks, each at a cost of $300. Finally, workers cleaned and planted

[31] Mosely, "Mulberry on Cooper River Noted for Its Great Beauty."

the fields. This marked the last step in a process that turned the once productive and profitable fields into a large waterfowling venue. The total cost: $29,300.[32]

Exactly when the Chapmans carried out subsequent improvements is unclear. Modest efforts to create a domestic setting for the main dwelling occurred immediately following the renovation of the main house, for in 1923 a visitor reported that the lawn in front "is kept mowed in such a way as to make one feel as if he was walking on a thick velvet carpet."[33] The Chapmans also planted camellias along a path to the inside of the rice fields, at the base of the bluff.[34] Later on, they rebuilt several outbuildings and erected a number of new structures. This may have occurred as early as about 1920, but some evidence suggests that it happened several years later. Whenever it actually took place, the effort created a new outbuilding complex. The Chapmans rehabilitated a superintendent's house, a servant's house (possibly a surviving slave cabin), and several barns and sheds. They also built a new stable and two cabins. The resulting complex lies a short distance downslope from the main house.[35]

The final phase of the grounds improvement campaign came in the early 1930s, when the Chapmans commissioned landscape architect Loutrel Briggs to prepare a formal landscape plan (see Figure 4.6). Briggs created a groomed setting for the main dwelling, a formal garden close alongside, and a network of pleasure walks leading to a large terraced garden along the bluff overlooking the river. His survey of the surviving landscape at Mulberry found no evidence of a historic garden, so he looked to the house for stylistic cues. Briggs's plan had three major components. Immediately in front of the house he created a small forecourt. This feature referenced English country estates and gave the Chapmans a designated space for greeting visitors. It also added formality and differentiated the main entrance from the service access in the rear. In creating the forecourt, Briggs reoriented the drive leading to the house, which had previously terminated behind the building. The forecourt

---

[32] "Mulberry Plantation, Pawley Plantation, Salt Point Plantation."

[33] Mosely, "Mulberry on Cooper River Noted for Its Great Beauty."

[34] When Briggs began working at Mulberry about 1930, he found that "a path had been rather recently made beside the river along which camellias were planted." See Briggs, *Charleston Gardens*, 138.

[35] Preservation Consultants, *Berkeley County Historical and Architectural Inventory* (1989), survey forms 110-0044.00–04, South Carolina Department of Archives and History, Columbia, SC; firsthand inspection of extant buildings and personal communication with Ben Miller, superintendent of Mulberry Plantation, Jan. 23, 2008.

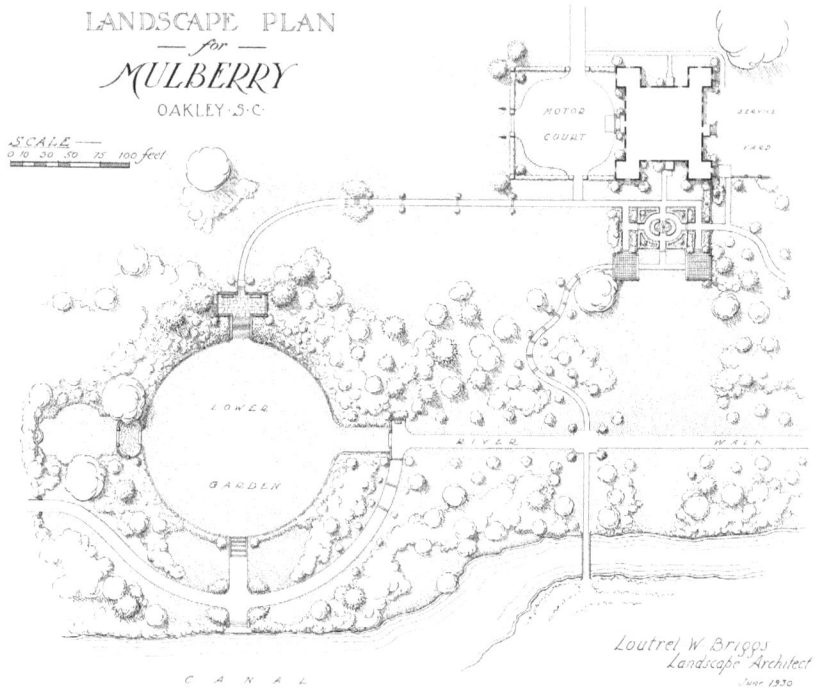

FIGURE 4.6 Landscape plan by Loutrel Briggs, 1930.
From the Loutrel Briggs Collection, South Carolina Historical Society, Charleston, SC.

encompasses a paved terrace enclosed by a low pittosporum hedge. Small cedar trees stand at each corner, and squared brick posts mark each of three entrances. Briggs stipulated that the hedge and cedars be kept carefully groomed. As a final touch, he placed two small cannons at the entrance to the lawn, thereby adding an element of dignified authority to the already elegant setting.[36]

Close by, along the east side of the building, Briggs laid out a formal garden. He described the site as an "obviously excellent" location: a small, nearly level area near the edge of the bluff. The garden centers on a small court formed of planting beds and walkways. Two square patios

---

[36] Briggs, *Charleston Gardens*, 138. The cannons were unearthed at the base of the bluff circa 1835. Although their provenance is unknown, some evidence suggests they saw use during the Battle of Charleston in 1780. See "Mulberry Sale to Summerville Man Negotiated," *News and Courier*, Nov. 8, 1946, 1; Warren Ripley to Tom Savage, Aug. 24, 1987, Historic Charleston Foundation Archives, Charleston, SC.

with wooden chairs and benches lie immediately adjacent, opposite the house. Brick screenwalls and wrought iron fences enclose the ensemble. Briggs populated the beds with small shrubs, flowers, and a few exotics and planted boxwoods along the edges. At the center of the court, a sundial rests on a sculpted pillar.[37]

With its stunning prospect and close proximity to the house, the garden afforded a charming and flexible venue for a host of activities, intimate entertaining in particular. Briggs's design established a close relationship to the dwelling. The sundial and brick screenwalls added scale and visual interest while accentuating contrasts between the formal landscape around the house and the more naturalistic features along the river and rice fields below.[38]

Further down along the edge of the bluff, at a distance of roughly 300 feet, Briggs placed the third piece of his composition: a large circular garden set into the hillside "somewhat like an amphitheater." This "lower garden," as Briggs called it, centered on a "solid circle of lawn" roughly 150 feet in diameter. Banks planted with azaleas, camellias, and other "early flowering shrubs" sloped gently down to its edges from the hillside above. A paved terrace at the top of the hill provides a dramatic view of the garden, the river, and the surrounding landscape. In the spring, when the azaleas and camellias reach full bloom, the lawn is ringed by a sea of pink, red, and white. The largest element of Brigg's plan, the lower garden offers a stunning counterpoint to the formal garden beside the house. Its scale dwarfs visitors and dramatizes views of the river and surrounding landscape. Whereas the formal garden affords intimacy and serves as an extension of the house, the lower garden strives for majesty. As a focal point in an already spectacular landscape, it inspires awe.

Throughout the gardens, Briggs used "statuary and ironwork" in a manner he considered appropriate to the Georgian features of the main dwelling and the age of the plantation. He also set "graceful little lead figures" atop the brick walls at the entrance to the lower garden. The upper and lower gardens feature wrought iron gates and fences with an elegant scroll pattern. Briggs used salvaged bricks for walkways, screenwalls, and corner posts. These possess a mottled, irregular appearance that supplies a deliberately crafted look of age. As an added touch, Briggs used "especially molded" bricks to dress the tops of the square posts of the forecourt. Sculpted to match the coping along the base of the

---

[37] Briggs, *Charleston Gardens*, 138–139.   [38] Ibid., 138–139.

main house, these bricks create visual continuity between the house and forecourt and suggest concurrent origins. In all, Briggs did a masterful job of giving Mulberry a set of brand-new gardens that appeared centuries old.[39]

Creating the gardens entailed substantial costs. Briggs's fees amounted to $3,500 and plant materials totaled at least $13,000 ($10,000 for camellias and another $3,000 for trees and shrubs). A crew of arborists, the "Davy Tree Men," worked at the plantation over a three-year period. Their services cost the Chapmans $6,000. To maintain the gardens and landscape, the Chapmans installed a $4,000 watering system. All told, the grounds beautification campaign cost at least $16,500. Construction of the outbuilding complex ran another $21,500.[40]

Development of the gardens, grounds, and outbuildings took important steps toward fulfilling the Chapmans' objectives. The gardens and grounds created a groomed landscape that highlighted the architecture of the main house and created a series of spaces for leisure and entertaining. Segmentation of spaces near the dwelling facilitated greeting and entertaining guests and allowed the landscape to be experienced sequentially, as an unfolding pageant of vistas and color. The gardens also provided an expansive setting for enjoying the countryside and mild weather. Meanwhile, the outbuildings supplied infrastructure for maintaining the estate and recreation. The barns and sheds provided storage for tools and equipment, the stables housed the Chapmans' horses, and kennels sheltered trained hunting dogs. Moreover, as a dedicated space of labor, the outbuilding complex removed maintenance activities from spaces close to the main dwelling.

Aesthetically, the Chapmans' version of Mulberry possessed a more refined and elegant appearance than its predecessors. Restoration of the main dwelling made a revered example of early eighteenth-century architecture the visual and functional center of a large estate. Its position at the crest of the bluff gave it a commanding presence; the gardens and lawns focused attention on it. Meanwhile, Briggs's landscaping gave a large portion of the estate a more formal appearance than it had previously possessed. Covering roughly twenty acres, the gardens and grounds formed a sequence of groomed settings that invited immersion in verdant

---

[39] Ibid., 141.
[40] "Mulberry Plantation, Pawley Plantation, Salt Point Plantation." According to this statement, the combined costs of renovating the house, rebuilding the rice fields, and creating the outbuildings and gardens totaled $112,300.

spaces, many with captivating views. The elegant styling of this landscape represented a dramatic change from the undeveloped, predominantly utilitarian spaces that had historically dominated Mulberry. The out-building complex also added refinement. Its neat and uniform appearance stood in contrast to the agricultural buildings and workyards that had historically been present.

The Chapmans' estate offered little evidence of the labor system and agricultural operations that had historically sustained plantations throughout the lowcountry. The loss of the rice mill, barns, and sheds at the far end of the bluff stripped Mulberry of structures associated with its role as an agricultural enterprise. Destruction of the slave street removed telling evidence of the same era. The loss of these buildings severed a powerful link to Mulberry's former use. Removal of the utili-tarian buildings that had historically stood near the main house also obscured portions of Mulberry's history. Finally, rehabilitation of the rice fields altered spaces essential to understanding the historical use and development of the plantation. By turning them into an elaborate water-fowling domain, the Chapmans made the central element of Mulberry's productive complex part of the recreational infrastructure of their estate. Although the fields remained present, the purposes they served, their appearance, and their relationship to other parts of the plantation differed fundamentally. Survival alone left them unable to convey a strong sense of their historical role.

The Chapmans' redevelopment of Mulberry turned the remnants of a well-ordered agricultural enterprise into an elaborate pleasure ground. By refiguring its material form, the Chapmans stripped away surviving evidence of slavery and agriculture and replaced it with a landscape dedicated to leisure and recreation. Within a few years of their pur-chase, Mulberry became dramatically different. Few signs of the labor and commodity-producing systems that had historically shaped and sustained the plantation remained. Neatly kept lawns and gardens occupied places where slaves had once lived and worked, and seem-ingly natural landscapes covered spaces once crucial to cultivation of staple crops. The Chapmans' Mulberry continued to be called a plantation, but its form and use identified it as a modern country place, a leisure retreat made possible by the fruits of industrial capitalism and a culture that recognized sharp distinctions between past and present. It made clear the decline of commercial agriculture in the lowcountry and the rise of dramatically different forms of land use and economic organization.

REPRESENTING MULBERRY

The Chapmans' efforts made Mulberry a focus of greater attention than ever before. The Charleston *News and Courier* applauded the couple for saving a revered dwelling and publicized the development of the grounds and gardens.[41] Tourist publications continued to highlight Mulberry's picturesque character and distinctive main dwelling. Architectural historians showed increased interest in the house. Fiske Kimball, one of the most prominent scholars of his generation, wrote about it at length in his *Domestic Architecture of the American Colonies and of the of the Early Republic* (1927). The Architects' Emergency Committee, a Depression-era initiative that cataloged important examples of early American architecture, included Mulberry in its survey of Georgian-style domestic buildings. In the early 1950s, Thomas T. Waterman, another pioneering architectural historian, featured the house in his *Dwellings of Colonial America* (1951).[42] Meanwhile, as Charleston's popularity as a tourist destination grew, local promoters began taking visitors to see Mulberry firsthand. Beginning in the early 1930s, tours of plantations and other sites along the Cooper allowed visitors to see some of the best-known plantations on the Carolina coast. Mulberry figured among several plantations featured on a 1934 tour of Cooper River plantations and remained open to the public on a limited basis in later years. By the late 1930s, it stood among the most revered plantations in the lowcountry.[43]

With new attention came new views of Mulberry's history. Representations of the plantation continued to emphasize many of the same qualities as before: Thomas Broughton, its role in the Yemassee War, and the architecture of the main dwelling. At the same time, representations became more consistent, narrower in scope, and more celebratory. Increasingly, histories of the plantation concentrated on a few owners and a few well-remembered events while omitting mention of Mulberry's historical purposes. Popular discourse reduced Mulberry to a grand house

---

[41] Martin, "Mulberry Castle, Built in Indian Days."

[42] Kimball, *Domestic Architecture of the American Colonies and of the Early Republic*, 67, 69, 72, 285; Architects' Emergency Committee, *Great Georgian Houses of America*, II: 49–54; Waterman, *The Dwellings of Colonial America*, 23–32.

[43] Margaret S. Middleton, "Old Plantations to Be Seen on Lowcountry Tour Today," *News and Courier*, March 20, 1934, 3. Loutrel Briggs wrote in 1951 that "the house and gardens have for several years been open to the public for an admission fee." See Briggs, *Charleston Gardens*, 141. After Lawrence A. Walker purchased Mulberry from Adelaide Chapman in 1946, he opened it for visitation.

in an idyllic setting, historically and at present. By the early 1930s, depictions of the plantation assumed an almost standardized form that varied little through repeated tellings.

In July 1931, the Charleston *News and Courier* published "Mulberry Castle, Built in Indian Days," an article by Chlotilde R. Martin that chronicled the history of the plantation and the Chapmans' efforts to reclaim its former splendor. Martin characterized Mulberry as "one of the most famous of the low-country plantations." The main house, she wrote, is "one of the most imposing and unusual of any of the homes in this entire section." Martin emphasized the building's early date of construction, Mulberry's association with the Broughton and Milliken families, and its role during the Yemassee War. Her history thus took a familiar form. Moreover, in presenting Mulberry as an icon of lowcountry heritage, Martin touted its early origins and longevity. In her estimation, Mulberry evidenced the fortitude and perseverance of early settlers and their determination to make Carolina a success.[44]

Martin credited the Chapmans with treating the dwelling in a manner appropriate for a revered edifice. "The house was in a bad state of disrepair when the Chapmans bought it," she reported. "Parts of the ceilings and floors were gone." Martin informed readers that the Chapmans had replaced these elements and installed modern bathrooms. At the same time, she said nothing about the other work that took place. Martin's account omitted mention of the extent of reconstruction Brendon had carried out, the new decorative features he installed, and the removal of the Victorian-era elements. Instead, Martin presented the house as little changed: refurbished and renewed but otherwise unaltered. "The mouldings and mantles are all original," she noted.[45]

Martin also expressed approval for the modifications Brendon made to accommodate contemporary living standards. She marveled, for example, at his decision to put a breakfast room in one of the towers and a gun room in another. Martin also praised the Chapmans' taste in decor. "The furnishings are exquisite and thoroughly in keeping with the atmosphere of the house itself," she declared.[46]

Martin devoted considerable attention to the gardens. She described them sequentially, in the same manner as a visitor might experience them. She began by characterizing the formal garden beside the house as an "ever-green garden of miniature shrubbery." Martin then described the

[44] Martin, "Mulberry Castle, Built in Indian Days."   [45] Ibid.   [46] Ibid.

plantings on the hillside below, which she called "a magnificent garden of rare beauty and charm." "Azaleas, japonicas, bulbs of various sorts, magnolias, beautifully shaped cedars, [and] shrubbery and live oaks" grew in "thick profusion," forming a picturesque mélange of color. Martin found the circular garden on the edge of the bluff equally impressive. She described it as having "the shape of a natural amphitheater." The handsome entrance gate, she noted, had a "grill brought from England." The garden included "an outdoor plaza" with "a tier of brick seats along one side." Planted in shrubbery, the plaza offered superb views of "the rice fields and the river."[47]

Martin's account possessed all of the features that would typify depictions of Mulberry for decades to come. By presenting a former slave plantation solely as a site of elite domesticity, Martin elided most of its history. By focusing on the dwelling and new gardens, Martin highlighted the elements of the estate most closely associated with white authority, historically and at present. By saying nothing about Mulberry's original purposes and the laborers who had sustained it, she omitted crucial dimensions of its past. In short, Martin offered a synoptic vision that told remarkably little about Mulberry's history. By portraying the plantation as a handsome residence set amid a well-developed landscape, she cast it as a site of peace and serenity and little else.

In the years that followed, other authors recounted Mulberry's history in similar fashion. The *News and Courier* published at least five articles during the 1940s and 1950s that followed the same basic scheme. Continuing interest among architectural historians made the house one of the best-known dwellings on the South Atlantic coast. In 1938, the cadre of architects and architectural historians who assembled *Plantations of the Carolina Low Country* included Mulberry in their catalog of valued plantations. By classifying the main dwelling as one of twenty-five buildings representative of the colonial and provincial period, they identified it as an exceptionally important vestige of Carolina's beginnings.

Accounts published in the 1930s and later decades consistently sounded several themes. One was the rarity of the main dwelling. In *Plantations of the Carolina Low Country*, Samuel Gaillard Stoney, Jr., described it as "the only remaining representative of a group of houses built just at the time when the colony of Carolina had come to be a well-founded success."[48] This description cast it as a rare example of a

---

[47] Ibid.   [48] Stoney et al., *Plantations of the Carolina Low Country*, 50.

dwelling from the early stages of colonial settlement. Stoney did not identify any of the houses that had been lost and overlooked the terror of the Yemassee War, which left many settlers uneasy about the colony's prospects immediately following the completion of the house at Mulberry. Nonetheless, his characterization encapsulated a view consistent with long-standing appreciation of the building and general nostalgia for the colonial era. Other writers followed suit. In 1950, Thomas T. Waterman described the dwelling at Mulberry as "a remarkable survival of an important plantation [house]" from the time when the early colonists "had overcome the initial difficulties of establishing themselves in the wilderness." He located its origins in the moment when colonists had "accumulated land and capital" and began "to build comfortable houses with architectural qualities that would recall their homeland."[49] Mulberry thus symbolized the passing of the worst hardships of the colonial experience and the beginnings of a stable plantation society.

Claims about authenticity figured prominently in many accounts. In 1947, the *News and Courier* characterized the house at Mulberry as a "well-preserved example of Lowcountry architecture."[50] When Loutrel Briggs wrote about the structure a short while later, he described it as "standing just as it was built by Thomas Broughton in 1714."[51] In touting the apparently unaltered state of the dwelling, these accounts ignored the eighteenth- and nineteenth-century alterations, the Chapmans' restoration, and accrued weathering and decay. They judged the restoration to have returned the house to its early or original appearance.

One especially florid account heralded Mulberry as an example of the grandeur that had characterized the lowcountry during the colonial and antebellum eras. In a newspaper feature that drew liberally from his *Plantations of the Carolina Low Country*, Samuel Gaillard Stoney, Jr., characterized Mulberry as a "monument" to what he called the "plantation era" – a two-century-long period that made Carolina the envy of other New World plantation societies. Recounting what had by then become a well-established perspective on the regional past, he described the lowcountry as "almost synonymous with plantations." Early settlers, he wrote, started a plantation soon after establishing the village of Charles Town in the spring of 1670. "Stability truly came to the colony,"

[49] Waterman, *The Dwellings of Colonial America*, 32.
[50] "Do You Know Your Lowcountry? Mulberry Plantation," *News and Courier*, Feb. 17, 1947, 10.
[51] Briggs, *Charleston Gardens*, 136.

he continued, "when rice, as a staple, first caused going plantations to spread along the rivers and creeks and over the islands of this coast." From these beginnings sprang an era of "long prosperity" that helped the "younger colony ... catch up with and then in many ways surpass its elder sisters to the north." Lowcountry planters enjoyed what Stoney called "an ample life." Surviving plantation dwellings and parish churches stood as "monuments of beauty to those happy times." As one of the most coveted of those dwellings, the house at Mulberry bore witness to "the culture that sprang directly from the plantations."[52]

Praise for the Chapmans' restoration appeared repeatedly. In *Plantations of the Carolina Low Country*, for example, Stoney contended that "much of Mulberry's charm was saved or enhanced by a most careful restoration made for its present owners, Mr. and Mrs. Clarence E. Chapman, by the late Charles Brendon of England."[53] In 1946, the *News and Courier* called the house "the most faithful restoration in this section of the country."[54] Assessments of this kind voiced approval for Brendon's skill as an architect, the work he carried out, and the Chapmans' apparent respect for Mulberry's history. Implicitly, these claims also cast the house as true to its early form and appearance.

Some accounts went so far as to credit the Chapmans' gardens with contributing to Mulberry's historical ambiance. A *News and Courier* article observed that "Mulberry's gardens," although "not as extensive as at other Lowcountry plantations," formed a "picture of Carolina history as true and nostalgic as any other." This remarkable claim posited recently created gardens as part of a "true" portrait of Carolina's early history. That post-rehabilitation commentators assumed the Chapmans' gardens to have been early or original features indicates how strongly popular views associated plantations with elegant architecture and formal landscaping. Characterizing the Chapmans' landscape as "true and nostalgic" misrepresented its origins and what survived from Mulberry's past.[55] As one of many misreadings that imposed contemporary expectations on extant features, it substituted assumptions about the historical appearance of plantations for fact.

Considered as a whole, the most striking characteristic of the accounts published after the Chapmans' rehabilitation is the limited perspective

[52] Samuel Gaillard Stoney, Jr., "Plantations Brought Riches to Lowcountry," unidentified newspaper clipping, MPVF.
[53] Stoney et al., *Plantations of the Carolina Low Country*, 50.
[54] "Mulberry Sale to Summerville Man Negotiated."
[55] "Do You Know Your Lowcountry? Mulberry Plantation."

they offered on Mulberry's physical extent and historical use. All focused closely on the main dwelling and the surrounding grounds. All focused on a few planter patriarchs and their families, a few noteworthy events, and the architecture of the main house. All focused on the Chapmans' rehabilitation and their use of the plantation. None mentioned slaves, toil, and hardship. None mentioned commercial enterprise. All cast Mulberry as a large house surrounded by a well-groomed landscape. By portraying these elements as essential features, post-rehabilitation accounts elided slavery, agriculture, and labor. They portrayed Mulberry in ways that clashed with its historical use and form.

The reasons for the perspective found in such accounts are not difficult to discern. Fascination with colonial-era architecture, idealization of the antebellum southern household, and interest in the activities of wealthy people from afar all played a part in delimiting the physical space of a property that had totaled 1,000 acres for most of its history and encompassed more than double that under the Chapmans' ownership. So, too, with the groomed landscape and gardens. These also drew attention. They encouraged a perspective centered on domestic spaces that ignored the tidal fields, upland fields and pastures, and forested land. Spaces associated with owners and their families thus rose in stature while those tied to productive activities and slaves and their descendants slipped from view. A selective focus went hand in hand with an incomplete reading of Mulberry's history.

Whether purposeful intent to diminish or erase slavery manifest itself in post-rehabilitation accounts is unclear. The general tone and orientation of published writings suggests that their form and content resulted more from inattention to dimensions of historical experience considered unimportant than deliberate omission. Pervasive belief in black inferiority proved sufficient to consign African Americans to the margins of the story or leave them out entirely. In an era when historical writing privileged elite actors and most whites regarded African Americans as primitive, childlike beings, the notion that slaves could have shaped the circumstances of their own lives, let alone the course of history, found few adherents.[56] The absence of African Americans in post-rehabilitation

---

[56] On early twentieth-century views of race, see especially Matthew Pratt Guterl, *The Color of Race in America, 1900–1940* (Cambridge: Harvard University Press, 2001); Matthew Frye Jacobson, *Whiteness of a Different Color: European Immigrants and the Alchemy of Race* (Cambridge: Harvard University Press, 1998); Grace E. Hale, *Making Whiteness: The Culture of Segregation in the South, 1890–1940* (New York: Pantheon Books, 1998; George M. Frederickson, *The Black Image in the White Mind: The Debate on Afro-American Character and Destiny, 1817–1914* (New York: Harper and Row, 1971),

accounts, then, is hardly surprising. Most contemporaries saw agency as something whites possessed and blacks did not.

The confluence of the Chapmans' activities and new histories of Mulberry offers telling commentary on the power of material remains to shape perceptions of the past. Material change and new forms of activity stirred interest in Mulberry's history while simultaneously obscuring large portions of it. Restoration of the main dwelling, rehabilitation of the grounds, creation of new gardens and landscaping, and repurposing of formerly productive lands constituted an expansive campaign of refigurement. These efforts not only attracted attention but, by implication, presented the spaces affected as historically important. At the same time, they drew attention away from other parts of the plantation. Selective attention to physical space informed narratives that celebrated only fragments of Mulberry's past.

That contemporaries saw the Chapmans' treatment and use of Mulberry as largely consistent with its past underscores the narrow boundaries of remembrance during the interwar era. Although narratives of regional history never failed to mention slavery and commercial agriculture, histories of individual plantations consistently ignored the activities that had sustained the rice-planting regime. By seeing strong commonalities between the Chapmans' estate and the plantation that had existed during the eighteenth and nineteenth centuries, contemporaries overlooked most of the latter's history. Seeing plantations, physically and historically, as lacking direct connections to slave labor and commercial enterprise disavowed the central dimension of the lowcountry's past. No matter what the exact reasons for such omissions, writers of the interwar era told Mulberry's history as the story of a plantation without slaves.

CONCLUSION

Although the Chapmans' efforts heralded new interest in vestiges of the lowcountry past, not for another decade did other northerners follow in their footsteps. In the years after the Chapmans restored the house at Mulberry, sportsmen and sportswomen created estates along the Cooper and in other parts of the lowcountry. In some cases, northerners showed concern for old buildings and landscapes, and some rehabilitated select

245–332; Joel Williamson, *The Crucible of Race: Black-White Relations in the American South since Emancipation* (New York: Oxford University Press, 1984), esp. 4–7, 327–522.

edifices in ways that privileged their early form and appearance. Examples of preservation in situ and rehabilitation without substantial alterations became common. Nowhere, however, did northerners show similar concern for authenticity as the Chapmans had. Brendon's painstaking approach represented a groundbreaking effort. Although Americans had "restored" early houses for roughly half a century, never before had anyone treated a building in the rural lowcountry in such fashion. In returning the house at Mulberry to its "historic" form, the Chapmans showed devoted care for a revered example of the region's architectural heritage. By surrounding the house with elegant gardens, they created a setting designed expressly for it. By using the house as their winter residence, the Chapmans gave it a privileged place in their lives. In all, the Chapmans gave the building new life.

During the estate-making boom of the late twenties and early thirties, other northerners created similar estates. Popularization of lowcountry history implicitly made nearly every surviving plantation structure worthy of reverent treatment, and northerners flocked to the region with visions of rehabilitating old houses and returning aging plantations to their former grandeur. At Harrietta Plantation on the Santee River, Horatio S. Shonnard and his wife returned an early nineteenth-century mansion to its original appearance. At the Wedge Plantation nearby, E. G. Chadwick renewed an exceptionally ornate dwelling and surrounded it with sumptuous gardens. On the Edisto River below Charleston, Owen Winston rehabilitated the house at the Grove Plantation, a handsome structure built about 1828.[57] These and other examples illustrated enthusiasm for returning early houses to their "historic" form and appearance. In each case, a deteriorated colonial or antebellum dwelling gained new life as the centerpiece of a new estate.

The Chapmans' efforts did not directly influence any of the estates that followed. Northerners arrived in the lowcountry with full knowledge of their options. They availed themselves of guidance from architects and other designers and proceeded with a focus trained on the lands they purchased and whatever stood on them. Although northerners took notice of other estates, nothing suggests that they looked to them for guidance. Regardless, the Chapmans created a pioneering example.

---

[57] Murray, "Harrietta on the Santee Blooms Again"; Chalmers S. Murray, "The Wedge on the South Santee," *News and Courier*, Sept. 20, 1931, A7; Isabella G. Leland, "Do You Know Your South Carolina? Edisto's Grove Plantation Was Built after Revolution," *News and Courier*, June 2, 1958, B1.

Their estate provided an early indication of what would become a major theme of northerners' efforts. At a time when views of the decaying landscape of rice slavery remained inchoate, the Chapmans saw a material heritage deserving of reverent care. Because they left no records recounting their views of Mulberry or what they hoped to achieve, it is impossible to know exactly how they saw their accomplishments. Nonetheless, their actions speak for themselves. In rehabilitating Mulberry, the Chapmans made it anew. The restoration of the main house celebrated a seemingly idyllic past, and the other improvements they made created a magnificent estate. Although most observers saw a plantation returned it to its former grandeur, in fact, the Chapmans created an estate that combined old and new in new ways.

The importance of the Chapmans' efforts has yet to be recognized. Their treatment of the main house signaled the advent of new sensibilities about the decaying landscape of plantation slavery. Although people within and without the lowcountry had seen select plantations as historically valuable since the late nineteenth century, the Chapmans led in seeking to "preserve" part of one. Their efforts cared for a celebrated edifice in ways that lowcountry people could not. In the same moment as lowcountry people awakened to the material heritage in their midst, the Chapmans showed that people from outside the region saw the built fabric of the early lowcountry in a similar manner – and would, in some cases, take steps to ensure its survival.

As an example of early twentieth-century Americans' enthusiasm for all things colonial, the Chapmans' version of Mulberry expressed an idealized view of the nation's past. Scholars of the Colonial Revival have long seen the movement as intensely nationalistic, socially and politically conservative, and motivated in part by concerns about European immigration and the potential for quintessentially American values to be lost amid rapid social and cultural change. Whatever else the Colonial Revival may have been, it claimed a noble, dignified Anglo-Saxon past as central to the nation's beginnings. In employing the idiom at Mulberry, the Chapmans revealed as much about their view of contemporary life as their aesthetic tastes. As members of a privileged elite, they endorsed the social hierarchies and traditions that the Colonial Revival symbolized.[58]

---

[58] On the ideological dimensions of the Colonial Revival, see William B. Rhoads, "The Colonial Revival and American Nationalism," *Journal of the Society of Architectural Historians* 35, no. 4 (Dec. 1976): 239–254; Alan Axelrod, ed., *The Colonial Revival in America* (New York: W. W. Norton and Co., 1985); Karal Ann Marling, *George*

The Chapmans' association with Mulberry lasted until the closing days of World War II. In January 1945, they sold the estate to Lawrence Walker, Jr., a resident of Summerville who used the plantation more or less as they had. Yet Walker also opened it for public tours, which gave Mulberry a role in the lowcountry's growing heritage tourism industry. By February 1947, visitors could stroll the grounds and tour the ground floor of the house daily. Walker also purchased all of the Chapmans' furnishings and household goods, which included antiques assembled over the course of nearly thirty years. The house thus became as much a shrine to their tenure as to any other period.[59]

In the decades since the Chapmans' time, portrayals of Mulberry crafted for popular audiences have continued to emphasize familiar themes. One recent coffee-table catalog of lowcountry estates depicts a grand plantation returned to its former glory by appreciative stewards of colonial-era architecture. Surveys of regional architecture continue to identify the main dwelling as an elegant example of Georgian design while ignoring the rest of the plantation and its history. Tourist publications routinely celebrate Mulberry's architecture without mentioning any of the activities that allowed Thomas Broughton to build such a substantial dwelling, let alone those that made Mulberry prosperous in later years.[60] Missing from these and other accounts are the activities that made Mulberry a plantation and sustained it for the better part of two centuries. Even after decades of scholarship emphasizing the central role of African slaves and Indians in the early lowcountry, popular accounts continue to lag. The discordance is less a product of competing visions of the past than failure to reconsider narratives that long ago became closely tied to physical vestiges of the past. Without reexamination of relationships between space, representation, and history, meanings ascribed to tangible remains continue to be rooted in outmoded portrayals, complete with all the biases and omissions they reflect.

Washington Slept Here: Colonial Revivals and American Culture, 1876–1986 (Cambridge: Harvard University Press, 1988); David Gebhard, "The American Colonial Revival in the 1930s," *Winterthur Portfolio* 22, nos. 2–3 (summer/autumn 1987): 109–148; Sarah L. Giffen and Kevin D. Murphy, eds., *A Noble and Dignified Stream: The Piscataqua Region in the Colonial Revival, 1860–1930* (York, ME: Old York Historical Society, 1992); Richard Guy Wilson, Shaun Eyring, and Kenny Marotta, eds., *Recreating the American Past: Essays on the Colonial Revival* (Charlottesville: University of Virginia Press, 2006).

[59] "Mulberry Sale to Summerville Man Negotiated"; "Do You Know Your South Carolina? Mulberry Plantation."

[60] See, for example, Baldwin, *Lowcountry Plantations Today*, 34–45.

In the years when the Chapmans made Mulberry anew, synoptic visions of its past and present became closely linked with a material edifice that represented a small portion of a once-sprawling estate. Each supplemented the other in interlocking fashion, giving shape and meaning to visions of a seemingly idyllic past and an equally idyllic present. Together they obscured the harsh realities of Mulberry's history. Representations of the plantation ignored what little evidence of racial slavery survived and instead depicted a country residence set amid pastoral surroundings. The Chapmans' Mulberry gave physical expression to an aestheticized, thoroughly sanitized past. Other dimensions of the plantation's history lay buried beneath layer upon layer of material refiguration and closely associated narratives. By the time the Chapmans' estate reached completion in the 1930s, it offered few clues about what Mulberry had been historically. Attempts to recover its history would need to rely on historical sources rather than firsthand analysis of the lands it had occupied and surviving remains. Although Mulberry constitutes but one example, it illustrates developments that occurred throughout the lowcountry. Where northerners remade old plantations, they did more than simply reshape material remains. Narratives and material form became intimately connected, with profound consequences for understandings of newly created places deemed "historic."

5

Medway Plantation

The Patina of Age

The new owners have done the interior of the house over, put in bathrooms
and otherwise modernized it without detracting from the charm of the
place. The grounds have also been beautified and made into a more
picturesque setting for the old house.

– Chlotilde R. Martin, 1931[1]

Oldest and best preserved plantation in South Carolina is Medway, which
dates from 1682.

– *Life*, 1938[2]

When Gertrude and Sidney Legendre purchased Medway Plantation in
June 1930, they became owners of a property with a storied history, an
unusual main dwelling, and an uncertain future. One of the oldest plantations in the lowcountry, Medway traces its roots to the late seventeenth
century. It passed through a succession of owners before becoming the
property of Theodore Samuel Marion, a nephew of Revolutionary War
hero Francis Marion, in 1827. Marion's purchase inaugurated a long
period of stability. Although he died within a year, his grandson, Thomas
Samuel DuBose, immediately took control of the plantation. DuBose
expanded its productive capacities. In 1833, he sold Medway to his
brother-in-law, Peter Gaillard Stoney. A rice planter on a large scale,
Stoney increased the plantation's agricultural output and established a

[1] Chlotilde R. Martin, "Medway, Historic Brick House," *News and Courier* (Charleston,
SC), Apr. 26, 1931, B6.
[2] "Life Goes to a Party with the Sidney Legendres on a Deer Hunt in South Carolina," *Life*,
Jan. 24, 1938, 54.

brickyard. Together, these developments made Medway one of the most successful plantations in the Goose Creek neighborhood.

Medway's fortunes plummeted after the Civil War. Brick production ended during the war years or soon thereafter, and rice production ended sometime later. By the early twentieth century, Medway lay in ruined condition, its infrastructure deteriorating and surviving buildings crumbling. Its condition underscored the fate of plantation agriculture on the Carolina coast. Weathered, ravaged by time, and overgrown, Medway told of a staggering decline.

The Legendres faced the same set of choices as other new owners. They had the option of renovating the main house, "restoring" it to the appearance of an earlier time, or turning it into a substantially new residence. Buyers of old plantations had taken these approaches elsewhere, and successful examples of each could be found throughout the lowcountry. Roughly the same set of options applied to the grounds and gardens. The Legendres might have "restored" them, rehabilitated and developed them with some respect for existing features, or created a new landscape. These formed the general range of options available.

Rather than following in others' footsteps, the Legendres took a different course. Instead of restoring the dwelling, they renovated and modernized it while maintaining its aged appearance. They refurbished the interior and added modern amenities but left the exterior weathered and worn. Elsewhere, they left signs of age undisturbed. They rehabilitated several outbuildings without changing their form and styling. For some new outbuildings and gardens, they used salvaged materials specifically for aesthetic reasons. For others, they employed plain styling that did nothing to detract from the appearance of the main house and surviving outbuildings. In making Medway anew, the Legendres strived to give it the appearance of a different time. By purposefully retaining material decay and selectively adding to it, they created a carefully crafted patina of age.

The Legendres' exact intentions are a mystery. Their efforts may have owed as much to practical considerations as aesthetic preferences. Refurbishing existing buildings saved time and money, and reusing building materials also limited costs. Moreover, fascination with romantic decay clearly motivated some northerners. The potential to use long-decaying remains for aesthetic effect supplied inspiration and compelling starting points. Whatever the Legendres' logic, the results spoke for themselves. Within about a decade, the Legendres turned Medway into a well-appointed estate that appeared centuries old. Although possessed of all the amenities needed for seasonal occupancy and sport, Medway appeared to belong to another era. Simultaneously part of the past and the present, Medway straddled two worlds, each radically different from the other.

The Legendres' activities illustrate a central theme of northerners' efforts. Sportsmen and sportswomen frequently integrated aged elements into the material environments of their "plantations." Some left deteriorating outbuildings standing as picturesque ruins. Others incorporated salvaged materials into new structures.[3] Both approaches made age conspicuous while refiguring the symbolism of the remains present. Purposeful retention of decay turned the results of cataclysmic upheavals into an aesthetic treatment. On the eve of the Legendres' arrival, Medway's condition marked the end result of the Civil War and emancipation. It told of the Confederacy's failed bid for nationhood, slavery's destruction, and the decline of a once-powerful plantation empire. Under the Legendres' care, decay became part of a material environment created for pleasure. It showed favor for picturesque decay, cultivation of a seemingly authentic past, and romanticization of plantations as exotic realms of endeavor.

The Legendres' treatment of Medway inspired narratives that emphasized age and remoteness but said little else. Rehabilitation of the plantation triggered a similar shift in representation as occurred with restoration of early houses. Brief accounts concerned with architecture, planter families, and well-remembered events became common; mentions of slavery and labor became rare. Histories of the estate became celebratory accounts concerned mainly with the appearance of a place with a long and complex history. By studiously avoiding discussion of Medway's relationship to

---

[3] Examples abound. At Clay Hall Plantation in Beaufort County, Geraldyn Redmond used bricks from "two old rice mills" to build a new hunting lodge. According to one visitor, the bricks gave the lodge a "time-worn appearance." Timber used in the interior came from "a house near Augusta [Georgia] ... over a hundred years old." See Chlotilde R. Martin, "Clay Hall, Redmond Mansion Lodge," *News and Courier*, Dec. 7, 1930, A11. At Castle Hill Plantation near Yemassee, John S. Williams took a similar approach. Like Redmond, he bought paneling and timbers from an old house in Augusta and used them in the interior of a new mansion. See Chlotilde R. Martin, "Lowcountry Gossip," *News and Courier*, Dec. 1, 1935, C1. At Hobcaw Barony, a house built for Bernard M. Baruch's daughter, Belle, "incorporated recycled nineteenth-century Charleston bricks." See Lee Brockington, *Plantation between the Waters: A Brief History of Hobcaw Barony* (Charleston: History Press, 2006), 107. For mentions of similar practices, see Chlotilde R. Martin, "Dominick Builds among Giant Oaks," *News and Courier*, Feb. 8, 1931; "Shooting Lodge in Carolina: 'Richmond,'" *Country Life* 65, no. 2 (Dec. 1933): 60–61; Chalmers S. Murray, "Waverly, Gem of the Waccamaw," *News and Courier*, July 26, 1931, A6.

the Civil War and emancipation, such histories portrayed the plantation as an iconic dwelling amid bucolic surroundings. In this fashion, a former slave plantation became a grand house in a pastoral setting, nothing more and nothing less.

DISCOVERING MEDWAY

Of the dozens of men and women who participated in the Second Yankee Invasion, few had as strong a zeal for adventure as Gertrude and Sidney Legendre. In an era when memorable exploits provided social capital among the upper classes, they surpassed virtually all of their peers. By the time the Legendres purchased Medway, they had traveled across Europe together and spent several months on safari in Africa. In later years they traveled to Indochina, Japan, Mexico, and back to Africa. As self-styled naturalist-explorers, the Legendres hunted big game in exotic locations around the world. Their adventures brought them into contact with African kings, European nobles, and primitive tribes in distant corners of the globe. As part of an elite who organized their lives around travel and exploration, the Legendres thrived on excursions to faraway destinations.[4]

In the early 1930s, Medway became the anchor of the Legendres' lives. Each fall, they returned to the plantation to relax, restore themselves, and enjoy their favorite pastimes. Medway offered superb opportunities for sport and leisure. Located on Back River, a tributary of the Cooper, Medway lies twelve miles east of Summerville and eighteen miles north of Charleston. It afforded the Legendres seclusion and isolation yet lay within easy reach of services, retail stores, and transportation connections to major cities.

In terms of personal background, the Legendres fit squarely within established norms for northern plantation owners. Gertrude and Sidney both hailed from distinguished families. Gertrude Sanford grew up

---

[4] On the Legendres' travels and big game expeditions, see Gertrude Sanford Legendre, *The Time of My Life* (Charleston: Wyrick, 1987), chaps. 3, 5, 7–9, and 11; "Gertrude Sanford, Explorer, to Marry," *New York Times* (New York, NY), Aug. 20, 1929, 38; "Girl Got 7 Monkeys for a Bottle of Brandy," *New York Times*, Aug. 21, 1929, 20; "All Animals, Even Fiercest, Run from Man, Berkeley Big Game Hunter Says," *News and Courier*, Feb. 2, 1936, C2. Sidney wrote two books about their expeditions, *Land of the White Parasol and the Million Elephants* (New York: Dodd, Mead, and Company, 1936) and *Okovango, Desert River* (New York: J. Messner, 1939).

in Amsterdam, New York, where her father ran a major carpet-manufacturing company. Her parents moved in exclusive social circles and divided their time between Amsterdam and New York City. The family routinely wintered in Aiken, South Carolina, a mecca for polo and other equestrian sports. Summertime destinations included Long Island, the English countryside, and select locations on the European continent. Although the Sanfords did not possess one of the great fortunes of the Gilded Age, they nonetheless belonged to the top echelon of American society. As members of the class of wealthy plutocrats that Americans simultaneously celebrated and reviled, they led leisured lives. By maintaining exclusive company, traveling extensively, and indulging in activities available to only a small elite, the Sanfords took full advantage of the opportunities their wealth provided.[5]

Sidney grew up in comparatively modest circumstances but nonetheless knew privilege and comfort. Born in 1900, he descended from French planters who had fled the slave revolt on Saint-Domingue in 1791–1804. Upon arriving in New Orleans, the Legendres turned to mercantile and professional pursuits. Sidney's paternal grandfather became a successful sugar planter and attorney, and his father also practiced law. By the time Sidney was born, the Legendres ranked among New Orleans's elite. He and his three brothers attended Princeton, where Sidney starred on the football team. Tall, handsome, and charming, he made friends easily and moved comfortably in exclusive settings. His schooling at Princeton opened up avenues of social access and opportunity that led to connections with northeastern elites.[6]

Although wealth and status placed Gertrude and Sidney in the mainstream of the Second Yankee Invasion, in other ways they stood apart. They spent more time at Medway than most of their counterparts did at their estates and threw themselves into Medway's upkeep and development. Although many northerners became deeply invested in their estates, the Legendres immersed themselves in the enterprise to an unusual degree. Even though they employed a superintendent, the Legendres managed workers, oversaw maintenance, and handled what Gertrude called "the endless paper work that the plantation created." Moreover,

---

[5] On the Sanford Family, see especially Alex M. Robb, *The Sanfords of Amsterdam: The Biography of a Family in America* (New York: William-Frederick Press, 1969).
[6] John Titford, *The Legendre Family of France, St. Domingue, and the United States of America* (Higham, UK: the author, 2006), 229–230, 247–259, 276, 309.

they talked about Medway constantly. For the Legendres, the estate became a never-ending project – a challenge comparable to their most ambitious adventures. At times it became an all-consuming passion. "We lived as if on an island, absorbed by daily events," Gertrude recounted. "We were both hopeless romantics when it came to Medway," she observed. "We had to be." Otherwise, "we would have been driven mad by its reality."[7]

The Legendres first encountered Medway during the winter of 1929–1930. After marrying in New York City on September 17, 1929, they traveled to California by train and then to the Cassiar Mountains of western Canada for a honeymoon that defied convention. Hiking across rugged terrain, they encountered trappers and prospectors, hunted sheep and goats in deep snow, and spent several days tent-bound in a blizzard. "It was glorious," Gertrude later recalled. "Our friends thought we were crazy," she noted. The Legendres found the experience exhilarating. Their return to civilization brought a shock. A sharp downturn in the stock market had thrown the world's financial markets into turmoil. The international crisis that would soon be known as the Great Depression had begun. "Fortunately, Father had very little money in stocks," Gertrude observed. "The misery of much of the world was unknown to us."[8]

Upon returning to New York, the Legendres set out on a driving tour of the East Coast. With Palm Beach, Florida, as their destination, they drove south through the Mid-Atlantic states and on into Virginia. In South Carolina, they stayed with Benjamin and Elizabeth Kittredge at Dean Hall Plantation. Gertrude's family had known the Kittredges for years. They moved in similar social circles and socialized in New York and at Palm Beach. The Kittredges welcomed the newlyweds. The Legendres' visit provided a chance to see close friends and an opportunity to introduce them to the countryside along the Cooper River.[9]

One morning, at the Kittredges' suggestion, the Legendres rode on horseback to a neighboring plantation. Gertrude and Sidney followed a sandy dirt road to Spring Grove Plantation and crossed the Back River. Then they followed game trails and worn paths until they came upon a hulking mansion. Their first glimpse left them virtually speechless.

[7] Legendre, *Time of My Life*, 75–76. The Legendres' involvement in running and developing the estate is detailed in Sidney J. Legendre, "Diary of Life at Medway Plantation, Mt. Holly, South Carolina," private collection (copy in author's possession).

[8] Legendre, *Time of My Life*, 69–70.

[9] Ibid., 70; Virginia C. Beach, *Medway* (Charleston: Wyrick and Co., 1999), 35.

"It stood derelict in a grove of live oaks several hundred years old," Gertrude wrote. Its brick walls had "faded to a soft pink." Although only a "shell," the building nonetheless inspired awe. Ivy climbed its walls, and every exposed surface seemed severely weathered. Thick clumps of Spanish moss hung from surrounding trees. Captivated, the Legendres stood and marveled. As an African American family ventured out of "their wooden houses," Gertrude and Sidney spread out a picnic lunch and sat down to eat. In that moment, beneath the warm sun of the Carolina coast, they "tried to imagine plantation life."[10]

For the Legendres, imagining "plantation life" meant contemplating what would be required to salvage the ruin before them and make it suitable for habitation. It also meant considering what would be needed to clear the surrounding grounds, renovate outbuildings, and maintain forests as hunting domains. The Kittredges' example supplied an obvious model. They had spent their winters at Dean Hall for twenty years and now possessed a comfortable, well-appointed estate. The Legendres may also have known of others. By the time of their visit, northerners owned several estates on the Cooper. George Bonbright had recently purchased Pimlico, and Clarence and Adelaide Chapman spent their winters at Mulberry Plantation. J. W. Johnson and his family owned Mepkin. These holdings varied in size, scale, and character, but all showed the potential of old plantations. As seasonal residences and venues for sport and leisure, they supplied examples to be emulated, replicated, and improved upon.[11]

The Legendres' encounter with Medway left an indelible impression. Although they continued on to Palm Beach, neither of them could stop thinking about what they had seen. "Something about it haunted us both," Gertrude recounted. As they drove to Florida, the Legendres considered their options. They eventually approached Gertrude's father for assistance. Although he viewed the Carolina coast as unhealthy and did not understand the couple's desire to live there, he recognized the strength of their determination. "Father didn't even try to argue,"

---

[10] Legendre, *Time of My Life*, 70. See also Beach, *Medway*, 35–37.

[11] Chlotilde R. Martin, "The Cypress Gardens, in Berkeley," *News and Courier*, undated newspaper clipping, Dean Hall Plantation Vertical File, South Carolina Historical Society, Charleston, SC (hereafter SCHS); Chlotilde R. Martin, "New Yorker Reclaims Pimlico," *News and Courier*, May 24, 1931, B2; Chlotilde R. Martin, "Mulberry Castle, Built in Indian Days," *News and Courier*, July 26, 1931, A5; Chlotilde R. Martin, "Mepkin, Henry Laurens Plantation," *News and Courier*, May 17, 1931, B14.

Gertrude explained. In the spring of 1930, the Legendres purchased Medway for $100,000.[12]

The house that had captured the Legendres' attention is one of the oldest and most unusual dwellings in South Carolina. The earliest portion dates to 1704. As late as 1738 it remained a one-story brick building with a footprint of about 36 by 26 feet. Over the next century, it gained a second story and a three-room addition. In 1855, Peter Gaillard Stoney erected a large ell in the rear. By 1875 the house had taken the form the Legendres saw. Tall, awkwardly proportioned, and lacking exterior ornamentation, its major features are a series of stepped gables and a façade tower that looks out toward the Back River. Viewed from the side, the house has the appearance of an oversized barn or mill, although the fenestration indicates domestic use. Put simply, the building looks nothing like the columnar mansions commonly associated with the Old South. As Gertrude noted, it "doesn't have the great Georgian portico of Tara fame."[13]

Medway had begun as a small stock farm in the late seventeenth century. In 1701, Edward and Elizabeth Hyrne, two English immigrants, bought it with aims of making a quick fortune and returning to their homeland. By then, Medway had a well-built dwelling that Hyrne described as "the best Brick-house in all the Country." Erected nine years earlier at a cost of £700, it measured "80 Foot long, 26 broad [and was] Cellar'd throughout." Fire destroyed the building in January 1704, leaving the Hyrnes homeless. "If it had not bin for some good people we must have perished," Elizabeth recounted. The Hyrnes immediately built a smaller dwelling that served as the plantation's main house for decades.[14]

By the late 1730s, Medway possessed a domestic complex that included barns, several sheds, and a well. Rice grew in inland fields and a rice mill stood along the Back River. The mill indicated intensive production of the staple that fueled the lowcountry economy. As planters began controlling water flows and using impoundments to increase output, machines for processing rough rice became common. Pounding mills relieved slaves of some of the backbreaking work involved in processing the annual crop.

[12] Legendre, *Time of My Life*, 70; Louisa G. Stoney et al. to Gertrude and Sidney Legendre, Deed Book C27, 871–875, Berkeley County Register of Deeds, Moncks Corner, SC (hereafter BCRD). See also Beach, *Medway*, 32.

[13] Legendre, *Time of My Life*, 70.

[14] Albert J. Schmidt, "Hyrne Family Letters," *South Carolina Historical Magazine* 63, no. 3 (July 1962): 150–157; Richard N. Côté, *Preserving the Legacy: Medway Plantation on Back River* (Mt. Pleasant, SC: the author, [1993?]), 9–19; Beach, *Medway*, 9–25.

The presence of such a machine at Medway indicated large-scale cultivation and adoption of current agricultural technology.[15]

Some time later, the center of rice cultivation at Medway shifted to tidal fields. During the mid-eighteenth century, lowcountry planters began reclaiming river swamp and creating impoundments that allowed for improved control of water levels and increased output. At Medway, slaves built dikes in the shallows of Back River to create a series of new fields. At the same time, slaves also adapted existing features for new purposes. Crain Pond, a spring-fed marsh located northwest of the main house, became a reserve – a reservoir used for irrigating the inland and tidal fields to the east. Later, in the early nineteenth century, one of Medway's owners created a second reserve immediately below the main dwelling. Called the "Home Reserve," this body of water formed part of a hydraulic system that ultimately spanned about three-quarters of the plantation.[16]

Increased production went hand in hand with the growth of Medway's labor force and development of a more complex built environment. By the end of the eighteenth century, a group of new buildings stood clustered around the main dwelling. Several standing off the west front were probably slave cabins. If so, they likely represented the full extent of slave housing on the plantation at the time. Several other buildings on the opposite side of the house showed greater variation in scale and form. Most likely, they included barns and sheds.[17]

A second set of buildings stood along a road on the east side of the plantation. Situated in between tidal and upland fields, they provided storage for tools and agricultural implements. During the annual harvest, initial processing of crops took place in workyards immediately adjacent.[18]

By the end of the eighteenth century, Medway encompassed 2,593 acres. Swamps accounted for over 1,550; high land occupied another 831; and rice grew on nearly 130. These numbers made Medway a plantation of respectable size – not as large as others in the immediate vicinity, but a formidable enterprise nonetheless. Medway also had room for expansion. River swamp could be turned into additional tidal fields, and upland terrain could be developed for row crops and pastures. Timber stands supplied firewood and building material. Moreover, at a site at the

---

[15] Côté, *Preserving the Legacy*, 4.
[16] Plat of Medway Plantation, 1792, SCHS; Gertrude S. Legendre, *Medway Plantation, 1686–1980* (Charleston: n.p., 1980), 3.
[17] Plat of Medway Plantation, 1792, SCHS.   [18] Ibid.

northeastern corner of the plantation near the Back River, brick manufacturing began sometime before 1792. This industry, common in select
parts of the lowcountry before the Civil War, gave many planters a
valuable source of supplemental income. After the Wando River basin,
the Cooper River became the most productive brick-producing area in
the lowcountry.[19]

During the late antebellum era, further changes took place. In the mid-
1830s, Peter Gaillard Stoney purchased Medway. His arrival marked a
turning point. It began a long period of stable ownership – he presided
over Medway until his death in 1884 and his heirs retained ownership for
another forty-five years. Stoney also initiated a series of improvements
that gave Medway a more refined appearance. By the late 1850s, it would
possess a well-developed domestic setting, a feature that made it one of a
small number of aesthetically pretentious plantations in the lowcountry.[20]

The form and condition of the main house at the time of Stoney's
purchase is uncertain. Most likely, it had become a two-story residence
with a lateral-gable roof and a three-bay front section with wings.[21] In
1855, Stoney added a large rear wing on the west (rear) elevation, which
enlarged the footprint and enclosed the original dwelling. The wing, or ell,

---

[19] Ibid. Some sources suggest that brickmaking at Medway began even earlier. See, for
example, John Beaufain Irving and Louisa Cheves Smythe Stoney, *A Day on Cooper
River*, 2nd ed. (Columbia: R. L. Bryan Company, 1932), 68. On the brickmaking indus-
try in the lowcountry in general, see Lucy B. Wayne, "'Burning Brick and Making a
Fortune at It Too': Landscape Archaeology and Lowcountry Brickmaking," in *Carolina's
Historical Landscapes: Archaeological Perspectives*, ed. Linda F. Stine et al. (Knoxville:
University of Tennessee Press, 1997), 97–111.

[20] Côté, *Preserving the Legacy*, 4–5. For a biographical sketch of Peter Gaillard Stoney, see
Chalmers Gaston Davidson, *The Last Foray: The South Carolina Planters of 1860, a
Sociological Study* (Columbia: South Carolina Tricentennial Commission by the Univer-
sity of South Carolina Press, 1971), 253.

[21] Côté and Stockton are cautious in interpreting the historical evidence. Both conclude
the second story and east front addition could have been added at any time between
1738 and 1875. Speaking strictly, they are correct. Documentary and physical evidence
shows the structure remained a small, one-story brick dwelling in 1738. Sometime before
1875, it assumed the form the Legendres saw. No evidence from the intervening years is
known to exist, save for information about the wing Stoney built in 1855.

When the development of the house is considered in relation to the general history of
the plantation, it seems likely that the second story and east-front addition had been built
by 1855, and probably by the time Stoney purchased the plantation in the mid-1830s.
The date of construction for the rear ell is known from family tradition. This tradition
likely encompasses the full extent of the work performed on the house by Peter Gaillard
Stoney. Had he built either of the other additions, they would have become part of the
family's recollections. Based on this reasoning, I believe the house probably had all of its
existing components, save for the rear ell, by about 1835.

stood two and a half stories tall with a dining room on the ground level. Two rooms and a hall occupied the second floor. The ell dramatically increased the size of the house and improved its proportions. Because of its haphazard evolution, the house never achieved a well-composed form, but the rear ell yielded greater balance and a more satisfying appearance.[22]

As construction of the rear wing proceeded, the Stoneys improved the grounds. Slaves working under the direction of Anna Porcher Stoney, the plantation mistress, planted a double avenue of oaks leading away from the west elevation. The avenue defined the inland approach to the house and complemented a grove of trees that had been planted around the building about twenty years earlier. In time, the avenue matured into a grand allée. Even when first planted, the trees added refinement and symmetry to an otherwise naturalistic landscape. By the time the Legendres saw the plantation, the avenue formed a towering, cathedral-like arch. Below lay a lawn stretching away from the house for more than 500 feet. The allée gave the landward approach a stunning appearance. Consistent with the monumentality and sequential organization of a landed estate, it introduced drama and grandeur to the plantation landscape.[23]

The changes that the Stoneys made demarcated an important shift in focus. As a self-conscious effort to display status and authority, they suggested how Peter Gaillard Stoney and his family conceived of themselves and their place in lowcountry society. Even before purchasing Medway, Stoney held an established position among the rice-planting elite. By the early 1830s, he owned two plantations, Calibogue on Hilton Head Island and another called Foot Point. These holdings supplied him with favorable returns. In purchasing Medway, Stoney significantly expanded his earning potential. It quickly became the most profitable of his plantations and proved more stable over the long run.[24]

The Stoneys' development of the main house and grounds gave Medway a more evolved form. The material environment increasingly

[22] Samuel Gaillard Stoney, *Plantations of the Carolina Low Country* (Charleston: Carolina Art Association, 1939), 46.

[23] Loutrel W. Briggs, *Charleston Gardens* (Columbia: University of South Carolina Press, 1951), 146; Stoney, *Plantations of the Carolina Low Country*, 46. The allée planted in 1855 apparently replaced a single avenue of water oaks that previously led up to the house. Whether these stood on the west or east side of the buildings is unclear.

[24] Peter Gaillard Stoney, List and Memorandum Book, 1824–1833, SCHS; Samuel G. Stoney, ed., "Recollections of John Stafford Stoney, Confederate Surgeon," *South Carolina Historical Magazine* 60, no. 4 (Oct. 1959): 208.

reflected the scale of agricultural production, Peter Gaillard Stoney's ambitions, and a diversified scope of activity. It also made clear the long and prosperous history of plantation agriculture in the lowcountry. The relative stability of rice production and the returns it yielded created conditions that encouraged investment in durable architecture. Even though planters generally spent only part of the year at their plantations, many chose to build large and stylish houses. Medway stood among the largest and best-appointed on the Cooper River.[25]

Medway's material environment embodied characteristics typical of tidal rice culture. Some time after 1792, the slave street moved from the main dwelling to a new location across the Home Reserve, about a quarter-mile to the southeast. This site proved consistent with patterns seen throughout the lowcountry. As Philip D. Morgan has noted, planters located slave villages two hundred yards or more away from main dwellings. The new location placed the Stoney slaves close to the river and rice fields, the brickyard, and several lowland fields. The tidal fields underscored the sophistication of rice cultivation on the South Atlantic coast. The product of massive investments in labor and engineering expertise, they told of the long hours slaves spent building embankments, sluice gates, and drains. Finally, the inland reserves and the array of associated water-control devices also demonstrated the scale of activity at Medway. As evidence of planters' ability to control water flows, they told of systems that sustained commercial rice production throughout the region.[26]

Medway produced handsome returns. In 1859, the plantation turned out 175,000 pounds of rough rice, a figure that made it the second most

[25] On social and economic conditions that encouraged investments in well-built dwellings, see especially Phillip D. Morgan, *Slave Counterpoint: Black Culture in the Eighteenth-Century Chesapeake and Lowcountry* (Chapel Hill: Omohundro Institute of Early American History and Culture by the University of North Carolina Press, 1998), 110. For examples of especially elaborate plantation dwellings, see especially Stoney, *Plantations of the Carolina Low Country*.

[26] Personal communication with Robert Horton, March 13, 2007. On spatial arrangements typical of rice plantations, see Morgan, *Slave Counterpoint*, 110–123. Development of the landscape of commercial rice production is ably detailed in S. Max Edelson, *Plantation Enterprise in Colonial South Carolina* (Cambridge: Harvard University Press, 2006); Morgan, *Slave Counterpoint*, 156; Mart A. Stewart, *"What Nature Suffers to Groe": Life, Labor, and Landscape on the Georgia Coast, 1680–1920* (Athens: University of Georgia Press, 1996), 87–116; Joyce E. Chaplin, *An Anxious Pursuit: Agricultural Innovation and Modernity in the Lower South, 1730–1815* (Chapel Hill: Institute of Early American History and Culture by the University of North Carolina Press, 1993), chap. 7.

productive rice plantation in St. James Goose Creek Parish.[27] Brick manufacturing also proved lucrative. In 1859, the brickyard at Medway produced 800,000 bricks, earning Stoney a return of $5,000. The yard itself represented a capital investment of $18,000.[28] It supplied builders in Charleston and the surrounding area with a steady flow of brown, gray, and red bricks. According to family tradition, bricks from Medway went into the construction of Fort Sumter, "numerous houses in Charleston," and the main house at Dean Hall Plantation.[29]

The 1860 federal census valued Medway at $20,000 and Stoney's personal estate at $75,000. These figures made him one of the richest planters in St. James Goose Creek and St. John's Berkeley parishes.[30] Stoney and his family lived in style and comfort. They spent their winters at Medway and summered in Charleston or at the Barrows, an upland retreat near Moncks Corner.[31] Like other members of the planter class, they feared the diseases that plagued lowland swamps during the summer months and departed for the relative safety and convivial social atmosphere of upland retreats as soon as the weather turned warm each spring.

The enslaved men, women, and children who sustained Stoney's enterprises knew no such comforts. They lived in meager conditions, had few

---

[27] Federal Census of Agriculture, 1860, St. James Goose Creek Parish, MS, line 8, p. 9, South Carolina Department of Archives and History, Columbia, SC (hereafter SCDAH). The 175,000 pounds of rough rice produced in 1859 would have placed Medway in the eighty-fourth percentile of rice-producing plantations in St. Johns Berkeley. It is important to note that this represented the bottom end of large producers in the parish. As in other parts of the lowcountry, rice production became increasingly concentrated over time, and by the late antebellum years, large plantations produced the vast majority of the annual crop. On the general course of rice production during this period, see especially William Dusinberre, *Them Dark Days: Slavery in the American Rice Swamps* (New York: Oxford University Press, 1996), 387–407.

[28] Federal Census of Industry, 1860, St. James Goose Creek Parish, MS, p. 1, SCDAH. Daily accounts of brick production at Medway appear in Overseer's Day Book, 1852, folder 2, box 18B, Stoney Family Papers, SCHS.

[29] Samuel Gaillard Stoney, "Where History, Tradition Architecture and Gardening Unite with Mellow Beauty to Provide a Charming Place," newspaper clipping cited as *News and Courier*, Apr. 8, 1948, in Medway Plantation File, South Carolina Vertical File Collection, Charleston County Public Library, Charleston, SC (hereafter SCVFC). Stoney's investment in the brickyard grew by 250 percent during the 1850s. The 1850 census listed its value as $4,500. Output also increased; in 1850, the yard turned out 600,000 bricks. See Federal Census of Industry, 1850, St. James Goose Creek Parish, MS, SCDAH. Dean Hall is located on the western branch of the Cooper River, about three miles to the northeast.

[30] Federal Census of Agriculture, 1860, MS, St. James Goose Creek Parish, line 8, p. 338, SCDAH; Federal Census of Population, 1860, M653, roll 1216, 133.

[31] Davidson, *The Last Foray*, 253.

material possessions, and suffered constant hardship. Nonetheless, they powerfully influenced Medway's material environment. In the late ante-bellum era, about 120 slaves lived and worked at Medway. Most belonged to or descended from a group of seventy-two that Stoney had purchased from his brother-in-law at the same time he bought Medway. These men, women, and children spent their days carrying out tasks that yielded profits for Stoney, maintained Medway, and continually developed its infrastructure. They gave shape and meaning to the plantation as a place of habitation and as an economic enterprise. Through their energies, Medway fulfilled its intended purposes year after year.[32]

Save for a few house slaves, the world of Stoney's human chattel centered on the crude dwellings that formed the slave street and the sites of labor where they spent the better part of their days. The principal zones of labor lay along the eastern edge of the plantation, where the brickyard, tidal fields, and several lowland fields were located. Slaves also worked fields near Crain Pond and in the immediate vicinity of the main house. Cutting and hauling wood and road work took some farther afield. How much freedom of mobility the Stoney slaves enjoyed is unclear, but they undoubt-edly moved among other plantations in the Goose Creek neighborhood, traveling along worn paths, the Back River, and other waterways to visit family members, perform errands, and take on extra work when allowed.[33]

One site at Medway probably meant more to the Stoney slaves than any other. Atop a small rise about a quarter-mile northeast of the main house lies the burial ground where slaves interred their dead. No records of slave funerals at Medway survive, but they likely adhered to practices common throughout the lowcountry. Slaves buried their dead in graves without permanent markings. Following African traditions, they often left small pieces of crockery and other items at gravesites. As scholars such as Charles Joyner have observed, slaves celebrated death as the climax of life and the moment of reunion with ancestors. For people who knew lives of unrelent-ing hardship, funerary rituals provided solace, hope, and a sense of salva-tion. They also provided momentary respite from the rigors of daily life.[34]

[32] Bill of Sale, Theodore S. DuBose to Peter Gaillard Stoney, Nov. 2, 1833, in Arithmetic Book of C. M. Stoney and Plantation Notes, 1832–1860, Stoney and Porcher Family Papers (microfilm), Southern Historical Collection, University of North Carolina, Chapel Hill, NC.
[33] Plat of Medway Plantation, 1792, SCHS; Overseer's Day Book, 1852, Stoney Family Papers, SCHS.
[34] Charles Joyner, *Down by the Riverside: A South Carolina Slave Community* (Urbana: University of Illinois Press, 1984), 138–139.

Stoney and six of his sons fought for the Confederacy in the Civil War.[35] Like nearly all of the lowcountry elite, they heeded the call of duty when the dual blows of John Brown's raid and the election of 1860 left them believing that they had but one choice. Their views on the war have been lost to history. They may have fought for principle, for home and family, or to save the only way of life they had ever known. Or, most likely, they fought for all three. No matter what beliefs sustained them, the Stoney men undoubtedly held Medway close at hand. As their home-place and a symbol of the world that slavery had made, it served as a beacon of their cause. Confederate success ensured Medway would remain unchanged; failure portended a different outcome.

LIFE AFTER SLAVERY AT MEDWAY

Emancipation inaugurated a series of dramatic upheavals at Medway. The immediate effects included the loss of Stoney's ability to maintain and operate the plantation as before and greater autonomy for Stoney's former slaves. Once stripped of the coercive force of the lash, slave-holders' efforts to direct the labor of their former chattel met with frustration. In the lowcountry, traditions forged under the task system and uncommonly high levels of solidarity among freedpeople challenged planters' aims. Freedpeoples' determination to control the pace of pro-duction and to shape the circumstances of their lives ran counter to planters' desire to effect as close a return to the antebellum order as possible. Struggles over land, labor, and wages ultimately produced tenu-ous arrangements that accorded planters some authority while giving freedpeople significant autonomy and a combination of land and wages in return for work performed on planters' crops. Arrangements of this kind became prevalent throughout the lowcountry by the late 1870s. Although effective in mediating sharply divergent interests, they sealed the fate of commercial rice production on the South Atlantic coast. By resuming production at a greatly reduced scale and with low levels of efficiency, the demise of the crop became all but certain.[36]

---

[35] Beach, *Medway*, 25.

[36] John Scott Strickland, "'No More Mud Work': The Struggle for Control of Labor and Production in Low Country South Carolina, 1863–1880," in *The Southern Enigma: Essays on Race, Class, and Folk Culture*, ed. Walter J. Fraser, Jr., and Winfred B. Moore, Jr. (Westport, CT: Greenwood Press, 1983), 42–62; John Scott Strickland, "Traditional Culture and Moral Economy: Social and Economic Change in the South Carolina Low Country, 1865–1910," in *The Countryside in the Age of Capitalist Transformation:*

Freedom undermined Medway's physical order even before its full effects became known. In the final days of the war, Union forces scavenged plantations near Goose Creek in search of food and fodder. On March 1, 1865, Union officer Henry O. Marcy visited Medway while his men foraged nearby. He described it as "a fine place." He found eighteen members of the Stoney clan present, all of them women save for a "lame son in law." Also at Medway were about 150 slaves. The "col[ored] people," Marcy wrote in his diary, "were of all ages and sizes, clothes in all sorts of garments, from the ordinary homespun to whole suits made of old carpets and blankets. Nearly all stood with heads and feet bare and very few were clothed sufficiently warm to prevent them from suffering." Marcy characterized the Stoney family as "intelligent but bitter rebels." He noted that "the ladies plainly spoke of their sentiments and like gloried in their struggle" but "were polite." At the Stoneys' invitation, Marcy and at least one other officer ate dinner with the family. Although Marcy found the Stoneys hospitable, he noted "sharp table talk" in his diary.[37]

Marcy's visit coincided with intense debates about the meaning of freedom in the lowcountry. After dinner, Anna Stoney requested that Marcy address her slaves. Days earlier, an officer from a Union gunboat had come ashore at Medway and told the slaves "to remain and obey their mistress as before." Marcy surmised that the officer must have been "overcome by the charm of the ladies, or by some other motive." The slaves had different expectations. According to Marcy, the officer's guidance "was not [what] they had heard Massa Linkin had promised

Essays in the Social History of Rural America, ed. Steven Hahn and Jonathan Prude (Chapel Hill: University of North Carolina Press, 1985), 141–178. Peter Coclanis has argued that structural changes in international markets sounded the death knell for rice production in the Carolina and Georgia lowcountry before the Civil War. Although his contentions have not gone unchallenged, they cast doubt on the potential for any labor arrangement to have sustained commercial production over the long term. This is an important point, but it does not diminish the argument advanced herein, which emphasizes slavery's importance in shaping and maintaining the physical landscape of large-scale rice production. The intensity of postbellum labor struggles and the arrangements they produced suggests that the order and control that had historically characterized the rice landscape virtually demanded slave labor. Only at enormous cost would planters have been able to achieve similar results. For Coclanis's arguments, see Peter A. Coclanis, *The Shadow of a Dream: Economic Life and Death in the South Carolina Low Country, 1670–1920* (New York: Oxford University Press, 1989); Peter A. Coclanis, "Distant Thunder: The Creation of a World Market in Rice and the Transformations It Wrought," *American Historical Review* 98, no. 4 (Oct. 1993): 1050–1078. A recent challenge is William Dusinberre, *Them Dark Days*, 387–396.

37 Henry O. Marcy, "Diary of a Surgeon: U.S. Army, 1864–1899," entry for March 1, 1865, SCHS.

them." Understandably, they "couldn't exactly see how they were free if they must do just as before." According to Marcy, the slaves had "talked it over among themselves and concluded they wouldn't just do as mistress told them."[38]

The resulting impasse left Mrs. Stoney distressed. She told Marcy that "the niggers had become unruly." Fearing "trouble," she begged that he "advise them to be quiet, obedient, & c." Marcy agreed to speak with the slaves, but the message he conveyed disappointed the Stoneys. From the "veranda" of the main house – probably the porch of the rear ell – Marcy addressed a gathering of the entire population. He spoke for "10 or 15 minutes." Although he failed to record the content of his speech, it apparently pleased most of the slaves present, for he recounted in his dairy that "they all rushed up when I was through and must shake me by the hand." Those closest to him during the speech, he noted, "seemed appreciative and happy." When Marcy finished his remarks, the slaves "thanked God for freedom and promised to make good use of it." The Stoneys reacted differently. According to Marcy, they "appreciated my eloquence less and in silence returned to the parlor."[39]

Marcy's visit to Medway offers a glimpse into a moment of radical upheaval: the undoing of a centuries-old social order and the beginning of new arrangements based on free labor. It also marked the beginning of a new era in the material condition of the plantation. The difficulty Anna Stoney noted with her former chattel underscored conflicts that would dominate the postwar era. The condition of the plantation at the time is unclear. Most likely, the war had undermined its order and upkeep. In the years that followed, Medway's condition deteriorated. Marcy's visit foretold of developments that would leave the plantation disheveled and in decline. Ultimately, they would seal Medway's fate and open the way for new possibilities.

Limited sources make it difficult to judge the rate of deterioration. Production of staple crops resumed slowly after the war. In the 1870s, many of the Stoney's former slaves worked on the plantation as wage laborers.[40] Rice cultivation at Medway apparently ended sooner than at many plantations on the Cooper River, although the exact date is uncertain. Brickmaking also stopped. The shattered economy reduced construction to a bare minimum, and the loss of slave labor dramatically

---

[38] Ibid., entry for March 1, 1865.   [39] Ibid., entry for March 1, 1865.
[40] Medway Account Book, 1872, folder 3, box 188, Stoney Family Papers; Stoney Family, General Accounts with Laborers, 1873, both at SCHS.

increased production costs. Limited demand for building materials and competition from mechanized producers in the Piedmont virtually destroyed brick manufacturing in the lowcountry. The brickyard at Medway probably lay idle by the closing days of the war. If not, it ceased to operate soon thereafter.[41]

By the end of the nineteenth century, the main dwelling at Medway stood vacant and decaying. The earthquake of 1886 had sent the stepped gables tumbling to the ground. Weeds and brush covered the grounds. Visitors who saw the plantation in these years remarked on its age and condition and the melancholy mood that seemed to hang in the air. When William Whitson hunted at Medway in 1899, he described the house as "an old-time mansion, and in its day a magnificent one." The surrounding landscape, he wrote, had become "a wilderness inhabited by wild beasts." Whitson's comments overlooked the dozens of African Americans who farmed small plots on those lands, but his words nevertheless conveyed a certain truth. Like other plantations, Medway told a story of ruin and decline.[42]

As productive activities ended, domestic and recreational uses grew more prominent. By about 1880, the Stoney family used the plantation principally as a hunting retreat and a site for family gatherings. The lands around Medway offered excellent deer hunting, and the Stoneys frequently took friends and acquaintances to the plantation for sport. In the fall of 1902, three of Peter Gaillard Stoney's sons and four other men formed the Back River Hunting Club, an organization that used Medway and neighboring lands as hunting domains. Records of the group's outings reveal consistent use of the plantation for sport throughout the 1900s and 1910s.[43]

As agriculture declined, the Stoneys' lives became centered in Charleston, and Medway assumed the role of a country retreat. From the beginning of the twentieth century onward, the Stoneys congregated at Medway at Thanksgiving and Christmas and occasionally on other holidays. These

---

[41] Wayne notes that brickmaking was "essentially abandoned" in the lowcountry after 1865. See Wayne, "'Burning Brick and Making a Large Fortune at It Too,'" 102.

[42] William Whitson, "A South Carolina Hunt," *Forest and Stream*, Sept. 30, 1899, 264–265.

[43] For accounts of hunting at Medway, see Whitson, "A South Carolina Hunt," 264–265; "Joseph Jones" [Dr. J. William Folk], "Hunters' Paradise Is Down at Medway," unidentified newspaper clipping, Stoney Family Papers, SCHS. On the Back River Hunting Club, see Record Book of the Oaks Hunting Club and the Medway Hunts, folder 4, box 188, Stoney Family Papers, SCHS.

gatherings used Medway as a refuge. They supplied opportunities to reminisce, to enjoy peace and quiet, and to renew ties to an ancestral homestead. They made the plantation a space of specialized activity that reflected new social and economic relations, new priorities, and large-scale agricultural decline.[44]

Despite Medway's continuing importance in the Stoneys' lives, recreation and holiday use evidenced the decline of a once-prosperous enterprise. These activities showed that Medway no longer served as a site of economic production, produced profits, or fulfilled its original purposes. The condition of the plantation contrasted sharply with the prewar era and highlighted growing uncertainty. The collapse of rice production and planters' failure to find new uses for their lands cast doubt over the future of commercial agriculture. Plantations already lying idle told of a near-complete lack of commercially viable uses for lowcountry lands. As the years passed, material distress signaled the difficulty of adapting to radically changed circumstances. Plantations increasingly appeared as anachronisms – as remnants of a bygone era and a disappearing way of life.[45]

Visitors to Medway continued to comment on its deteriorating condition, especially the state of the main house. About 1920, for example, Henry Lowndes described the building as "badly in need of repairs." He returned several years later to find conditions unchanged. As he rode away, Lowndes "glanced back at the old house and thought how majestic it looked and what a shame it was that it was left to go to ruin." He judged the chances of seeing it "restored to its original splendor" virtually nil (see Figure 5.1).[46]

Lowndes's comments could have applied to any number of houses along the coast. Many of the largest and most refined dwellings in the region stood in similarly decrepit condition. In the eyes of many lowcountry people, these structures told a tragic story. As much as they underscored the wealth and grandeur that Carolina planters had enjoyed in their heyday, they told of an astonishing downfall. As vestiges of a different era, the plantation dwellings of the old elite marked time.

---

[44] Augustine T. S. Stoney to Dear People [Stoney family members], Nov. 22, 1917, folder 7, box 18A, Stoney Family Papers; Stoney Family Notebook, 1888–1916, both at SCHS.

[45] On planters' efforts to adapt to changing conditions, see James H. Tuten, *Lowcountry Time and Tide: The Fall of the South Carolina Rice Kingdom* (Columbia: University of South Carolina Press, 2010), 37–38, 46–50, 56–74.

[46] Henry Lowndes quoted in Beach, *Medway*, 30–32.

FIGURE 5.1 Main house at Medway Plantation, circa 1920.
Courtesy of the Charleston Museum, Charleston, SC.

Increasingly, they appeared as a history written in material form – a physical record of what the lowcountry had been, what it had become, and the gulf between the two (see Figures 5.2 and 5.3).

NEW BEGINNINGS: THE LEGENDRE ERA AT MEDWAY

The Legendres' arrival signaled the beginning of extensive changes. The couple initially focused on rehabilitation of the main house, several outbuildings, and portions of the surrounding grounds. In the years that followed, they continued creating gardens and landscaping, adding outbuildings, and making other changes as they saw fit. Almost immediately, Medway assumed a new and markedly different appearance. The Legendres' efforts announced Medway's new role as a site of leisure and recreation. Although the Stoneys had previously used Medway for these purposes, the Legendres introduced new priorities, new social conditions, and more intensive use. Under their ownership, Medway became a different place than before.

FIGURE 5.2 African American workers and an unidentified white male on the porch of the main house at Medway Plantation, circa 1920. The latter figure is probably one of Peter Galliard and Anna Porcher Stoney's sons.
Courtesy of the South Caroliniana Library, University of South Carolina, Columbia, SC.

At the time of the Legendres' purchase, Medway encompassed 2,530 acres, with boundaries that differed little from the acreage Thomas Smith had assembled in the 1680s. The shape of the parcel approximated the form of a squat arrow lying horizontally, with the tip pointing westward. Back River formed the eastern boundary; the right-of-way of the Seaboard Air Line Railroad lay on the west. Most of the tract lay covered in pine forest. The main house and its associated outbuildings stood near the center of the property, immediately north of the Home Reserve. Along the eastern edge of the plantation, ninety acres of old tidal fields remained intact. Over 1,200 acres of marshlands crisscrossed the plantation, the majority of them in three large concentrations: one at Crain Pond, another in the southwest corner, and the third along the stream between the Home Reserve and the river. Only a handful of outbuildings remained standing near the main house. These included a small stable, a smokehouse, the schoolhouse where Stoney's children had studied, and an unknown number of dwellings occupied by African Americans. Across the Home Reserve, the slave street of the antebellum era had collapsed, and none of the buildings that had stood along the road leading

FIGURE 5.3 Residence of David Gourdine and family, circa 1920. Gourdine was one of the Stoneys' favored servants. He may have been born a slave; if not, he was born soon after the war. His grave is among the few marked burials in the African American cemetery at Medway.
Courtesy of the South Caroliniana Library, University of South Carolina, Columbia, SC.

to the brickyard survived. Nearby, the rice mill remained standing and faint outlines of the brickyard remained visible. Only a few open fields survived.[47]

In purely physical terms, Medway amounted to a sizable tract of land covered with decaying remains and rapidly advancing vegetative growth. The main house stood in isolation, in greatly diminished circumstances. Elsewhere, the basic infrastructure of rice production and brick manufacturing remained intact, but erosion and decay had taken a heavy toll. Although some elements of Medway's earlier form survived, a great deal had been lost, and most of what remained lay in poor condition. Natural processes had become the dominant force in shaping the plantation's physical space.

The Legendres immediately set to work on making Medway their own. They began by hiring a Charleston architect, Albert Simons, to renovate

[47] J. P. Gaillard, "Map of Part of Medway Plantation, 1930," Plat Book B, 122, BCRD.

and modernize the main dwelling. A Stoney family relation by marriage, Simons had studied architecture in France under Ernest Hébrard and later worked for Lawrence Hall Fowler in Baltimore. After serving in World War I, he returned to Charleston and formed a partnership with Samuel Lapham, another lowcountry native. Together, the pair restored Charleston townhouses and plantation dwellings for wealthy northerners. After decades of limited maintenance, the house at Medway needed extensive work. The Legendres spent the winter of 1930–1931 in Charleston while the renovation began. According to Gertrude, they met daily "with lawyers, contractors and architects" and frequently trekked to Medway to monitor the progress of construction. Under Simons's direction, workers rebuilt the roof and windows and made extensive masonry repairs. The interior of the house received a thorough refurbishing, and workers installed indoor plumbing, modern bathrooms, and a central heating system. At the same time, crews left the exterior essentially unchanged, as per the Legendres' wishes (see Figure 5.4).[48]

As the renovation of the main house proceeded, work on the grounds got under way. Workers cleared underbrush, reclaimed overgrown pasture, pruned trees, and planted ryegrass. They also added ornamental plantings and flowering shrubs in several locations. Beneath the oak allée off the west front, workers rehabilitated and enlarged a series of terraced gardens that Louisa Smythe Stoney had created between about 1906 and 1920. These lined the verdant terrace between the oak avenues with flowers that billowed multiple shades of red, pink, and white for several weeks each spring. While in bloom, the terraced gardens formed a spectacular vista that greeted visitors on the approach to the house (see Figure 5.5). The Legendres also rehabilitated and enlarged a small flower garden. For this project they enlisted the services of Ellen Biddle Shipman, a pioneering landscape architect and close friend of Gertrude's mother. A thick layer of overgrowth, weeds, and accumulated detritus obscured all but the garden's general location off the southwest corner of the house. Shipman carefully excavated the old beds and walking paths and redefined the garden's overall form. Then she selected flowers, trees, and shrubs and supervised their planting. The revived garden, probably complete by the mid-1930s, became a floral showcase and setting for quiet reflection and intimate conversation. Occupying roughly a third of an

---

[48] Legendre quoted in Beach, *Medway*, 42. See also Legendre, *Medway Plantation*, 3. On Simons, see Stephanie E. Yuhl, *A Golden Haze of Memory: The Making of Historic Charleston* (Chapel Hill: University of North Carolina Press, 2005), 8–9, 36–52.

FIGURE 5.4 Main house at Medway Plantation, 1940. Photograph by
C. O. Greene.
Courtesy Historic American Buildings Survey, Washington, DC.

FIGURE 5.5 Allée and gardens, 1940. Photograph by C. O. Greene.
Courtesy Historic American Buildings Survey, Washington, DC.

acre, it lies adjacent to a lawn on the south side of the house and at the upper end of the allée.[49]

As Shipman carried out her work, the Legendres also renovated several surviving outbuildings. They turned the old schoolhouse and the summer kitchen into guest cottages and created an office out of another structure. An old smokehouse nearby received minor repairs. Together, these buildings formed a row of wood-frame structures off the southeast corner of the main dwelling.[50]

In April 1931, the Charleston *News and Courier* reported on the Legendres' efforts. In "Medway, Historic Brick House," Chlotilde R. Martin surveyed the history of the plantation and described the rehabilitation of the house and grounds. She informed readers that "the new owners have done the interior of the house over, put in bathrooms and otherwise modernized it without detracting from the charm of the place." She also credited the Legendres with beautifying the grounds and creating a "more picturesque setting for the old house."[51]

In later years, the Legendres continued making improvements. In the spring of 1937, they renovated the ground-floor rooms of the house once again. Workers removed cypress paneling from a dwelling on a neighboring plantation and installed it in the living room, the library, and the front hall. This gave the entry hall and principal entertaining rooms a richer, more sumptuous appearance. The paneling is a deep golden color that radiates warmth and a look of age. Gertrude described the post-renovation interior as "cozy, warm and familiar, with fireplaces and simple wood mantles" and the feeling of "an English country house."[52]

The single largest landscape feature the Legendres created lies on the north side of the house. In the late 1930s, the couple decided to expand a vegetable garden dating to the Stoneys' era and build facilities for maintenance of the gardens and grounds. The Legendres hired Ides Van Der Gracht, a Dutch architect practicing in New York City, to design a walled garden and greenhouse complex. Van Der Gracht fashioned a design that echoed the architecture of the main house and supplied the Legendres with the additional buildings they wanted. Two greenhouses adjoining a small pavilion form the north wall. Serpentine walls define the west and south sides, while a wall with a series of inset arches stands on the east.

---

[49] Beach, *Medway*, 26–27, 44–45; Briggs, *Charleston Gardens*, 146–147.
[50] Legendre, *Medway Plantation*, 3.   [51] Martin, "Medway, Historic Brick House."
[52] Legendre, *Time of My Life*, 70.

At one of the corners closest to the house is a small pavilion with a hipped roof. Similar pavilions form the outer ends of the greenhouse wing.[53]

The garden and greenhouse complex references the main house in two ways. First, Van Der Gracht placed stepped gables on the greenhouse pavilion to replicate the most distinctive feature of the main dwelling. The result is a combination of visual continuity and the suggestion of contemporaneous origins. Second, Van Der Gracht used salvaged bricks to give the complex an aged appearance. Van Der Gracht did not replicate the stuccoed exterior of the main dwelling but instead left the brick unfinished to highlight its worn and mottled appearance. The resulting difference introduced a subtle contrast but does not undermine the intended effect. The house and garden complex convey a strong sense of age and appear as a unified whole.[54]

Construction of the garden complex began in February 1940. Materials came from the plantation rice mill. The Legendres hired the Charleston Construction Company to dismantle this large, heavily built structure and salvage the timber framing, floorboards, bricks, and roof slate. Female laborers stripped the bricks of mortar, earning $1.50 per thousand. Construction proceeded at a moderate pace – not fast enough to satisfy Sidney, who complained on several occasions about the slow rate of progress, but still swiftly enough for the complex to be completed by the early summer.[55]

Immediately beyond the garden to the east, the Legendres erected a series of outbuildings that include a kitchen, laundry, garage, stables, tack house, and large barn. These structures all stand one story tall and are of frame or brick-masonry construction. Instead of giving them an aged appearance, the Legendres specified white paint and green trim, resulting in plain, utilitarian styling typical of agricultural buildings of the period. Set close together off the northeast corner of the main dwelling, these buildings supported upkeep of the estate and recreational activities. They made possible Medway's operation as a sporting estate.[56]

Beyond the domestic complex, the Legendres made modest changes. They maintained open pastures to the north and south in order to provide grazing land for horses and, for a time in the early 1940s, a herd of cattle. Farther afield, the Legendres instituted other changes. Like all northern owners, they adopted land-management practices aimed at improving hunting conditions. In the old fields along Back River, they planted small

[53] Beach, *Medway*, 44; "Diary of Life at Medway Plantation," 206.
[54] Beach, *Medway*, 44, 47.   [55] "Diary of Life at Medway Plantation," 108, 206.
[56] Ibid., 206.

quantities of rice as feed for ducks. On upland terrain, the Legendres planted several crops to attract quail. They also rehabilitated the water-control systems associated with the inland reserves and tidal fields. The Legendres rebuilt trunks, dikes, drains, and sluicegates throughout the entire system in order to regulate water levels, which aided their efforts to grow rice and attract ducks.[57]

The changes the Legendres made to the larger landscape marked the final stage in the process that turned a ruined plantation into a private leisure retreat. Coupled with redevelopment of the domestic complex, these measures took Medway beyond its earlier form. More than simply rehabilitating surviving remains, the Legendres literally made Medway anew. Their estate incorporated vestiges of past eras, new buildings and landscape features, and stylistic elements developed for the sake of aesthetics. More had been lost than added, and the resulting whole differed profoundly from its antecedents. If Medway remained a "planta-tion," it was only as a plantation of a dramatically different kind.

A RUIN REMADE

In carrying out one of the most elaborate transformations of the era, the Legendres not only turned Medway into a handsome country estate, they refigured its physical space. Development of the domestic complex and concentration of activities in immediately proximate areas went hand in hand with decreased use of lands once dedicated to agriculture and brickmaking. Under the Legendres' ownership, the domestic complex became central to sporting and leisure pursuits and hosting friends and visitors. The barns and sheds buzzed daily with workers caring for horses and dogs, readying wagons, cleaning freshly killed game, and maintaining the grounds and gardens. Meanwhile, the remainder of the estate became a vast recreational domain, used for hunting, horseback riding, picnick-ing, and a host of other activities. An undetermined number of African American tenants also lived on and farmed these lands, which meant that portions of the estate remained in productive use. Still, as compared with the era of slavery, Medway became a site of diminished activity. Whereas agricultural operations and brick manufacturing had once maintained

---

[57] The pastures are shown on Galliard, "Map of Part of Medway Plantation, 1930," and also in a photo that appears in "Life Goes to a Party with the Sidney Legendres," 54. The Legendres kept a herd of cattle as one of many attempts to offset operating costs. See "Diary of Life at Medway Plantation," 203–205.

a broad footprint, the Legendres' plantation assumed the two-part organization typical of northern-owned estates: a domestic core centered on the main dwelling and a less intensively used landscape reserved for recreation.

The changes the Legendres carried out came on the heels of roughly half a century of decline. In the years when the Stoneys had used Medway for hunting and holiday gatherings, the landscape of rice culture fell into ruin. Croplands became overgrown, outbuildings deteriorated, and the slave street and associated domestic spaces virtually disappeared. By the beginning of the twentieth century, large portions of the infrastructure that once sustained the estate had suffered heavy losses. In the years that followed, decay continued to advance. By the early 1920s, the better part of the plantation's historical landscape had perished.

The Legendres did not set out to strip Medway of physical evidence of slavery and agricultural production. Their aims focused on creating a winter residence and private hunting retreat, and they worked to achieve their goals in the same way that dozens of other northerners did. Yet in reshaping the material environment at Medway, the Legendres dealt powerful blows to what survived. To be sure, some elements escaped largely unscathed. The Legendres left the tidal fields and inland reserves in recognizable if somewhat altered condition, and the handful of surviving antebellum buildings also retained their basic character. These edifices, however, represented only a small portion of what had existed historically. The loss of the slave street, the brickyard, most of the barns and sheds, and the upland fields took a huge toll on Medway's surviving form. Without these elements, the landscape told relatively little about the plantation's history.

The changes the Legendres carried out refigured Medway in several ways. First, they created a more elegant and elaborate domestic complex than had historically existed. Although the Stoneys had substantially improved the appearance of the grounds, the Legendres went much further. By giving Medway a well-developed ensemble of gardens, lawns, and outbuildings, the Legendres demonstrated the extent of their resources, their ambitions, and the scale of activity that took place during their sojourns. Visually and symbolically, the landscape highlighted the architecture and authority of the main dwelling. Meanwhile, spaces in the immediate vicinity proved crucial to Medway's operations. The gardens and lawns accommodated outdoor social and recreational activity and the outbuildings and workyards facilitated maintenance and upkeep.

Overall, the domestic complex provided essential infrastructure for a "plantation" as conceived by people of the 1930s.

Second, the Legendres achieved a sweeping consolidation of physical space. The two-part organization placed the domestic complex in relative isolation amid several thousand acres of forest. In effect, it became a park-like setting surrounded by a sprawling expanse of undeveloped land. The new pattern contrasted sharply with Medway's historical organization. Before the Civil War, Medway had teemed with activity. Slaves had labored on the better part of its lands and continual activity around the main house had underscored the plantation's role in an expansive world of commerce and trade. By the Legendres' era, Medway lay in seclusion and isolation. Occupied seasonally by its owners and farmed year-round by tenants, its condition reflected low-intensity land use and economic decline.

The relative absence of activity had powerful implications for perceptions of the Legendres' estate. The idealized image of plantations as serene enclaves of agrarian virtue and aristocratic authority found seemingly authentic counterparts in estates such as Medway, where social and material conditions suggested isolation and self-sufficiency. Although not an intentional byproduct of the Legendres' efforts, Medway's character echoed the mythic plantation world of Old South lore. As yet another quality that suggested continuity with earlier periods, limited activity validated the myth of a romantic past while obscuring patterns of dramatic change.

A third consequence of the Legendres' efforts resulted from purposeful cultivation of ruination and decay. By making age the central feature of the domestic complex, the couple gave Medway an ostensibly timeless aura. The weathered and worn appearance of the main dwelling and walled garden and the maturity of the surrounding plantings made Medway appear as though it had stood for ages. Although the plantation ranked among the oldest in the lowcountry, its material form had recent origins. The Legendres' estate resulted from careful retention of select elements and the addition of new features, some styled to look old, others plain. The resulting material environment proved profoundly ahistorical, even as it appeared otherwise. Created specifically to meet the Legendres' tastes, Medway articulated the needs and priorities of its owners while telling little about its past.

Although the Legendres and others continued to refer to Medway as a plantation, the meaning of the word differed dramatically from before. A hundred years earlier, Medway had been a site of toil and suffering for more than a hundred enslaved men, women, and children. Commercial imperatives had shaped its operations and material form. Now Medway served as a place of leisure for people of wealth and privilege. In between the two lay vast differences. In part, those differences reflected changes in use, purpose, and social relations. Yet they also reflected extraordinary material changes. Not only had slavery ended, but most of its physical vestiges had disappeared. Remnants survived, but only in scattered and diminished form.

REPRESENTATIONS OF A PLANTATION REMADE

Material change played a crucial role in shaping popular views of the plantation and the lowcountry past. In the years following the Legendres' arrival, Medway attracted attention as never before. The Charleston *News and Courier* featured the plantation in an article published in April 1931, and *Life* magazine ran a photo essay on Medway in January 1938. It also mentioned the plantation in another spread published in December 1939. *Town and Country* ran a feature on Medway in March 1949. Local writers and historians also showed interest. Samuel G. Stoney, Jr., and his collaborators featured the main dwelling in *Plantations of the Carolina Low Country*, a compendium of colonial- and antebellum-era plantation houses. Moreover, by the mid-1930s, lowcountry people recognized Medway as one of a small number of plantations that northerners had "restored" to its former grandeur. In this role, it exemplified the veneration of regional traditions after years of neglect and decay.[58]

With new attention came new claims about Medway's past. For decades, observers near and far had recognized Medway's age and historical associations. As early as the mid-1870s, popular writers had identified the plantation as a remnant of the colonial era. Most accounts focused solely on the main dwelling. In 1875, Constance Woolson described Medway as belonging to a group of "old houses and gardens" in a neighborhood "rich in colonial memories and Revolutionary legends." She noted its association with Landgrave Smith and identified the dwelling as "the first brick house in Carolina, still standing."[59] When William

---

[58] Martin, "Medway, Historic Brick House"; "Life Goes to a Party with the Sidney Legendres"; "The Low Country," *Life*, Dec. 25, 1939, 38–45; "Medway Plantation," *Town and Country* 103, no. 4318 (March 1949): 76–79; Stoney et al., *Plantations of the Carolina Low Country*, 46–47, 86–89. For mention of Medway as "restored," see, for example, "Life Goes to a Party with the Sidney Legendres," 54.

[59] Constance Woolson, "Up the Ashley and Cooper," *Harper's New Monthly Magazine* 52 (Dec. 1875): 16, 22.

Whitson visited Medway in 1889, he described the house as "an old-time mansion."[60] Two decades later, another visitor identified the plantation as the "ancestral estate" of the Stoney family, owners "for many generations." These and other accounts characterized Medway as old and archaic but said little else. Viewing the plantation principally as a novelty, they emphasized its age, its distressed appearance, and its early origins but made no judgments about its historical importance and meaning.[61]

The Legendres' arrival made Medway's historical value plainly apparent. For the first time, commentators described the plantation as "historic" – a term that contemporaries saw as signifying origins before the Civil War. In "Medway, Historic Brick House," Chlotilde R. Martin characterized the main dwelling as "the first brick house in South Carolina built out[side] of Charleston." She praised the Legendres' treatment of the building and surrounding grounds.[62] Subsequent accounts followed in a similar vein. In 1938, *Life* magazine characterized Medway as the "oldest and best-preserved plantation in South Carolina."[63] This statement echoed Martin's accolades while crediting the Legendres with reverent care of the entire estate. "Oldest" suggested historical import; "preserved" indicated that *Life* saw Medway as largely unchanged from its historical form. Together, these terms identified Medway as a rare and important vestige of South Carolina's past.

Representations of physical space changed as well. The tendency to focus solely on the main dwelling became even more pronounced. In describing Medway as representative of "South Carolina's great plantation houses," *Life* highlighted the majesty of the Legendre's residence, the towering oaks around it, and the beauty of the countryside. It depicted Medway as a great mansion with a park-like setting and enviable isolation. In effect, *Life* depicted the entire plantation as akin to a great country house. Moreover, the magazine said nothing about slavery, commerce, labor, or spaces where productive activities had historically taken place. In fact, *Life* said nothing about Medway's past, save for passing mention that the plantation "was once used for rice growing."[64] Readers unfamiliar with American history would have been hard-pressed to

[60] William Whitson, "A South Carolina Hunt," *Forest and Stream*, Sept. 30, 1899, 264.
[61] Harriette Kershaw Leiding, *Historic Houses of South Carolina* (Philadelphia: J. B. Lippincott and Co., 1921), 30. On first-generation plantation architecture, see especially Shelley Elizabeth Smith, "The Plantations of Colonial South Carolina: Transmission and Transformation in Provincial Culture" (PhD diss., Columbia University, 1999), chap. 1.
[62] Martin, "Medway, Historic Brick House."
[63] "Life Goes to a Party with the Sidney Legendres," 54.   [64] Ibid., 54.

deduce that plantations had historically operated as commercial enterprises worked by slave labor.

In characterizing Medway as "historic," Martin's account crossed a threshold. It revealed growing acceptance of plantations as valued pieces of lowcountry heritage. That histories of Medway said nothing about large portions of its past apparently mattered little. Narratives centered on planter families echoed popular and scholarly views of the era, which paid little attention to non-elite actors and tended toward triumphalism. "Historic" identified Medway as exceptional. Implicitly, it also indicated belonging to a distant, thoroughly different time. No longer simply dated, Medway had slipped fully into the past. "Historic" thus represented more than an arbitrary judgment; it expressed sentiments that had grown over time and crystallized with the Legendres' activities. In using the term, Martin expressed what others had already thought and felt.[65]

By categorizing Medway as a tangible piece of lowcountry history, Martin and others made several important judgments. First, they emphasized continuity with earlier periods, a perspective that downplayed recent changes and presented the Legendres' estate as having evolved slowly over time. Second, Martin and other commentators cast Medway as a historical artifact, a status that emphasized authenticity. Third, "historic" included Medway in a developing schema that ranked plantations according to perceived importance. By the mid-1930s, most observers saw Medway as a rare example of a plantation from the earliest period of South Carolina history. Like about two dozen other plantations accorded similar acclaim, Medway stood as an iconic example of regional heritage. Plantations in this category validated sweeping narratives of the regional past. One account after another cast them as evidence of the grandeur of the antebellum era and the accomplishments of the rice-planting elite. "Historic" plantations made an imagined past a material reality. By grounding exclusionary narratives in space and time, plantations such as Medway implicitly legitimated romanticized visions of a sordid and still-contested history.[66]

Recognizing Medway as historically valuable did not inspire new investigations of its past. In fact, recognition of historical import may have dissuaded further study. In the years following the Legendres' remaking, observers continued to tell Medway's history in the same

---

[65] On the cultural processes involved in historical valuation, see David Lowenthal, *The Past Is a Foreign Country* (Cambridge: Cambridge University Press, 1985), chap. 4.

[66] See especially Stoney, *Plantations of the Carolina Low Country*.

way as before: as a lineage of ownership centered on a handful of planter-patriarchs and their families. Histories of the era portrayed the plantation as consisting solely of a revered house with beautiful surroundings. Landgrave Smith and the Stoneys figured prominently and other owners received occasional mention. Many accounts also noted the main dwelling's status as the oldest brick structure outside Charleston. Otherwise, histories of the era told little. Slavery, labor, brick manufacture, commerce, and the lives of enslaved men and women are conspicuous omissions. Recognition, celebration, and veneration apparently cast Medway's history as well understood and therefore not needing further examination. If anyone conducted additional investigations, their work is absent from the historical record.[67]

Despite their limitations, histories published during the era appeared in well-regarded publications and thus possessed a guise of credibility. The rigidly exclusionary public culture of the segregated South provided additional authority. Without countering views to contest the vision presented, an idealized view of Medway's past became the sole perspective voiced. Consistency and frequency of expression gave it a quasi-official guise. African Americans undoubtedly viewed Medway differently but focused their energies elsewhere. In an era when African Americans faced daunting barriers, contesting the history of a single plantation – or, for that matter, many – took a back seat to more pressing needs. Representing the past required resources of a kind that lowcountry blacks simply did not possess. Nor did it figure among their immediate concerns. African Americans instead focused on access to jobs, housing, and equal treatment under the law. Histories of Medway thus went unchallenged for the time.[68]

Just as the absence of competing claims accorded legitimacy to depictions of the plantation, so did its physical form. Narrative and material fabric developed a mutually reinforcing relationship. The Legendres' Medway implicitly validated historical accounts focused mainly or exclusively on the main dwelling. Without spaces of labor and commercial enterprise, the Legendres' estate implicitly sanctioned histories focused on planter families, architecture, and natural beauty. The reverse also proved

---

[67] See, for example, Martin, "Medway, Historic Brick House."

[68] On black priorities of the era, see Walter J. Fraser, Jr., *Charleston! Charleston!: The History of a Southern City* (Columbia: University of South Carolina Press, 1989), 359–364; Yuhl, *Golden Haze of Memory*, 15.

true. Narratives of Medway corresponded neatly to a "plantation" comprised mainly of a handsome building complex and pastoral soundings. The result was a self-validating pair.[69]

The groundswell of attention triggered by the Legendres' remaking made an exclusionary narrative a staple of popular memory. Histories published in the 1930s and later decades recounted a saga of white elites and a curiously styled building that had withstood wind, war, and repeated upheavals. In effect, they reduced a long and complicated history to a few details. Not only did such histories elide much of Medway's past and hundreds of historical actors, they devalued what did receive mention. Medway had always encompassed more than a large mansion set deep in the woods. Any meaningful history could not ignore the full extent of its past, the experiences of the many people associated with it, and the spaces where they had lived, worked, loved, and played. Synoptic accounts offered little of value. Only by eliding the historical roots of contemporary problems did writers of the era produce histories of plantations without slaves and continuity across time.

CONCLUSION

Nearly a century after the Legendres first imagined "plantation life," Medway remains one of the best-known plantations in the lowcountry. In part, its status reflects continuing interest in the region's colonial and antebellum heritage. With a well-developed tourism industry promoting iconic buildings and sites throughout the region, plantations such as Medway receive ample attention. Like Mulberry, Chicora Wood, Middleton Place, and a few others, Medway is routinely identified as a plantation with early colonial origins. Although set in a secluded location and not open for public tours, Medway nonetheless figures prominently in public discourse. As one of the most celebrated of all lowcountry plantations, it holds a revered place in memory.

Medway's prominence also reflects Gertrude Legendre's legacy and the strength of her personality. Sidney died suddenly in 1948 at the relatively youthful age of forty-eight. After his passing, Gertrude continued wintering at Medway. Over time, she gained acceptance among

---

[69] Also important is the fact that the Legendres' estate occupied the same physical space as earlier iterations of Medway and incorporated vestigial remains. These also made Medway seem an authentic remnant of the lowcountry past.

Charleston's notoriously exclusionary elite and became a stalwart advocate of land and wildlife conservation. In her later years, Gertrude wrote about her experiences at Medway and her love of the lowcountry. A memoir published in 1987 recounts some of her adventures and her life with Sidney. Other publications have chronicled her association with Medway and efforts to ensure its preservation. These have not only raised awareness of Medway; they have cast light on the Second Yankee Invasion. By calling attention to wealthy sportsmen's and sportswomen's role in rehabilitating old plantations, such accounts have revealed important dimensions of a little-understood phenomenon. The Legendres consistently figure among the most frequently invoked examples of "rich Yankees" who turned old plantations into winter havens.[70]

Recent accounts have portrayed Medway in the same manner as newspaper and magazines features of the 1930s and 1940s. In *Lowcountry Plantations Today*, a coffee-table compendium intended as a successor to *Plantations of the Carolina Low Country*, William P. Baldwin describes Medway in roughly the same way as *Life* did on the eve of World War II. Stunning photographs depict a majestic house surrounded by lavish gardens, towering oaks, and miles of forest. Interior views reveal elegant décor, sumptuous furnishings, and beautiful antiques. The walled garden, outbuildings, and main dwelling form a stunning ensemble. Emphasis on Medway's timeless beauty suggests continuity with earlier periods. Meanwhile, the absence of remains associated with slavery, labor, or commercial activity is striking – and consistent with Baldwin's inattention to these subjects. The "plantation" is depicted simply as a grand house surrounded by resplendent landscaping in a thoroughly familiar representational idiom.[71]

Baldwin's portrayal reflects more than willful disregard for the findings of decades of scholarship and growing acceptance of more inclusive perspectives on the American past. It reflects more than awe of the Legendres' house and its immediate environs. Further, it shows more than a tendency to see planter families as somehow more deserving of attention than the enslaved men and women who worked their lands. Baldwin's portrait reveals inattention to physical space – to the material form

---

[70] See, for example, Beach, *Medway*; Carola Kittredge, "Charleston's Grandest Dame," *Town and Country* 149, no. 5179 (Apr. 1995): 118–121. Legendre's memoir is *The Time of My Life*.

[71] William P. Baldwin, *Lowcountry Plantations Today* (Greensboro, NC: Legacy Publications, 2002), 22–33.

Medway once possessed, its present-day configuration, and the profound changes that took place over time. Baldwin's portrayal insists on viewing a plantation as a large house and nothing else. Medway, of course, never had such a form, nor did other plantations. Even in the Legendres' era, Medway encompassed more than the domestic complex. For the better part of Medway's history, the rice fields, brickyard, and upland fields constituted central zones of activity. These spaces figured as Medway's most developed features and clearly conveyed its purposes. They anchored its productive capacities and shaped the lives of the people who worked its lands. To ignore all that Medway historically encompassed invariably yields an incomplete view of its history.

In the years when the Legendres remade Medway, narrowly focused views of past and present became closely tied to a place that exuded complexity, spatially, socially, and economically. The changes the Legendres carried out stimulated unprecedented interest in a plantation long seen as archaic but otherwise unremarkable. New discourses quickly produced judgments about Medway's historical importance. These lashed exclusionary narratives to a small part of a plantation. New histories foregrounded select elements of Medway's past and a small part of its physical space. In the process, activities central to the plantation's origins and development became obscured. So, too, with the spaces most familiar to the hundreds of men, women, and children who had lived and labored at Medway as slaves. Nowhere did the shifting relationship between historical memory and material form manifest itself more clearly. By the 1940s, narratives of Medway's past and the core of the Legendres' estate formed a sumptuous image of rural gentility. Medway may have remained a "plantation" of a kind, but it differed fundamentally from those that had dominated the lowcountry landscape before the Civil War. What Medway meant and the purposes it served differed profoundly from those of its predecessors.

6

Representing a New Plantation World

> The plantation hum of happy purposes, as a beehive, is gone. The industry is closed these twenty years. Remains the beauty and grace, the golden days in a lovely climate. What is the next chapter?
>
> – Susan Lowndes Allston, 1929[1]

Northerners' estates captured the attention of people far and wide. As sportsmen and sportswomen remade old plantations, newspaper and magazine articles chronicled the changes taking place. By comparing new estates with historical plantations, contemporary accounts explained what plantations had once been and what some had become. By describing northerners' activities and interests, they identified the purpose of new estates. By highlighting commonalities and contrasts with plantations that remained in agricultural use, newspapers and magazines underscored the growing significance of leisure. As productive uses lost viability, northerners identified alternatives. They revealed possibilities, suggested untapped options, and underscored the lowcountry's shifting roles within the South and the nation.

Newspaper and magazine reporting formed part of a wave of attention that placed the lowcountry at the center of national consciousness. In the late 1920s, lowcountry artists, writers, and promoters began celebrating the region in new ways. Calling attention to its rich heritage and age-old traditions, they cast it as a land unto itself. Dubose Heyward's *Porgy* and Julia Peterkin's *Black April* and *Scarlet Sister Mary* introduced readers to

---

[1] Susan Lowndes Allston, "Windsor Place Spot of Beauty," *News and Courier* (Charleston, SC), Dec. 22, 1929, A3.

a timeworn landscape, colorful characters, and the "ancient" city of Charleston. Daniel Elliot Huger Smith and Alice Ravenel Huger Smith's *Dwelling Houses of Charleston* and Albert Simons and Samuel Lapham's *Charleston, South Carolina* highlighted the elegance and grandeur of Charleston in its heyday. Ambrose Gonzales's *The Black Border* and Samuel Gaillard Stoney, Jr., and Gertrude Shelby Matthews's *Black Genesis* described the curious dialect and enchanting folkways of "Gullah negroes." The "Charleston," the dance that became a popular sensation in 1926, attracted further attention.[2] In the 1920s, the lowcountry became widely recognized as a land unto itself, quintessentially southern yet unique. No longer a forlorn region topped by repeated traumas and tragedies, it became a place where, as Stephanie E. Yuhl has written, "a glorious past lived on."[3]

Interest in northern-owned estates figured among the major trends of the era. Lowcountry people viewed the activities of wealthy sporting enthusiasts as an index of regional progress. The remaking of old plantations and concomitant creation of new estates heralded renewal on an almost unimaginable scale. As long-stagnant districts gained new life, lowcountry people looked on with interest and anticipation. Journalists followed suit. Eager to record the unfolding drama, they paid close attention to northerners' efforts. Meanwhile, sportsmen and sportswomen and throngs of other middle- and upper-class Americans viewed the new estates as heralding the rise of a recreational destination. Although devoted sport hunters had prized the lowcountry for decades, the advent of a well-developed sporting scene, replete with ample socializing and a full range of leisure pursuits, gave the region new significance.

---

[2] Du Bose Heyward, *Porgy* (New York: George H. Doran Co., 1925); Julia Mood Peterkin, *Black April: A Novel* (Indianapolis: Bobbs-Merrill Co., 1927); Julia Mood Peterkin, *Scarlet Sister Mary* (Indianapolis: Bobbs-Merrill Co., 1928); Alice Ravenel Huger Smith and Daniel Elliott Huger Smith, *The Dwelling Houses of Charleston, South Carolina* (Philadelphia: J. B. Lippincott Co., 1917); Albert Simons and Samuel Lapham, *Charleston, South Carolina* (New York: Press of the American Institute of Architects, 1927); Ambrose Elliott Gonzales, *The Black Border: Gullah Stories of the Carolina Coast* (Columbia: The State Co., 1922); Samuel Gaillard Stoney and Gertrude Shelby Matthews, *Black Genesis: A Chronicle* (New York: Macmillan Co., 1930). On the "Charleston," see Stephanie E. Yuhl, *A Golden Haze of Memory: The Making of Historic Charleston* (Chapel Hill: University of North Carolina Press, 2005), 172–173; Walter J. Fraser, Jr., *Charleston! Charleston!: The History of a Southern City* (Columbia: University of South Carolina Press, 1989), 368.

[3] Yuhl, *Golden Haze of Memory*, 6.

As attention to northerners' estates grew, three modes of representation developed. One cast northerners as heirs to the planters of the antebellum era. This interpretation grew out of the writings of journalists Chlotilde R. Martin and Chalmers S. Murray. Martin wrote more than thirty articles on northerners' estates for the Charleston *News and Courier*. Murray penned more than twenty. These pieces introduced northerners and their "plantations" to lowcountry audiences. They described northerners' backgrounds, the history of their estates, and choices regarding architecture and landscapes. By describing existing conditions and offering glimpses of past eras, Martin and Murray viewed the new estates in a historical perspective. Both writers depicted them as the latest phase in a long sequence of plantation development. Rather than seeing radical changes, Martin and Murray emphasized continuity. Although neither downplayed the scale and extent of the changes taking place, both saw a revival of old ways. Instead of heralding the rise of a new order, Martin and Murray saw restoration on a grand scale.

A second mode of representation depicted northerners' estates as merging a romantic past with the best features of modern life. Articles in magazines such as *Country Life*, *House Beautiful*, and *Town and Country* portrayed northerners as combining time-honored traditions with modern comforts and conveniences. In this vision, colonial-era architecture and centuries-old gardens mingled with well-appointed kitchens, elegant furnishings, and machine-age amenities. Northerners' estates thus became settings where a gracious past blended seamlessly with everything that people of wealth and taste favored in 1930s-era America. In turn, northerners became modern-day purveyors of the renowned hospitality and graciousness commonly associated with "the South."

A third mode of representation advanced new claims about the significance of plantations in lowcountry history. Although plantations had long been a part of the lowcountry landscape, not until the 1930s did people explicitly identify them as central to the history of the region. New claims about plantations did not pertain exclusively to northerners' estates; they also applied to other plantations. Nonetheless, they resulted in part from northerners' activities. The remaking of old plantations prompted devotees of lowcountry history to celebrate valued remains and interpret the changes taking place. By highlighting early buildings and asserting their significance, avocational historians narrated the transformations under way. Their efforts responded directly to the destabilizing force of northerners' activities. Although recognition of plantations as historically valuable predated the development of large estates, an outpouring of interest in the 1930s demonstrated the concern sparked by northerners' actions. Bold, seemingly authoritative claims about the significance of plantations revealed growing interest in the lowcountry and detailed study of early plantation houses. Self-appointed guardians of lowcountry history proffered their version of events before others did, motivated partly by pride and partly by a desire to affirm existing social hierarchies.

Collectively, newspaper and magazine reporting told lowcountry residents, tourists, and distant observers about the changes taking place. By judging northerners' actions, journalists and other writers narrated the origins of the new estates and explained their relationship to historical antecedents. By emphasizing continuities, they presented a misshapen vision. Rather than acknowledging the breadth and extent of the changes under way, these commentators emphasized renewal. By casting northerners' estates as old plantations reborn, they left vestiges of slavery unmentioned or depicted them as picturesque elements of an endlessly enchanting landscape. As northerners transformed old plantations, commentators near and far downplayed the consequences of their actions. Representations in different forums crafted a portrait of a resurgent lowcountry, not a region reshaped by the swift and decisive actions of wealthy outsiders. In turn, slavery's debilitating legacy slipped away, marginalized by a wall of portrayals that looked past it or saw only an unproblematic piece of a thoroughly captivating history.

CHRONICLING THE ESTATE-MAKING BOOM, ONE PLANTATION AT A TIME

In October 1930, William Watts Ball, the editor of the Charleston *News and Courier*, hired Chlotilde R. Martin to write a series of stories on northerners' estates. Martin, a native of Aiken, had worked as a reporter since the late 1910s. She initially found employment at *The State* newspaper in Columbia, where Ball had held his first editorship. Later she worked for a group of small-town papers owned by Allendale publisher Eugene B. McSweeney. Martin apparently wrote to Ball in search of work. She had lost her husband to Hodgkin's disease almost four years earlier and had two young children to raise. With the international financial crisis a year old and unemployment still at unprecedented levels, Martin looked for paying jobs wherever she thought they might be found.

Although still writing for McSweeney's conglomerate, she eagerly took on freelance assignments as her time allowed.[4]

Ball responded to Martin's initial inquiry hesitantly. "It is possible that The News and Courier might offer you temporary work," he wrote. He had "a series of stories" in mind that he felt certain she could do, although he was "not yet prepared to say that we shall undertake to obtain them." Nor did he identify the subject matter. He first wanted to know if she had access to a car and a camera. He concluded with a cautionary note: "Please do not regard this letter as particularly encouraging. The work would be temporary at best."[5]

Martin responded immediately. "I have my own automobile and can take good, clear pictures with my kodak," she told Ball. "The work which you suggest sounds very interesting," she added. Hopeful that Ball would be able to make the project "materialize," Martin concluded by writing, "Please let me know as soon as possible."[6]

Despite the chilly tone of his initial letter, Ball hired Martin at once. The same day he received her letter, Ball responded by outlining the terms of her employment and explaining the assignment. "The plan of The News and Courier," he wrote, "is to publish a series of illustrated stories about the estates in coastal South Carolina purchased and improved on by wealthy men, from the Savannah river to Georgetown." You will need "to visit the places and obtain the facts," Ball advised. Stories about the estates would be "the main object," but she could write "any other incidental stories connected with the region," provided they did not interfere with the main task. The pay would be sixty dollars per week, inclusive of expenses, save for the cost of developing photographs. Ball thought the whole project could be completed quickly, probably within two or three months. Whether Martin would earn a profit remained to be seen. But even if she did not, Ball predicted that the work would still "be interesting."[7]

---

[4] William Watts Ball to Chlotilde R. Martin, Oct. 29, 1930, folder 9, box 377, News and Courier Records (hereafter N and C Records), South Carolina Historical Society, Charleston, SC (hereafter SCHS); Robert B. Cuthbert and Stephen G. Hoffius, eds., *Northern Money, Southern Land: The Lowcountry Plantation Sketches of Chlotilde R. Martin* (Columbia: University of South Carolina Press, 2009), xix.

[5] William Watts Ball to Chlotilde R. Martin, Oct. 29, 1930, folder 9, box 377, N and C Records.

[6] Chlotilde R. Martin to William Watts Ball, Oct. 30, 1930, folder 9, box 377, N and C Records.

[7] William Watts Ball to Chlotilde R. Martin, Oct. 31, 1930, folder 9, box 377, N and C Records.

Martin set to work at once. She started with "a strenuous week in Beaufort," where she visited twenty northern-owned estates and learned of others near Bluffton, Hardeeville, and in Jasper County. She immediately began writing articles about those she had visited and making plans to travel to "four or five estates in Hampton county." By the end of the month she had been to at least forty-two. During each visit, she interviewed the owner or superintendent, took photographs, and gathered as much information as she could. Research in county records and conversations with local informants rounded out her sources. For historical information she consulted Harriette Kershaw Leiding's *Historic Houses of South Carolina*, the most detailed account of lowcountry domestic architecture available. For general background on northerners' lands she spoke intermittently with Neils Christensen, a Beaufort realtor. Christensen's firm handled inquiries from well-to-do northern sportsmen searching for hunting preserves and plantations in the southeast corner of the state. Probably the single-most knowledgeable authority on northerners' activities in the area, Christensen supplied Martin with information about northerners' estates and the new breed of plantation owners.[8]

By the end of the first week of December, Martin had written at least thirty articles and had four more nearly ready, pending additional information "from several sources." The following week she visited several plantations in Colleton County. On December 13 she wrote to Ball with an update. The week spent traveling the quiet roads of Colleton County had been productive, she told him, but another trip would be necessary. She had been unable to gain access to two plantations, both surely worth writing about. One, she told him, "is not yet completed – a half-million dollar affair."[9]

As Martin continued her work, the *News and Courier* published the first of her sketches, "Bingham House in Beaufort," on December 2. It described Cotton Hall Plantation, a 1,300-acre estate owned by Harry Payne Bingham of New York. Bingham and his wife had just completed a twenty-six-room house that Martin described as "magnificent." They reportedly spent about $200,000 creating their estate.[10] Five days later, the newspaper published "Clay Hall, Redmond Mansion Lodge."

[8] William Watts Ball to Chlotilde R. Martin, Nov. 4, 1930, folder 9, box 377, N and C Records.

[9] Chlotilde R. Martin to William Watts Ball, Dec. 7 and 13, 1930, folder 9, box 377, N and C Records.

[10] Chlotilde R. Martin, "20-Acre Lawn to Surround Bingham House in Beaufort," *News and Courier*, Dec. 2, 1930, 5.

It related the sad story of Geraldyn L. Redmond, a Long Island resident who had purchased several Beaufort-county plantations, only to pass away a short while later. In what would become a hallmark of her writing, Martin related the comments of an African American resident who praised Redmond and his largess. "It was jest like slavery time when Mr. Redmond was livin'," said Becky Frazier, an elderly woman whom Martin described as an "ancient black crone." "Dis place ain't the same since Jedus done come and tuck Mr. Redmond up to Heaven," Frazier declared.[11] On December 9, the newspaper published an article on several northern-owned tracts at the north end of Laurel Bay. Readers learned about J. K. Hollins's 1,000-acre estate on Brays Island, a neighboring tract owned by Phillip A. Carroll, and Warren H. Corning's estate near Gray's Hill. Martin characterized these holdings as supplying superb shooting and ample opportunities for relaxation.[12]

In the week leading up to Christmas, Martin wrote frantically. Eager to reserve the holiday for her family, she strived to finish as many sketches as she could. By the evening of December 23 she had completed ten and come close to finishing another three. She spent Christmas day in Camden and then drove down to the coast. On December 29 she met with Ball in Charleston, where she delivered two stories and outlined ideas for several more.[13]

Although Ball had predicted only few weeks' work, reporting on northerners' estates kept Martin busy much longer. She apparently completed her fieldwork and writing sometime early in 1931. Her sketches continued to appear into July of that year. Thereafter, she wrote a handful of articles for the paper: a piece on two Hilton Head Island estates, another on preparations for the 1932–1933 season, and a third on low-country rivers. These articles shared Martin's knowledge of northerners' estates and made use of information gathered during her research.[14]

---

[11] Chlotilde R. Martin, "Clay Hall, Redmond Mansion Lodge," *News and Courier*, Dec. 7, 1930, A11.

[12] Chlotilde R. Martin, "Largest Tree along Coast Found on Beaufort Island," *News and Courier*, Dec. 9, 1930, 14.

[13] Chlotilde R. Martin to William Watts Ball, Dec. 29, 1930, folder 9, box 377, N and C Records.

[14] Chlotilde R. Martin, "Hilton Head Island Estates Merged," *News and Courier*, Feb. 7, 1932, 3; Chlotilde R. Martin, "Low-Country Plantations Stir as Air Presages Coming of New Season," *News and Courier*, Oct. 16, 1932, 6; Chlotilde R. Martin, "Indian-Named Rivers Contribute Breath of Life to Carolina Low-Country," *News and Courier*, Nov. 7, 1932, 3.

Meanwhile, Martin sold Ball on the idea of a weekly column. "Low-country Gossip," a chronicle of lowcountry folklife, began in February 1932 and ran for several decades. It featured portraits of local characters, news about seasonal happenings, and detailed descriptions of settings throughout the region. Mentions of northerners' estates and activities appeared intermittently. Like Martin's other writings, "Lowcountry Gossip" became a vehicle for informing readers about northerners' comings and goings, their favored pastimes, and the legions of "old-time Negroes" who populated the rural lowcountry.[15]

Ball initially hoped that Martin would produce sketches of all northern-owned estates. As he told her in December 1930, "of course we expect you to continue this service through Berkeley and up the coast including Georgetown."[16] Martin's response is not known, but she apparently declined. By then she may have realized that the modest pay made it difficult for her to earn a profit. The added expense of traveling farther from home surely would have made the work even less remunerative. Whatever the case, Ball turned to Chalmers S. Murray, an Edisto Island native and journalist. Like Martin, Murray desperately needed work. He began writing articles on estates north of Charleston and published his first sketch in March 1931. The *News and Courier* continued publishing others throughout the summer and the fall. By the time "Wee-Nee, Black River Lodge," appeared on October 4, the newspaper had run twenty-one.[17]

Martin and Murray's articles told readers about changes taking place throughout the countryside. They immediately became the single most accessible source of information about winter colonists and their estates. Martin and Murray produced detailed, readable sketches about successful businessmen, heirs to industrial fortunes, and the places where these and other figures passed the winter months. As lowcountry people took notice of the transformations in progress, Martin and Murray provided an authoritative accounting of northerners' activities.

Martin and Murray's articles established a representational idiom that would shape popular views of the Second Yankee Invasion for decades. Both writers recognized the scale and extent of the changes under way.

[15] Cuthbert and Hoffius, eds., *Northern Money, Southern Land*, xix–xx.

[16] William Watts Ball to Chlotilde R. Martin, Dec. 15, 1930, folder 9, box 377, N and C Records.

[17] Chalmers S. Murray, "Wee-Nee, Black River Lodge," *News and Courier*, Oct. 4, 1931, 11.

They reported northerners' activities honestly and frankly, often in copious detail. Readers learned of the "improvements" new owners made, the changes they carried out, and how they refigured the landscape and architecture of their estates. In some cases, both writers also described the personal lives and backgrounds of estate owners. Readers learned the source of northerners' wealth, their familial lineage, and their most important accomplishments. Moreover, Martin and Murray's writings told a great deal about owners' recreational pursuits. How northerners spent their time in the lowcountry became a mainstay of the series.

At the same time, neither Martin nor Murray described the historical use of plantations. Neither mentioned slavery, labor, commerce, or the profit-seeking motives of the planter class. Neither mentioned the conflicts that had historically characterized life on lowcountry plantations or those that erupted after emancipation. Nor did Martin and Murray write about any of the other activities that had historically taken place on plantations. Rather, both writers depicted plantations simply as sites of upper-class habitation. By focusing on planter's dwellings, grounds and gardens, and owners, Martin and Murray provided an incomplete view of the past and present. By overlooking crucial dimensions of history and contemporary life, they misrepresented historical plantations and northerners' estates.

Martin and Murray's perspective manifest itself in several ways. One lay in emphasis on continuities with past eras. In delineating the relationship between historical plantations and new estates, Martin and Murray portrayed northerners as reviving spaces that had nearly faded into oblivion. Both writers portrayed northerners' actions as rehabilitative. Terms such as "renewal" and "restoration" made this clear. Construing northerners' actions in this fashion portrayed the new estates as, if not exactly like plantations of the past, closer to them than not. In Martin and Murray's eyes, the changes northerners carried out fell short of effecting wholesale transformations.[18]

Martin and Murray depicted social relations in similar fashion. By casting sportsmen and sportswomen as heirs to the old planter class, they

---

[18] See, for example, Chalmers S. Murray, "Mansfield, Ancestral Parker Home," *News and Courier*, March 1, 1931, A9; Chalmers S. Murray, "Waverly, Gem of the Waccamaw," *News and Courier*, July 26, 1931, A6; Chalmers S. Murray, "Springfield on the Peedee," *News and Courier*, July 27, 1931, 3; Chalmers S. Murray, "Holliday Winters in Peedee," *News and Courier*, Aug. 30, 1931, B11; Chlotilde R. Martin, "Tomotley, Brewton Plantation, Bindon, Castle Hill – Beaufort Restorations," *News and Courier*, Nov. 23, 1930, A12. For an instance of similar language, see "Interest in Reviving Old Plantations," *New York Times*, Jan. 22, 1933, real estate section, 2.

saw a return of the poise and self-confidence popularly associated with the antebellum southern elite. Northerners, in their view, possessed the same self-assuredness and noblesse oblige commonly associated with the old planters. Northerners' commitment to aristocratic field sports and favor for refined surroundings added to the similarities. In multiple ways, Martin and Murray portrayed northerners as following in the footsteps of the antebellum elite.[19]

Both writers also saw northerners' relations with African Americans as echoing antebellum practices. Martin made this point repeatedly with portrayals of elderly, seemingly docile blacks who welcomed the new-comers with open arms. Becky Frazier is one example. Others abound. At Myrtle Grove Plantation in Colleton County, Martin encountered "ancient black Jim Frazier." Frazier worked as a watchman on the estate. He reputedly dreamed of going to New York with his employer, Joseph S. Stephens. Frazier told Martin that Stevens, a retired stockbroker, had promised to show him "how dey makes money." Martin quoted Frazier as saying, "dey piles de gold up and one pile so and de silber up in anudder pile dere. I sho does wanta see dat much money."[20] At Gravel Hill, Martin noted a "picturesque aged negress" who spent her days sweeping the grounds with a "brush broom."[21] At Dean Hall Plantation, Martin marveled at a butler whose family had lived and worked on the estate for four generations and noted that descendants of slaves still occupied several slave cabins.[22]

Murray, for his part, did not include such portrayals in his sketches but did in other writings. A noted authority on "Gullah negroes," Murray published extensively on the folklore, language, and behaviors of low-country blacks, especially residents of the Sea Islands. Several articles on Gullah culture appeared in the *News and Courier* during the same era as the plantation series. Meanwhile, other white lowcountry authors also called attention to the cultural and social distinctiveness of lowcountry

---

[19] See, for example, Chlotilde R. Martin, "Crane House Reflects Old South," *News and Courier*, March 1, 1931, A6; Chlotilde R. Martin, "The Copps and Spring Island," *News and Courier*, Jan. 4, 1931, A6.

[20] "Myrtle Grove Plantation in Colleton," *News and Courier*, May 31, 1931, A7.

[21] Martin, "Gravel Hill, R. P. Huntington Home," *News and Courier*, Apr. 5, 1931, B3. For a similar portrayal, see Martin's comments about "Daddy Scripto" in "The Hutton Estate of 16,000 Acres," *News and Courier*, Jan. 25, 1931, B3.

[22] Chlotilde R. Martin, "The Cypress Gardens, in Berkeley," *News and Courier*, undated newspaper clipping, Dean Hall Plantation Vertical File, SCHS. For another portrayal of an "ancient darkey," see Chlotilde R. Martin, "Lowcountry Gossip," *News and Courier*, Dec. 3, 1933, A7.

blacks. Collectively, portrayals of the 1930s and 1940s identified "Gullah negroes" and "Sea Island negroes" as a distinctive type. Implicitly, this literature cast lowcountry blacks as contented, subservient, childlike creatures with few cares in the world. In the eyes of white commentators, lowcountry blacks' jovial demeanor and superstitions provided entertainment and amusement, and their apparent contentment indicated acceptance of their place on the bottom rung of the social ladder. Martin, Murray, and other writers thus cast northerners' estates as rekindling the supposedly harmonious race relations of the antebellum era.[23]

Martin and Murray's plantation articles made for informative, entertaining reading. Both writers covered a wide range of subjects and showed genuine interest in the places and people they wrote about. In general, their articles conveyed a lively, upbeat tone. The basic intent of the series is partly responsible. Ball viewed northerners' estates as beneficial to the lowcountry. Seeing them as the most promising development to take place in decades, Ball sought to build support for northerners' activities. He instructed Martin and Murray to portray the newcomers and their estates in a favorable light.[24] Both writers complied. Martin and Murray wrote approvingly of northerners and their activities. They heaped praise upon northerners' estates and touted the newcomers' neighborliness toward lowcountry people. Martin and Murray also lauded northerners' treatment of old houses and long-dilapidated landscapes. By casting northerners as saviors of a nearly lost world, they portrayed the newcomers with resolute favor.

At the same time, Martin's and Murray's writings did considerably more. By describing sportsmen's and sportswomen's ties to the lowcountry, they delineated emergent relationships between people and place. By reporting northerners' views of the lowcountry and its people, they related estate-owners' perspectives on the region. By judging the quality of the buildings that northerners erected and the renovations they carried out, Martin and Murray viewed the newcomers' efforts critically. In short, both writers did more than merely report. Their writings mediated northerners' relationship to the lowcountry and its people.

[23] See, for example, Chalmers S. Murray, "Negroes of South Carolina Low-Country Inhabit Ether with All Manner of Ghosts," *News and Courier*, Feb. 16, 1930, A7; "Gullahs Evolve Code for Whites," *News and Courier*, Oct. 22, 1933, B1; and "Only Negroes Can 'Feel' Spirituals," *News and Courier*, Oct. 29, 1933, A1.

[24] William Watts Ball, general letter of introduction for Chlotilde R. Martin, n.d. [Nov. 1930], folder 9, box 377, N and C Records.

Several themes quickly became hallmarks of the series. Both writers devoted considerable attention to plantation environments. Martin and Murray described prominent buildings, landscape features, and terrain in vivid detail. Readers learned of stately houses, well-kept outbuildings, "ancient" ruins, handsome lawns, and carefully maintained gardens. They learned about northerners' efforts to care for their estates and what they prized about them. Readers learned about captivating vistas and scenery and northerners' development of gardens and grounds. They learned about the lavish sums northerners spent on new houses and outbuildings, sumptuous decor, and extensive plantings. Readers also learned about rehabilitation of old rice fields and development of new shooting grounds. In sum, Martin and Murray produced rich portraits of northerners' estates. Their descriptions informed readers about the settings where northerners passed their time, how northerners merged old and new, and the results of such efforts.[25]

Sportsmen's and sportswomen's ties to the lowcountry also assumed prominent roles. In many cases, Martin and Murray simply explained how estate owners had discovered the region and what they liked about it. In others, they revealed more complicated stories. Familial relations, intersectional marriages, and self-proclaimed affiliations revealed the limitations of the term "Yankee." In several cases, couples with roots on opposing sides of the Mason-Dixon Line owned large estates. The Kittredges of Dean Hall fit this model. So too did Nicholas and Emily Roosevelt of Gippy Plantation and Horatio and Sophie Shonnard of Harrietta Plantation. The Kittredges' union occurred when Benjamin R. Kittredge, scion of a distinguished New England family, had married Elizabeth Maynard of Charleston. Nicholas Roosevelt, a distant cousin of President Theodore Roosevelt and native of New York, had married lowcountry native Emily Wharton Sinkler in 1916. Horatio S. Shonnard also hailed from New York, while his wife, the former Sophie A. Meldrim, was a native of Savannah. Mentions of these couples challenged conceptions of all winter colonists as "northerners." Details about their lives

---

[25] See, for example, Martin, "Crane House Reflects Old South"; Chlotilde R. Martin, "Buckfield, Kress Narcissus Farm," *News and Courier*, Apr. 12, 1931, B14; Chlotilde R. Martin, "New Yorker Reclaims Pimlico," *News and Courier*, May 24, 1931, B2; Chlotilde R. Martin, "Henry W. Corning's Place," *News and Courier*, Jan. 24, 1932, A11; Chalmers S. Murray, "Huntington House Reminds One of Africa," *News and Courier*, June 7, 1931, A8; Chalmers S. Murray, "LaFayette House on Edisto Restored," *News and Courier*, July 13, 1931, B3.

forced readers to consider the development of intersectional relationships and what they meant for familiar social categories.[26]

In other cases, Martin and Murray's investigations gave estate owners opportunities to rebuff their presumed status as "Yankees." As Robert P. Huntington, the owner of Gravel Hill Plantation, told Martin, "They insist on making me a Yankee down here and I resent it. I'm a Southerner and one side of my family came from South Carolina." By the time Martin interviewed Huntington in the spring of 1931, he had visited the lowcountry annually for thirty years. Irritated by locals' persistence in viewing him as an outsider, Huntington touted his ancestral ties to the South.[27]

Although Huntington's comments surely made for entertaining reading, they proved somewhat unusual in the context of Martin and Murray's reporting. Most sketches offered more succinct information about owners' backgrounds. Murray told readers about E. G. Chadwick, a vice president of a large real estate firm whom he described simply as a "New York man"; Jesse Metcalf, a "New York sportsmen" and a nephew of the US Senator from Rhode Island of the same name; "New York multimillionaire and sportsmen" Thomas Yawkey, owner of the Boston Red Sox; and Henry M. Sage, a land and timber baron from Albany, New York. Martin introduced readers to such figures as Z. Marshall Crane, head of the "famous Crane Paper Company"; Hans K. Hudson, "a New York man and member of the stock exchange"; Bayard Dominick, a "New York broker"; and J. Fritz Frank, a "New York publisher."[28] In a few cases, more detailed descriptions appeared. In her sketch of Belfair Plantation, for example, Martin wrote at length about W. Mosely Swain and his two daughters. Swain's ancestors included a founder of the

---

[26] On the Kittredges, see Louisa Cheves Stoney, ed. and comp., *A Day on Cooper River*, 2nd ed. (Columbia: R. L. Bryan Co., 1932), 28. On the Roosevelt-Sinkler marriage, see "Nicholas Roosevelt to Take a Bride," *New York Times*, Feb. 17. 1916, 11. On the Shonnards, see "Harrietta Plantation," *Georgetown Times*, Dec. 27, 1929, 6; John M. Lofton, Jr., "Do You Know Your Charleston? Harrietta Plantation," *News and Courier*, Apr. 25, 1938, 10. For mention of another trans-sectional marriage, see Chlotilde R. Martin, "Tomotley, Brewton Plantation, Bindon, Castle Hill – Beaufort Plantations," *News and Courier*, Nov. 23, 1930, A12.

[27] Martin, "Gravel Hill, R. P. Huntington Home." For a similar example, see Laura C. Hemingway, "Williamsburg Boasts Many Game Preserves for Hunt," *News and Courier*, Jan. 11, 1937, 3.

[28] Chalmers S. Murray, "The Wedge on the South Santee," *News and Courier*, Sept. 20, 1931, A7; Chalmers S. Murray, "Metcalf Builds on Peedee Bluff," *News and Courier*, March 22, 1931, B14; [Chalmers S. Murray], "South Island, Where Cleveland Hunted," *News and Courier*, Feb. 8, 1931, B9; Chalmers S. Murray, "New Yorker Lives in Coastal

Baltimore *Sun* and Philadelphia *Ledger* newspapers. An artist, Swain had designed the house at Belfair, a monumental, two-story dwelling built of tabby, and decorated it with his paintings. He and his daughters spent their days hunting, fishing, and enjoying the outdoors. The eldest daughter, Phyllis, had recently graduated from Smith College. "Eager, responsive, likeable people are the Swains," Martin declared. She considered them an asset to the coastal region.[29]

Plantation histories constituted a third area of emphasis. Descriptions of individual estates usually included information about historical owners and notable events. Some accounts mentioned early houses and other features of significance. Because Martin and Murray relied on informants and amateurish histories for information, some of their accounts lacked authority and presented more lore and legend than credible information. Moreover, Martin and Murray presented fragmentary views, not coherent narratives. Both writers used historical information to fill out articles concerned principally with new developments. Neither aimed to relate well-developed histories. As a result, their sketches tended to supply piecemeal details – interesting tidbits and anecdotes but little else.[30]

Although Martin and Murray reported on a broad range of subjects, they paid close attention to northerners' treatment of old plantations. Anything that created a sense of age and authenticity elicited praise. Restored houses provide a convenient example. Describing the house at Isaac Emerson's Arcadia estate, Murray observed that "the colonial dwelling" had been "restored to its former splendor." Additions designed

Garden," *News and Courier*, May 24, 1931, B6; Martin, "Crane House Reflects Old South"; Chlotilde R. Martin, "The Delta – Home of H. K. Hudson," *News and Courier*, Jan. 20, 1931, 7; Chlotilde R. Martin, "Dominick Builds among Giant Oaks," *News and Courier*, Feb. 8, 1931, B14; [Chlotilde R. Martin], "Bluff Plantation Overlooks Cooper," *News and Courier*, June 14, 1931, A2. On Chadwick, see also "E. G. Chadwick, 63, Realty Ex-Leader," *New York Times*, March 24, 1945, 17; on Yawkey, see "Yawkey Ambition Realized by Deal," *New York Times*, Feb. 26, 1933, sec. 3, p. 6; on Sage, see "H. M. Sage is Dead; Ex-State Senator," *New York Times*, Sept. 26, 1933, 26.

[29] Chlotilde R. Martin, "Belfair – Designed by Artist Owner," *News and Courier*, Jan. 18, 1931, B9. For additional commentary on the Swains, see Chlotilde R. Martin, "Low-country Gossip," *News and Courier*, Jan. 23, 1938, 13. Murray showed similar praise for Jesse Metcalf and his wife in his sketch of their Hasty Point Plantation. See Murray, "Metcalf Builds on Peedee Bluff."

[30] For examples, see Martin, "Bluff Plantation Overlooks Cooper"; Martin, "New Yorker Reclaims Pimlico"; Martin, "The Hutton Estate of 16,000 Acres"; Martin, "Buckfield, Kress Narcissus Farm"; Murray, "Huntington House Reminds One of Africa"; Murray, "LaFayette House on Edisto Restored"; Murray, "The Wedge on the South Santee." For an example that Martin may have written but is not formally attributed to her, see "Frelinghuysen Builds at Rice Hope," *News and Courier*, Aug. 9, 1931, B12.

in the style of the original block had made it "doubly attractive." Emerson's attention to maintenance also struck Murray as fitting. "No structure in rural America is better maintained," Murray wrote. Workers "renewed" the "resplendent coat of white paint" when "the first blemish appears."[31] Murray made similar comments about the Shonnards' Harrietta. "In all the Santee region there is hardly an estate that can hold a candle to Harrietta," he opined. "It has everything that goes to make up for a perfect whole – a noble old house, now restored to its pristine splendor; an imposing avenue of oaks and magnolias ... [and] wide flung fields and virgin forests." Murray labeled the restoration "a remarkable thing." "The new has been harmoniously blended with the old in such a way," Murray wrote, "that the spectator is led to believe that he is looking at the house as it stood in the days of the wealthy rice planters."[32]

Other treatments also achieved satisfactory results. Even when speaking about new houses, Martin and Murray praised measures that created a sense of age. Describing the recently completed house at Henry W. Corning's estate in Beaufort County, Martin observed that the building appeared to have stood "for many years." The whitewashed brick exterior lent "a mellow air of age."[33] Both writers lauded new houses that paid homage to earlier structures. Describing Friendfield Plantation in August 1931, Murray credited its owner, Radcliffe Cheston of Philadelphia, with building a new house that captured the "atmosphere" of the original dwelling. Murray observed that Cheston had sought to "recreate as nearly as possible" the form and features of the original structure, an early nineteenth-century building that had burned several years earlier. The results showed the wisdom of Cheston's approach. "Unlike most new houses," Murray wrote, the mansion appeared "as though it is already a necessary part of its surroundings."[34] In one particularly florid account, Martin credited a northern sportsman with capturing the "spirit of the old south." Describing Z. Marshall Crane's Hope Plantation in Colleton County, Martin observed that Crane had incorporated an existing building – the "ugly hulk of a house which he

[31] Chalmers S. Murray, "Arcadia, Where LaFayette Stopped," *News and Courier*, June 21, 1931, B5.

[32] Chalmers S. Murray, "Harrietta, on Santee Blooms Again," *News and Courier*, Sept. 13, 1931, A7. For a similar account, see Chlotilde R. Martin, "Wappoolah, Berkeley Country Estate," *News and Courier*, March 22, 1931, B9.

[33] Chlotilde R. Martin, "Lowcountry Gossip," *News and Courier*, March 31, 1935, C10.

[34] Chalmers S. Murray, "Cheston Builds on Old Plantation," *News and Courier*, Aug. 2, 1931, B2.

found on his plantation" – into a new dwelling. The resulting structure stood two stories tall with a columnar façade. Martin viewed it as "not entirely new except as to appearance." She exclaimed: "Why this is one of those real, old plantation homes. Not a new imitation – the real thing!" In her view, Crane had "created beauty and charm and dignity where there were none."[35]

Consistent with their enthusiasm for age and authenticity, Martin and Murray wrote approvingly about plantations that seemed more a part of the past than the present. Murray depicted the Wedge Plantation on the South Santee River, for example, as having been "reclaimed by the past." The entire estate, he observed, "speaks of a different age." He credited the owner, E. G. Chadwick of New York, with having "re-captured the atmosphere of Colonial America."[36] Murray found similar conditions at Chicora Wood Plantation on the Pee Dee River. Visitors felt as though they had entered "another world," he explained. "Free from the rush of traffic," Chicora Wood and several neighboring estates spoke of "the days when wealthy rice planters lived in isolated splendor on their extensive estates."[37]

When Martin and Murray encountered new mansions, they became more circumspect but nonetheless spoke favorably about what they saw. In many cases, they supplied even greater detail than usual. Sketches of this kind provided well-developed portraits of northern-owned estates. Martin's sketch of Delta Plantation, an estate owned by New York stockbroker Hans K. Hudson, provides an instructive example. Martin characterized the estate as almost unparalleled in "magnificence." She described its major features sequentially, as a visitor would encounter them. A pair of "great gates" framed the entrance. A short distance beyond stood a stable, which Martin characterized as exuding "dignity and ornateness." Next came the servants' quarters, a building that Martin judged "something to be exclaimed at." To the left stood a grove of "beautiful oaks" and a "wide sweep of velvety green lawn." In the center of the grove stood Hudson's mansion, a structure "fashioned upon long spacious lines." Neoclassical features and a red brick exterior gave it a characteristically colonial appearance. According to Martin, the architect had modeled the design on "a very old and famous colonial

[35] Martin, "Crane House Reflects Old South."
[36] Murray, "The Wedge on the South Santee."
[37] Chalmers S. Murray, "Chicora Wood on Georgetown Peedee," *News and Courier*, Aug. 9, 1931, A5.

mansion." The surroundings enveloped the structure in a sea of green. Views of the lowcountry landscape vied with the building group for attention. Martin considered the vista looking toward the old rice fields along the river especially charming. A sunken garden near Hudson's house, she noted, looked "luxuriant and as green as though it had been there for years."[38]

Martin's portrayal of Delta Plantation is more than an example of a sketch brimming with descriptive detail. It also exemplifies an important quality of her and Murray's articles. Few lowcountry people saw the new sporting estates firsthand. The majority of those who did saw them as carpenters, tradesmen, and delivery drivers, not as members of the elite who recreated with owners. Even then, those who saw northern-owned estates likely caught glimpses of one or two, not many. The number of lowcountry people who saw more than a handful remained small, even as time passed. For most residents of the region, northerners' domains occupied an exclusive realm that lay close at hand yet removed from view, fully off limits.

Martin and Murray's sketches thus provided a kind of vicarious experience that informed while emphasizing exclusivity. Copious detail and vivid descriptions allowed readers to imagine seeing and visiting new estates for themselves. Carefully crafted portraits of remade plantations offered glimpses of a world that most readers could know only at a distance. Although a poor substitute for firsthand experience, Martin and Murray's articles nonetheless supplied the best available view of northerners' holdings. Their insights informed, captivated, and entertained. By relating information about elegant spaces and the people who owned them, Martin and Murray stimulated interest in the new estates and their role in a region undergoing dramatic change.

Martin and Murray's sketches established a model that the *News and Courier* relied upon throughout the 1930s. Each of their articles occupied about a third of a page. Most had one or two accompanying photographs. As Martin and Murray's reports appeared, the *News and Courier* began publishing similar features by other writers. Virtually all followed the same general approach. Most paid close attention to new owners and their backgrounds, to the physical environment of northerners' estates, and to notable historical events. These accounts related information about recent improvements and steps taken by new

---

[38] Martin, "The Delta – Home of H. K. Hudson."

owners to accommodate their needs. They also noted remnants of past eras and their place in the physical space of new estates. Consistently attentive to the influence of earlier epochs, *News and Courier* reporters eagerly seized on opportunities to mention linkages to the lowcountry of long ago.[39]

In addition to sketches of individual plantations, the *News and Courier* also published short reports on northerners' estates and surveys of northern-owned landholdings throughout the lowcountry. The former announced plantation sales, construction of new houses, and the arrival of notable guests. These articles tended to be brief, usually a few paragraphs or less, and rarely included photos. At the opposite end of the spectrum, surveys of estates and other holdings provided general overviews of the winter colonists' world. These articles ran half a page or more and featured multiple photos. Focused on broad trends, they provided perspectives not found in sketches of individual plantations.[40]

Although the *News and Courier* published news about northerners' estates year-round, the bulk of its reporting coincided with the height of sportsmen's and sportswomen's activities. Plantation sketches and other features appeared mainly in the fall and the early spring, at the beginning of hunting season and during the peak of garden season, respectively. Thus, the *News and Courier* focused attention on northerners' estates when in full operation, with northerners and their guests present. Overlap with the height of the tourist season in the spring, when major newspapers and magazines celebrated Charleston attractions, virtually ensured strong attention.

The volume, depth, and variety of the *News and Courier*'s reporting suggests the importance lowcountry people ascribed to northerners' activities. Not for decades had developments of comparable significance taken place. Lowcountry people saw the new estates as evidence of a sweeping

---

[39] See, for example, Susan Lowndes Allston, "Fairfield Plantation on the Waccamaw," *News and Courier*, Jan. 26, 1930, A3; Thomas R. Waring, Jr., "Do You Know Your Charleston? Mepkin Plantation," *News and Courier*, Jan. 24, 1938, 10; J. V. N., Jr., "Do You Know Your Lowcountry? Prospect Hill," *News and Courier*, May 9, 1939, 10.

[40] "Black Border Plantations Lure Wealthy Northerners," *News and Courier*, Apr. 21, 1929, A9; Murray, "Attracted by Climate Northerners Become Land Barons of Carolina Coast," *News and Courier*, Dec. 15, 1929, A3; Martin, "Low-Country Plantations Stir as Air Presages Coming of New Season"; Laura C. Hemingway, "First of Winter Colonists Arrive in Williamsburg for Season," *News and Courier*, Oct. 22, 1933, B2.

revival. New buildings, new gardens, and large acreages heralded large-scale rehabilitation. As northerners' activities transformed the country-side, the *News and Courier* emphasized the significance of the events taking place. By raising awareness of northerners' estates and casting them as central to the lowcountry, the newspaper identified the new "plantations" as indispensable to the region's developing image as a destination for outdoor recreation and heritage-inflected leisure.

Although the *News and Courier* sought mainly to inform readers about northerners' activities, it also aided sportsmen's and sportswomen's interests in several ways. First, it portrayed new estates largely as owners desired. By depicting sporting plantations as elegant country seats, the newspaper affirmed the vision of old plantations turned into settings for leisure and recreation. Consistent emphasis on architecture, landscaping, and scenic beauty accentuated the design elements that best reflected northerners' efforts. Extolling northerners' role in redeveloping old plantations sanctioned the general approach they took and encouraged further efforts of the kind. Moreover, by portraying northerners as central to a revived lowcountry, the newspaper encouraged readers to see them as permanent and necessary. Martin, Murray, and other writers cast sportsmen and sportswomen as vital to the region and its future.

Second, the *News and Courier*'s reporting discouraged unfavorable commentary on northerners and their activities. Although public criticism of northerners' estates is virtually absent from the historical record, privately, some lowcountry people voiced strong disapproval. Complaints about the loss of prime lands, preferred hunting domains, and cherished houses appeared in personal correspondence. Occasional rumblings about "Wall Street planters" expressed disdain for the newcomers. That few disapproving comments appeared in public discourse suggests the power of the *News and Courier*'s reporting. The favorable portrayals found in Martin and Murray's articles may have suppressed dissenting views. At a minimum, the newspaper showed acceptance of northerners and their estates and left counter-perspectives unrecognized.[41]

---

[41] See, for example, Chalmers S. Murray to Herbert R. Sass, n.d. [Jan. 1934?], folder 14, box 101, Herbert Ravenel Sass Papers, SCHS; Yuhl, *Golden Haze of Memory*, 182–183; Jennifer Betsworth, "Reviving and Restoring Southern Ruins: Reshaping Plantation Architecture and Landscapes in Georgetown County, South Carolina," in *Leisure, Plantations, and the Making of a New South: The Sporting Plantations of the South Carolina Lowcountry and Red Hills Region, 1900–1940*, ed. Julia Brock and Daniel Vivian (Lanham, MD: Lexington Books, 2015), 72–75.

Third, portraying the new estates as rooted in plantations of earlier periods subtly aided northerners' aims. Although sportsmen and sportswomen did not actively seek to align themselves with the old planter class, neither did they resist the comparison. In fact, northerners tacitly accepted the parallels that others identified, eager for whatever status might be gained. Comparisons to the aristocratic planters of the antebellum era supplied potent indicators of belonging to a true elite. Estate owners looked upon the *News and Courier*'s portrayals with favor, eager for the authority they connoted.

Ultimately, the *News and Courier*'s reporting shaped perspectives on new estates and historical plantations. Articles by Martin, Murray, and other writers described both selectively, in ways that ignored much of what plantations had been historically and what some had become. By portraying plantations principally as sites of elite residency, the newspaper obscured slavery, labor, and agricultural enterprise. By ignoring the hardships that slaves had endured and the still-marginalized status of African Americans, the *News and Courier* ignored the tragic outcome of Reconstruction and contemporary inequalities. By leaving the profound differences between northerners' estates and their predecessors unmentioned, the newspaper misreported the origins of the new estates and their relationship to earlier plantations. In sum, the *News and Courier* fundamentally distorted the historical use, purpose, and experience of plantations. Its portrayals cast an imagined era of grandeur and racial harmony as a seemingly "natural" precursor to northerners' activities. Although plantations remained part of the lowcountry landscape, dozens had become "plantations" of an entirely new kind. Yet by viewing recent developments through a lens shaped by myth and historical amnesia, the newspaper misread the changes taking place.

GRACIOUS TRADITIONS, MODERN LIVES

As Martin and Murray's articles provided snapshot views of northerners' estates, other writers viewed sportsmen's and sportswomen's activities somewhat differently. Magazines such as *Country Life*, *House Beautiful*, *Home and Field*, *Town and Country*, and *Life* offered intimate views of select estates. Articles in these magazines echoed Martin and Murray's portrayals while offering new and novel interpretations. Like the *News and Courier* series, they praised northerners' care of old plantations and efforts to return long-distressed houses to their former grandeur. Yet they also cast northerners as attaining a style of life accessible only to a privileged few. Made possible by more than mere wealth, that lifestyle displayed taste, cultivated knowledge, and appreciation of finery from past eras and the present.

In December 1931, a six-page feature in *House Beautiful* chronicled the rehabilitation of Harrietta Plantation, a Santee River estate owned by Horatio and Sophie Shonnard of New York City. The author, Sara Furman, praised the Shonnards' efforts. "Harrietta," she observed, "stands to-day as a perfect and beautiful example of the domestic architecture and decoration of the late eighteenth century." The Shonnards, she explained, had restored the house to its eighteenth-century form with "sympathetic judgment and taste." "Perfection of architectural detail is everywhere apparent," Furman added. Inside, the decor exemplified the "beauty of design associated with the names of the Adam brothers, Chippendale, and Sheraton." In Furman's view, the owners had shown "rare" taste. The results qualified as the "finest local adaptation of the beauty of design."[42]

Furman's profile continued in a manner that would soon become common. She moved systematically through the house, describing its principal rooms and features and paying close attention to architecture, aesthetics, and decor. Throughout the account, Furman commingled references to regional design traditions with praise for the Shonnards' taste. She described the architecture of the house as "delicate and beautiful." The stairways and piazzas, she noted, expressed "the graciousness of the South." A number of antique furnishings supplied special charm. Some had belonged to Harrietta's "original owners," Furman noted. Meanwhile, she considered the "crystal chandeliers and fixtures in the dining- and drawing-rooms" and two "Aubusson carpets" especially beautiful. "Connoisseurs," Furman opined, would find them worthy of careful study. So, too, with "eight or ten" original Adam-style mantles that Furman judged to be "among the best examples in the South."[43]

Furman praised the Shonnards' design sensibilities and inspired use of local influences. "Antique mirrors and family portraits," she noted, supplied the main rooms with "proper dignity." The library, she added, housed a "fine collection of old books." Furman found Sophie Shonnard's decision to use local flowers as inspiration for the decor of the bedrooms "delightful." The "Magnolia Room," Furman observed, "is all white and yellow and old wood." The woodwork displayed "the patina

---

[42] Sara Furman, "Harrietta: An Old Plantation House on the Santee River," *House Beautiful* 70, no. 6 (Dec. 1931), 475–480.

[43] Ibid., 477.

that comes only from many years of gentle care." Draperies with magnolia blossoms complemented mahogany furnishings. Several mirrors
caught and extended the "mellow sunshine" that flooded the room during
the morning hours. The "Wild-Flower Room" had a different personality.
Wallpaper reproduced from a design originally used in the house featured
flower bouquets of light blue and rose on a blue background. A black rug
and rose-colored chintz curtains provided rich contrasts. Another bedroom took the Cherokee rose as its namesake. Oyster-white walls and
curtains "with pale blue blossoms trailing over a brown background"
produced a "pleasing color combination."[44]

As Furman lauded the decor, she also praised the spirit felt throughout
the house. "The restoration of life to an old house is indeed a joy
comparable to nothing else," she wrote. The Shonnards' restoration had
supplied "a feeling of satisfied gratitude," she added. Harrietta, Furman
wrote, possessed none of the "self-consciousness" sometimes found with
"relics" placed on display. To the contrary, it brimmed with vitality. In
Furman's view, Harrietta exuded the "charm" the South had possessed
during "the height of its past glory."[45]

Furman did not ignore Harrietta's slaveholding past, but neither did
she consider it particularly important. Rather, she viewed slavery simply
as a part of history and a source of the lowcountry's charm. Instead of
seeing remnants of the institution as troubling or disturbing, she considered them picturesque and enchanting. Speaking of lowcountry blacks,
Furman wrote, "there is probably no place in the country where the race
is as pure or its expression as spontaneous." The region's "Gullah"
population, in her view, gave plantations such as Harrietta much of their
"old-time atmosphere." According to Furman, visitors could not help but
be captivated by the "primitive singing and dancing" of the "colored
people." Furman especially marveled at Sophie Shonnard's rapport with
African Americans. Harrietta's "mistress," Furman noted, took "great
interest in the life of the 'quarters.'" "Having been brought up in the
southern tradition," Shonnard seemed "to live and entertain in the
manner of the romantic past."[46]

Of course, Furman left a great deal unstated. Her paternalistic description of Harrietta's African American population went hand in hand with
inattention to the realities of life under slavery. Furman made no mention
of Harrietta's original purposes or the conditions that had historically

[44] Ibid., 477–480.   [45] Ibid., 477, 480.   [46] Ibid., 475, 477.

characterized Santee River plantations. As one of the largest, most productive plantations in the area, Harrietta had undoubtedly been a site of unrelenting toil and trauma for hundreds of enslaved men, women, and children. When Furman depicted Harrietta as a sort of wonderland – an elegant retreat steeped in the charm of time – she presented a deeply distorted view of the past. Rather than offering anything close to a complete account, Furman provided a biased, highly selective glimpse that overlooked at least as much as it conveyed.

Furman's account established a model that would be replicated by other writers. As interest in northerners' estates grew, other magazines turned their attention to the lowcountry. Most catered to middle- and upper-class readers with interests in architecture and interior design. *Country Life, House Beautiful, Town and Country,* and *Home and Field* fit this category. Eventually, major news magazines also devoted attention to the new estates. During the 1930s, *Fortune* and *Life* published articles that emphasized the merging of old traditions with new comforts and conveniences.

Reporting on northerners' estates proved remarkably consistent. Celebration of northerners' role in renewing old plantations became a dominant theme. *Town and Country* credited Clarence and Adelaide Chapman with "a very intelligent and cautious restoration" of the house at Mulberry Plantation.[47] *Country Life* heralded a "renaissance" of lowcountry plantations. "Scores of [old plantations] have been bought and rejuvenated by non-residents, most of them Northerners," the magazine announced. If the trend continued, the road "from Georgetown to Beaufort, something more than a hundred miles," would again be "famous for its wealth and its beautiful homes."[48]

Along with the theme of renewal, several articles emphasized the lingering influence of the past. One feature in *Country Life* summarized the effects of northerners' activities by declaring, "Memories of the Old South ... live again." Alongside hand-drawn sketches of elegant mansions surrounded by lush foliage, the magazine claimed, "the art of gracious living reached its apogee in this country on southern plantations in the spacious days before the Civil War." Now, "more than half a century later," the "long departed glory of those days" had returned. At old plantations restored to their former grandeur, out-of-state residents

---

[47] "South Carolina Plantations," *Town and Country,* Jan. 15, 1935, 30.
[48] James C. Derieux, "The Renaissance of the Plantation," *Country Life* 61, no. 3 (Jan. 1932): 35, 37.

served up "openhanded hospitality" to a continually changing cast of guests. For *Country Life*, southern plantations had again become places of elegance.[49]

In the eyes of many observers, the continuing influence of southern traditions infused life in the present with rare qualities. Elizabeth Lounsbery described Gayer G. Dominick's estate on Bulls Island as "a world of romance and adventure." Dominick's estate occupied about 5,000 acres. Lounsbery portrayed it as providing seclusion, serenity, and a lush landscape. Forests teeming with wildlife provided ample opportunities for hunting and outdoor recreation. The main dwelling, a handsome structure "carried out along Colonial lines," gave Dominick and his wife a prime venue for intimate entertaining. Lounsbery viewed the estate as "an environment that offers complete relaxation surrounded by all the elements of social enjoyment."[50]

Whatever northerners' estates gained from old traditions, they also possessed modern comforts and conveniences. One feature after another highlighted amenities that reflected owners' insistence on contemporary standards. Well-appointed kitchens and bathrooms, closets, and sturdy, attractive furnishings proved common. "Gun rooms" supplied a clear indication of the reasons northerners came to the lowcountry. At Dominick's Bulls Island estate, a room lined with gun cases, "racks for fishing rods," "old maps," and taxidermied animal heads opened to a porch in the rear of the house. An adjacent "cold storage room" allowed for game to be "properly hung and cared for."[51] At Richmond Plantation, George Ellis, Jr., erected a kennel with sleeping quarters and grooming facilities for his "Springer Spaniels and bird dogs," a room filled with "trophies and records," and a kitchen where workers prepared meals for the canines.[52]

By the late 1930s, the themes that Furman had first identified became a consistent part of reporting on northerners' estates. A photo essay published in the December 25, 1939, issue of *Life* offers an apt illustration. Alongside images of northern-owned "plantations," Charleston streetscapes, and lush vistas, a brief essay recounted the history of the Carolina coast and its rise as a destination for tourism and outdoor

---

[49] "Memories of the Old South," *Country Life* 57, no. 3 (Jan. 1930): 60–61.

[50] Elizabeth Lounsbery, "Home for the Hunter," *House and Field*, Dec. 1932, 60–61, 69.

[51] Lounsbery, "Home for the Hunter," 69. See also "Do You Know Your Charleston? Dominick Estate," *News and Courier*, Dec. 5, 1932, 10. On the gun room at Harrietta Plantation, see Furman, "Harrietta," 479.

[52] "Shooting Lodge in Carolina: 'Richmond,'" *Country Life* 65, no. 2 (Dec. 1933): 61.

recreation. Before the Civil War, *Life* noted, crops such as "indigo, rice and long staple cotton" had supplied the region with great wealth. Now, however, the lowcountry economy had different underpinnings. "Today the big cash crop is rich Northerners," *Life* explained. These men and women had "come South and bought the plantations where Carolina aristocrats once lived and ruled." The "'Yankees,'" *Life* continued, come "to hunt, to relax and to enjoy the feudal feeling of property which owning thousands of plantation acres gives them." Accompanying photographs depicted shimmering rice fields, an elegant plantation house, brick slave cabins, and a group of hunters riding on horseback. In words and images, *Life* depicted northerners' estates as places of leisure, beauty, and elegance. Moreover, the magazine emphasized the lingering influence of the past. "Many historians," *Life* noted, "believe that the plantation culture of the Old South reached its apogee in the Low Country." Although aware of manifold "inequalities," *Life* contended that antebellum southerners had enjoyed a rarified existence. "Life was dignified and gracious and satisfying," it noted.[53]

Portrayals of northerners' estates demonstrated journalists' eagerness to celebrate a romantic history. By downplaying contrasts with historical plantations, writers such as Furman and Lounsbery emphasized the resurgence of old traditions. Their reading of the changes taking place owed as much to the mythology of the Old South as contemporary practices. Only by ignoring the harsh realities of slavery and its aftermath could they and other journalists see strong parallels between old and new. Only by ignoring the historical purpose of plantations could they view northerners' estates as roughly similar. Yet by establishing an interpretive framework for sportsmen's and sportswomen's actions, Furman, Lounsbery, and other writers ascribed meaning to the changes taking place. By informing interested onlookers about the processes under way, they publicized northerners' activities and the role of new estates in a resurgent lowcountry. For select portions of the magazine-reading public, their writings shaped beliefs about plantations of the past and the present. These same portrayals also played a crucial role in turning old plantations into venerable landmarks and new estates into reincarnations of a noble past. Newspaper and magazine reporting blended old and new seamlessly, erasing differences across time and space. Behind depictions of grand houses and sumptuous decor, plantations ceased to be sites

---

[53] "The Low Country," *Life*, Dec. 25, 1939, 38–45.

of suffering and hardship. Instead, they became elegant estates, now and long ago.

The estate-making boom coincided with new attention to the role of plantations in lowcountry history. Recognition of select plantations as historically valuable raised questions about the significance of planta-tions across time and the importance of examples with colonial and antebellum origins. During the 1930s, new narratives of lowcountry history placed plantations at the center of the region's story. Herbert Ravenel Sass's 1931 essay "The Low Country" and the introductory essay to *Plantations of the Carolina Low-Country*, an oversized volume produced by Samuel Gaillard Stoney, Jr., Albert Simons, and Samuel Lapham in conjunction with photographers Frances Benjamin Johnston and Ben Jehudah Lubschez, proved especially influential. Both publica-tions reached large audiences, won critical acclaim, and informed popular and scholarly accounts. As the lowcountry became a focus of national attention, Sass's essay became a semi-official history of the region. *Planta-tions* became the authoritative text on the region's most revered plantations. Together, these publications shaped popular views of the lowcountry and its past.[54]

Historians have failed to give Sass's essay and *Plantations* the attention they deserve. Sass is one of the lesser-known figures of the Charleston Renaissance. Although noted for his involvement in organizations such as the Society for the Preservation of Negro Spirituals (SPNS) and his collaborations with his cousin, artist Alice Ravenel Huger Smith, his role as the leading historian of the lowcountry during the 1930s, 1940s, and 1950s has yet to be appreciated. As for *Plantations*, Gene Waddell has recognized that its authors intended the book to be published as part of the short-lived Octagon Library of Early American Architecture, but its content and influence have yet to be examined. Its role in highlighting

---

[54] Herbert Ravenel Sass, "The Low-Country," in Augustine T. Smythe et al., *The Carolina Low-Country* (New York: Macmillan Co., 1931): 3–29; Samuel Gaillard Stoney, *Planta-tions of the Carolina Low Country* (Charleston: Carolina Art Association, 1938). For critical assessments of *The Carolina Low-Country*, see "Carolina Low-Country" and "Costal Civilization's Story Told Graphically," *News and Courier*, Nov. 22, 1931, B1 and B7, respectively; Charles McD. Puckette, "Charleston and the Carolina Low Coun-try," *New York Times Book Review*, Dec. 13, 1931, 4.

a small number of plantations and shaping popular views of the past has yet to be understood.[55]

Sass's essay and *Plantations* are products of their time. Both reflect the sensibilities, prejudices, and concerns of the moment. Neither viewed the lowcountry critically. Both are rife with romance, nostalgia, and selective memory. Eager to portray the antebellum era as a halcyon age, they offer nostalgic accounts of a planter class that lived as cultured aristocrats and treated slaves with benevolent care. Viewed in retrospect, Sass's essay and *Plantations* are immediately recognizable as products of mythmaking and historical amnesia. Their influence among contemporaries, however, should not be overlooked. In an era when most Americans recalled slavery as a civilizing institution, Sass's essay and *Plantations* affirmed well-established myths. Both celebrated the lowcountry past while identifying surviving remains as evidence of an era of lost grandeur. Both presented histories that echoed the mythology of the Old South while drawing important contrasts. Both portrayed the lowcountry as exceptional yet representative of "the South." Sass's essay and *Plantations* set the region at the center of American history at a time of surging interest in the southern states.

Sass's essay appeared in *The Carolina Low-Country*, a collection prepared by the SPNS and published by the prestigious Macmillan Company of New York. The SPNS grew out of concerns that the traditional music of African American plantation laborers might be lost as the last generation born into slavery passed on. Founded in Charleston in 1922, the SPNS dedicated itself to collecting, transcribing, and recording African American spirituals. Its membership consisted mainly of descendants of slaveholders, most of them born after Reconstruction. Throughout the 1920s and the 1930s, the SPNS collected and recorded spirituals and performed for audiences in Georgia and South Carolina. Tours in 1929, 1930, and 1935 took the SPNS to northern cities. SPNS performances typically involved fifteen to twenty members dressed in costume singing spirituals in Gullah. Members sang, shouted in the tradition of lowcountry blacks, and commented on the origin and meaning of spirituals and the society that had produced them. SPNS members viewed spirituals as a living link to a nearly vanished era of racial harmony and social stability.

---

[55] Yuhl, *Golden Haze of Memory*, 64, 69, 71–72, 133, 154; Gene Waddell, "The Only Volume in the Octagon Library: *The Early Architecture of Charleston*," in *Renaissance in Charleston: Art and Life in the Carolina Low Country, 1900–1940*, ed. James M. Hutchisson and Harlan Greene (Athens: University of Georgia Press, 2003), 124.

As a remnant of the plantation culture of the antebellum era, spirituals seemed to demonstrate the religiosity and "primitiveness" of lowcountry blacks and faithful bonds between slaves and masters.[56]

In the late 1920s, the SPNS decided to publish a collection of spirituals. Members initially envisioned a book that would publicize their activities and share some of the music they had collected. The project quickly grew in scale, however, and before long, members decided to include poetry, essays, and illustrations. The group's efforts ultimately produced *The Carolina Low-Country*, a mammoth celebration of regional history and culture. Published in 1931, the book presents a thoroughly nostalgic view of the antebellum era. A powerful statement on lowcountry heritage, *The Carolina Low-Country* laid the foundation for similar portrayals that appeared during the 1930s.[57]

Sass's essay occupies a crucial place in the collection. As the first of eight essays, it provides a general overview of lowcountry history and outlines major themes. Sass cast the story of the region as a grand drama, replete with elements of tragedy, triumph, and intrigue. Like his collaborators, Sass viewed the region's past as charming, captivating, and deeply appealing. He viewed it as demonstrating the grandeur and importance of the lowcountry.

Sass began his history by recounting the region's distant origins. The lowcountry began, he explained, with the coming of "Spaniards, Frenchmen and buckskin-clad hunters, pirates and privateersmen." Later came "redcoats marching through the streets of Charleston [and] mysterious cavalry riding along secret paths in the swamps at night." "The Negro," Sass noted, arrived early, "in increasing numbers through many years." He brought "an unexpected gift, a gift of rhythm," Sass observed, apparently in reference to African Americans' renowned love of song and dance. "By the sweat of [the Negro's] brow the great plantations" took shape. "The fruit of the white man's brain and the black man's sinews" created "thousands of acres of waving rice, of indigo and sea island cotton." From these fields flowed great wealth. With development of a plantation economy, the "Golden Age of the Low-Country" began.[58]

Sass depicted that "Golden Age" as synonymous with the colonial and antebellum eras. He viewed it as an era of prosperity, social harmony, and unparalleled accomplishment. "For many decades," Sass wrote, "a plantation civilization" flourished. In status and sophistication, it matched

[56] Yuhl, *Golden Haze of Memory*, chap. 4.   [57] Ibid., 154.
[58] Sass, "The Low-Country," 4–5.

those of Tidewater Virginia and the lower Mississippi Valley. "This was the civilization of the great Low-Country rice and cotton plantations," he explained. Not only did it produce great wealth, it created "a society probably as distinctive as any that America has produced." According to Sass, it produced "men and women cast in a very fine mould; it contributed many honorable names to American history; it achieved great feats of agriculture and of agricultural engineering" and "built houses whose dignity is not often equaled today." Ultimately, "it gave to the land itself character, atmosphere – almost a soul." Then, "suddenly, tragically, it perished."[59]

Sass viewed the Civil War as the great divide in lowcountry history. "When General Sherman's victorious Army of the West ... turned northward into South Carolina," the region's "Golden Age" swiftly ended. "The old happy life of the plantations was over," Sass explained. "The tramp-tramp-tramp of Sherman's infantry ... sounded the death-knell" for the lowcountry's plantation "civilization."[60]

Sass said little about the aftermath of the war or Reconstruction, an era that most Americans viewed as a misguided experiment gone terribly wrong. Intent on avoiding controversy, he glossed over the decades after Appomattox. Picking the story back up in his time, Sass emphasized surviving vestiges of the "Golden Era." Visitors to the lowcountry, he claimed, could "not go anywhere ... without feeling the glamour of its rich past." "Everywhere and on every hand are the visible signs of an old order," Sass wrote. "The old plantation houses, the old parish churches ... the old gardens, the stately avenues of hoary live oaks, the wide rice fields abandoned now ... all these whisper tales of great days that once were lived here. All these are memorials, monuments, of that Golden Age." As touchstones of "memories and visions," these and other remains called "historic associations" to mind.[61]

Sass's emphasis on tangible remains rooted his narrative in places and spaces that readers could visit for themselves or at least imagine as they read. The "old houses," the "old rice fields," the "old reserves," "ruins of fine columned mansions" – Sass saw these as underscoring the glamour of the region's history. A countryside "abounding in the landmarks of a vanished order" supplied evidence of the "great feats" and "dignity" of the "older civilization." That so many remnants of earlier periods survived showed how little the lowcountry had changed. According to Sass,

[59] Ibid., 3, 5.   [60] Ibid., 3, 28–29.   [61] Ibid., 3–4, 6.

the transformative effects of "modern industrialism" had passed the region by. Although Sass admitted "there are many other regions of America as rich in history and legend," he believed that few retained more than the "old tales." Sass portrayed the lowcountry as a land apart, virtually unmatched for historical ambiance.[62]

Plantations figured at the center of the region's history. "In the old days the plantations *were* the Low-Country," Sass declared. "Even now," he added, "one can not think of the Low-Country, one can not describe its physical features, except in terms of the plantations." According to Sass, plantations had "shaped the very face of the land itself. They changed marshes into rice fields, they converted swamps into lagoons, they created groves and stately avenues of live oaks and magnolias, [and] they made gardens," some of them "beautiful and famous today." Moreover, they had "moulded the traditions, the manners and customs of the people." Sass portrayed plantations as the essence of the lowcountry's past, the region's most prominent features, and the wellspring of its distinctive way of life.[63]

Even with the "Golden Age" long passed, plantations remained a powerful influence. "The plantations," Sass wrote, "are still in a real sense the Low-Country." Spread across the region "from end to end," he believed they had "left their mark." Although Sass noted that "many of the old houses have vanished," those that remained, he insisted, told the story of the lowcountry of long ago. They conveyed the grandeur of the "great plantation civilization" and its legacy. For Sass, plantations "helped to make the Low-Country what it is."[64]

Sass's account turned the myth of a genteel slave society into a regionally specific narrative of origins and development. The lowcountry, in Sass's view, did not stand apart from the South but exemplified its finest qualities. Like Tidewater Virginia and the lower Mississippi Valley, the lowcountry had exerted a powerful influence on the South and its history. As interest in the southern states swelled and Americans took note of contrasts among them, Sass's essay situated the lowcountry in relation to the larger whole.[65]

Sass's essay had immediate influence. *The Carolina Low-Country* sparked enthusiasm throughout the lowcountry and won critical acclaim. A review published in the *News and Courier* offered unrestrained praise.

[62] Ibid., 5–6, 8, 28–29.　　[63] Ibid., 14.　　[64] Ibid., 3, 5–16.
[65] Francis Pendleton Gaines, *The Southern Plantation: A Study in the Development and the Accuracy of a Tradition* (New York: Columbia University Press, 1924), 143–144.

Calling Sass's essay "a beautiful and highly emotional description" of the Carolina coast, it touted the "emotional effect" produced by Sass's "use of names," particularly the "long lists of plantations" he included. These, the newspaper observed, "spell romance to all of us."[66] Writing in the *New York Times Review of Books*, Charles Puckett offered similar praise. Puckett, a *Times* staffer with familial ties to Edisto Island, called Sass's essay a "picturesque description of the Low Country and its rice-and-cotton-plantation backgrounds." Seeing it as the most important essay in the volume, Puckett devoted about a third of the review to summarizing Sass's history. He related the narrative uncritically, taking Sass's claims without question and adding embellishments of his own. Puckett extolled the region's "rare, distinctive charm" and the lasting "imprint" of a "precious civilization." "Even the most critical observers of Southern plantation life," he wrote, recognized that "a manner of living which had gentleness, simplicity, beauty and integrity" had once existed. Although the South of yesteryear had inspired "much romantic nonsense," Puckett placed the lowcountry in a different category. "Here was a genuine dignity and grace," Puckett insisted.[67]

Additional influence came in the form of emulation. Sass's narrative quickly became a basis for sketches of the lowcountry and informed several more ambitious accounts. By the mid-1930s, it became the best-known history of the region. Following the success of *The Carolina Low-Country*, Sass continued to concentrate on historical writing. In 1931 he published his first novel, *Look Back to Glory*, a drama set in the lowcountry during the Civil War. Sass also continued to develop his narrative of lowcountry history. He published new, somewhat different versions of it in popular magazines and other forums. The most famous appeared in 1936 as the historical essay in Alice Ravenel Huger Smith's *A Carolina Rice Plantation of the Fifties*. A collection of thirty watercolors depicting the "Golden Era" of rice culture, Smith's book is among the best-known publications of the Charleston Renaissance. Smith's paintings depict a serene world of racial harmony and pastoral beauty. The prose portion of the book consists of Sass's essay, "The Rice Coast: Its Story and Meaning," and "A Plantation Boyhood," a memoir by Smith's father, Daniel Elliot Huger Smith. In combination with Smith's art, both essays offer a nostalgic, thoroughly paternalistic view of the lowcountry of the late

---

[66] "The Carolina Low-Country," *News and Courier*, Nov. 22, 1931, B1.
[67] Puckette, "Charleston and the Carolina Low Country."

antebellum era. As Stephanie E. Yuhl has written, the book sought to create "a visual record of the historical importance and beauty of an extinct culture."[68]

In 1939, three years after the publication of Smith's volume, the Carolina Art Society released *Plantations of the Carolina Lowcountry*. A decade in the making, this book fused a celebratory narrative of regional history with documentary photographs, measured drawings, and historical sketches of plantation houses. The author, Samuel Gaillard Stoney, Jr., and his two editors and collaborators, Albert Simons and Samuel Lapham, descended from patrician families. As members of the generation that dedicated itself to memorializing their ancestors, the trio looked back on the antebellum era in roughly the same way as Sass did. As devotees of lowcountry history and architecture, they saw the preservation of colonial and antebellum-era buildings as paramount. In *Plantations*, Stoney, Simons, and Lapham highlighted the "dignity and beauty" of early houses. Preparation of the volume took place as ties to the antebellum era grew increasingly tenuous. The group worked earnestly to record houses they considered important, well aware that further decay or a strong storm might cost them their chance. As Simons's wife, Harriet Porcher Simons, related:

Sunday after Sunday saw two cars full of men, women and children set out on our exploring and measuring parties. We all worked – the men with grandiloquent gestures and machêtes clearing underbrush grown second story high in the ruins; the women pressing their soon-toughened thumbs against the bricks to hold the end of the tape lines; the children clearing the trash and débris so that buried corners and steps might be found. Houses still standing and houses only a pile of earthquake-shaken bricks, houses approached by cement highways and houses which had to be reached by picking our way through briars and rattlesnakes – all were carefully studied and their floor plans and gardens brought to life again.[69]

Plantations is a handsome, oversized volume printed on glossy stock. Two-hundred and forty-five pages long, it brims with illustrations. Johnston and Lubschez contributed 146 photographs, 86 of which appear as full-page images. Alongside these are more than 50 floorplans and drawings. Augustine T. S. Stoney, one of Samuel Gaillard Stoney,

---

[68] Herbert Ravenel Sass, *Look Back to Glory* (Indianapolis: Bobbs-Merrill Co., 1933); Alice Ravenel Huger Smith, Herbert Ravenel Sass, and Daniel Elliot Huger Smith, *A Carolina Rice Plantation of the Fifties* (New York: W. Morrow and Co., 1936); Yuhl, *Golden Haze of Memory*, 69–73 (quotation on 69).

[69] Stoney, *Plantations of the Carolina Low Country*, 7 (first quotation), 9 (second quotation).

Jr.'s brothers, drew three maps for the book, including one showing the locations of well-known plantations around Charleston that graces the endpapers. The book is divided into three sections. The first, "The Country and the People," is a thirty-four-page history of the lowcountry. The others, "The Colonial and Provincial Period" and "The Republican Period," group plantation dwellings by historical eras and associated architectural trends. Plantations are identified by name and described in short historical sketches. Like other historical writing of the era, the focus rests squarely on historical owners and their houses. Rather than acknowledging slavery and commercial agriculture, *Plantations* instead depicts the country residences of social and economic elites.[70]

"The Country and the People" shares strong similarities with Sass's historical essays. It employs a similar chronology and is equally insistent on the lowcountry's importance. Like Sass, Stoney began his narrative with the earliest stages of European settlement. He noted the French attempt to found a colony near the town of Port Royal, territorial struggles between the French and the Spanish, and the Lords Proprietors' plans for the colony of Carolina. He discussed the founding of Charles-Town, early efforts to attract colonists, conflicts with Native Americans, and the beginnings of rice culture. Yet, unlike his counterpart, Stoney gave greater attention to colonial politics, economic activities, and society. Whereas Sass favored sweeping claims over narrative detail, Stoney delved deeply into a wide range of topics. He discussed, for example, the arrival of settlers from the West Indies, the influx of Huguenot settlers after 1685, the establishment of the Anglican Church, the rise of the "Goose Creek Men" and their politics, the transition from proprietary to royal rule, and the development of early plantations. Stoney noted the importation of "crews of negroes" and slavery's role in creating plantations. He discussed topography, climate, and natural features and their influence on trade, settlement, and communication. Stoney also mentioned the development of a colonial government and the colony's role

---

[70] Stoney, *Plantations of the Carolina Low Country*. A somewhat similar study on the plantations of a particular parish is John R. Todd and Francis M. Huston, *Prince Williams Parish and Plantations* (Richmond: Garrett and Massie, 1935). On Johnson's life and career and her work in the South, see Bettina Berch, *The Woman behind the Lens: The Life and Work of Frances Benjamin Johnston, 1864–1952* (Charlottesville: University Press of Virginia, 2000), esp. chap. 6; Mary N. Woods, *Beyond the Architect's Eye: Photographs and the American Built Environment* (Philadelphia: University of Pennsylvania Press, 2009), chap. 2. On Lubschez, see Benjamin J. Lubschez Membership File, American Institute of Architects Archives, Washington, DC.

in a growing contest for power in the New World.[71] In short, Stoney's readers learned more about South Carolina's early history than Sass's did. Although Stoney did not strive to provide a comprehensive portrait of the state's beginnings, he related more information than his counterpart.

Stoney continued his narrative by surveying the social and economic development of the lowcountry and its rise within the British Empire. He discussed the growth of rice production, the advent of crops such as cotton and indigo, and the rise of a planter class. In tracing these developments, Stoney relied heavily on David Duncan Wallace's four-volume *The History of South Carolina*, then the most authoritative history of the state available, and Edward McCrady's histories of colonial and Revolutionary Era South Carolina.[72]

Stoney organized his narrative around several themes. After chronicling Carolina's early history, he turned to the built environment and its role in the colony's development. Limited sources, Stoney noted, made it difficult to say much about the beginnings of Charles-Town or early plantation houses. He recounted the comments of a visitor who saw Charles-Town in 1682 and characterized the settlement as having about a hundred houses, most of them built of wood. As for plantation dwellings, the record offered even less. In lieu of factual information, Stoney offed plausible suppositions and conjecture. He suggested, for example, that the earliest settlers had occupied structures built of "poles, brush and sod." He saw "the ubiquitous log-cabin of the Low Country," a prevalent regional type, as hinting at the form and features of early colonists' buildings. According to Stoney, its simple construction, "foundations ... of live oak or light wood," and "chimneys made of sticks set up pig-pen fashion and daubed over" likely echoed early houses. As for extant structures, Stoney discussed the houses at Medway, Middleburg, and Mulberry plantations, which he saw as markers of progress during the early stages of colonization. The "fairly capacious dwelling" at Medway, he argued, showed significant development a mere sixteen years after the colony's founding. Its "pretence to esthetics in style" and setting showed progress beyond the less-developed houses of earlier years. The house at

---

[71] Stoney, *Plantations of the Carolina Low Country*, 11–25.

[72] Ibid.; David Duncan Wallace, *The History of South Carolina*, 4 vols. (New York: American Historical Society, 1934); Edward McCrady, *The History of South Carolina under the Proprietary Government, 1670–1719* (New York: Macmillan Co., 1897); Edward McCrady, *The History of South Carolina under the Royal Government, 1719–1776* (New York: Macmillan Co., 1901); Edward McCrady, *The History of South Carolina in the Revolution, 1780–1783* (New York: Macmillan Co., 1902).

Middleburg, the lowcountry's oldest surviving frame building, showed "another jump of years and development." Instead of "a European dwelling set up on the middle of an American wilderness," it showed "considerable understanding of the needs of the climate" in "its shape, its planning, and its materials." Meanwhile, the dwelling at Mulberry showed "how in a half century the Low Country had 'arrived.'" Noting its sturdy construction and distinctive design, Stoney considered the house "very much developed." In his view, it evidenced the stability and order the colony achieved by the eve of the Yemassee War.[73]

Family ties among the developing planter elite provided another theme. Stoney characterized the lowcountry's early society as an amalgam of "West Indians, Huguenots, and Dissenters." Intermarriages proved common. An "aristocracy of birth" developed among families "who had arrived at political or commercial prominence and wealth." Stoney also saw marriages between the daughters of Carolina planters and "younger sons of the English nobility" as indicating "the rise of the Low Country in imperial estimation." Although he only identified one such union, he spoke of them as a growing trend. In describing colonial-era society, Stoney concentrated almost exclusively on elites. People outside the planter and merchant classes rarely received mention. Slaves and whites of lesser status played minor roles in his account.[74]

Plantations supplied the third theme. Like Sass, Stoney placed plantations at the center of lowcountry history. He viewed them as the basis of the region's historical wealth and power, its social order, and its overall importance. Moreover, Stoney saw them as virtually inevitable. Using dramatic language, Stoney depicted the "coastal region of South Carolina as foreordained to plantations." "The lay of the land, its sorts, its climate, even the way its tidal rivers run" suited plantations "peculiarly." According to Stoney, early settlers came to the region with "plans and ambitions" for "a spacious system." Their success in developing plantations showed the wisdom of their vision. Subsequent growth and development supplied additional confirmation. In Stoney's view, the success of rice culture demonstrated the suitability of the Carolina coast to large-scale plantation agriculture. With an unabashedly circular argument, Stoney claimed a natural relationship between the lowcountry and

---

[73] Stoney, *Plantations of the Carolina Low Country*, 23–24 (first through sixth quotations on 23; seventh through eleventh on 24).

[74] Ibid., 25 and passim.

plantation slavery and thus legitimated the regime that had made his ancestors wealthy at the expense of African lives.[75]

Like Sass, Stoney saw the Civil War as the central drama in lowcountry history. The struggle over slavery reduced the planter class to penury, left Charleston in ruins, and destroyed plantations throughout the region. Union raids and the early occupation of Port Royal turned the Sea Islands into "a sort of No-Man's-Land." Emphasizing architectural losses, Stoney noted that houses such as William Seabrook's on Edisto Island experienced "strange adventures" and "curious dilapidations at the hands of foraging parties." Worse still, Union General William Tecumseh Sherman's march northward in the spring of 1865 destroyed plantation houses in areas below the Edisto River. Fighting elsewhere caused similar damage. Then came "raiding bands of demoralized freedmen" who "tore down houses and carried away the material of which they had been built." According to Stoney, the end of the war "found many a big house turned into a communal dwelling by the plantation negroes and half wrecked in the process."[76]

The "grim poverty and wholesale demoralization" that set in after 1865 demonstrated the consequences of what Stoney called the "violent destruction" of the "natural era" of the plantation "system." Although he failed to specify the constituent elements of that era, his narrative left little doubt that he saw slavery as the natural counterpart of plantations. Once emancipation destroyed the only labor system the lowcountry had known, the region struggled to adapt. In recounting the saga of the postwar era, Stoney employed a well-worn tale. Rice and Sea Island cotton both rebounded, but not to the same levels as before. "Without the discipline of slavery," Stoney lamented, "negro labor fell off steadily in quality." Rice became a "chancy" crop. Growing competition from domestic producers and the storms of the 1890s and 1900s sealed its fate. Sea Island cotton hung on longer but eventually succumbed to the boll weevil. As it passed from the scene, the lowcountry became a "region of deserted fields growing up in forest, of ragged dying gardens and grim, cold, pathetic houses, solemnly awaiting their doom by fire or dilapidation."[77]

Yet even as commercial agriculture collapsed, plantations survived. Indeed, some even gained new life. Stoney credited wealthy sportsmen

[75] Ibid., 11, 41 (first through fifth quotes on 11, sixth on 41).

[76] Ibid., 41–42 (first through third quotes on 41, fourth through six on 42).

[77] Ibid., 11, 41–42 (second through fourth quotes on 11, first and fifth through eight quotes on 42).

with reviving long-decaying plantations. "The coming of the automobile and the good roads it demanded" led to a rediscovery of the lowcountry, and the wealth of the post–World War I era "set men to searching for game preserves and winter homes." "Ruined rice-fields and cotton lands" gave these men exactly what they desired. Old plantation houses supplied "homes already equipped with the charm of time." In recent years, "house after house that seemed *in extremis*" had been "raised from the dead." Now, Stoney reported, "many of the ruined fields have been cleared again, river banks built and rice replanted." "The plantations of the Low Country," he confidently declared, "are well on their way to new and long careers not only of beauty but of usefulness, and of an active life once more."[78]

Stoney thus saw northerners' estates in the same way as other people of the era: as old plantations given new life. He saw them as evolving out of a storied past, not new and different. Like other commentators, Stoney saw continuity in the face of radical change. His account misled for the same reasons others did. Selectivity, emphasis on beauty and refinement, inattention to slavery and labor, and a grandiose view of the antebellum era cast plantations as the country seats of an aristocratic class. In this sense, Stoney's history affirmed established views of the region and its history. Yet for other reasons, *Plantations of the Carolina Low Country* presented a more misleading view than any of its counterparts. A seemingly authoritative body of research, the appearance of comprehensiveness, and the power of visual documentation placed it in a league of its own. Moreover, a focus on plantations with extant houses produced a profoundly circumscribed view of colonial and antebellum plantations. Instead of surveying characteristics common to plantations before the Civil War or concentrating on representative examples, Stoney and his collaborators focused on unusual, truly exceptional plantations. Although they did not intend to deceive readers, they did. By taking plantations with surviving houses as representative and failing to consider others that had once existed, the authors presented a misleading view of the past and the role of plantations in lowcountry history.

Plantations of the Carolina Low Country contains historical sketches for forty-five plantations and nine parish churches. Some sketches are accompanied by floor plans. All of the plantations and churches mentioned are depicted in photographs and measured drawings. The photographs

[78] Ibid., 42.

show plantation houses and church buildings in their existing condition – some well maintained, some weathered, some in distress. The drawings focus on ornamental features, mainly paneled walls, mantels, and moldings. Text and visual information play mutually supporting roles. Together, they link an idealized narrative of lowcountry history to vestigial remains.[79]

Although the title of the book and its text suggest a complete inventory of lowcountry plantations, *Plantations* provides far less. The authors concerned themselves with a small portion of the lowcountry bounded by the Edisto River on the south, the Santee River on the north, and the interior towns of Eutawville, Pineville, and St. Stephen. This triangle-shaped territory, all of which lies within fifty miles of Charleston, encompasses the Stono, the Ashley, the Cooper, and the Wando rivers.[80] It forms part of what S. Max Edelson has identified as the "core settlement area" of colonial South Carolina. Eighteenth-century planters divided the Carolina coast into a "core plantation zone" and a "frontier zone." Plantations of modest size dominated the former; "huge plantations" dedicated to growing rice for export characterized frontier areas. Core zone plantations featured "increased material refinement" and included examples styled as country seats. Frontier plantations amounted to crude labor camps.[81] Thus, when Stoney and his collaborators cast the longest-settled portion of the coast as the "lowcountry," they employed a misleading sample. They cast an area with exceptionally refined dwellings and landscapes as typical of the region as a whole.

In addition to geographic selectivity, Stoney, Simons, and Lapham made other choices that produced further biases. By focusing on surviving examples with architecturally notable houses, for example, *Plantations* presented a small number of plantations as characteristic. The forty-five discussed in the book, however, represented nothing of the sort. They included many of the region's most ornate plantation dwellings. All featured unusually high levels of material development. Still, Stoney and his collaborators portrayed them as typical, as representative of plantations from before the Civil War. In this manner they presented the uncommon as commonplace and elided large portions of lowcountry history.[82]

---

[79] Ibid., 46–233.   [80] Ibid.

[81] S. Max Edelson, *Plantation Enterprise in Colonial South Carolina* (Cambridge: Harvard University Press, 2006), chap. 4.

[82] On the material form of lowcountry plantations, see Edelson, *Plantation Enterprise in Colonial South Carolina*; Joyce E. Chaplin, *An Anxious Pursuit: Agricultural Innovation*

In other ways, too, *Plantations* is more misleading than not. Throughout the book, Stoney and the other contributors wrote about plantations in an all-encompassing fashion. Although they did not specify the extent of their coverage, the title and text of the book suggest they knew of all lowcountry plantations that had historically existed and those that survived. Nowhere did they indicate a selective focus. Most readers likely assumed that *Plantations* cataloged all or most plantations from the colonial and antebellum eras, a representative sample, or at least most noteworthy examples. Certainly, nothing in the book suggests otherwise.[83]

Stoney and his collaborators also wrote about plantations in a way that focused attention on planter's dwellings, often to the exclusion of all else. In many cases, they used the term "plantation" to refer solely to main houses. Thus the sketch for Wappaoolah Plantation begins: "The older portion of this house was built in 1806." The sketch for the Bluff Plantation starts in similar fashion: "Major Isaac Child Harleston ... seems to have built this little house at his Bluff Plantation ... shortly after the [Revolutionary] war." In other cases, however, "plantation" refers to the full landholding, as in the sketch for the Wedge Plantation: "This plantation gets its name from its peculiar shape, for from a bearing on the big-road that gives barely the space for a couple of gates it widens to a broad expanse of rice fields on the Santee." Inconsistencies notwithstanding, the effect is for "plantation" to be equated with "plantation house" much of the time. Rather than making clear distinctions, Stoney and his collaborators conflated the two terms, thus developing a rhetorical convention that gave "plantation" dual meaning yet emphasized a vision focused on big houses and inattentive to other features.[84]

---

and *Modernity in the Lower South, 1730–1815* (Chapel Hill: University of North Carolina Press for the Institute of Early American History and Culture, 1993), chap. 7; Phillip D. Morgan, *Slave Counterpoint: Black Culture in the Eighteenth-Century Chesapeake and Lowcountry* (Chapel Hill: University of North Carolina Press, 1998), 104–107, 110, 112–124; James L. Michie, *Richmond Hill Plantation, 1810–1860: The Discovery of Antebellum Life on a Waccamaw Rice Plantation* (Spartanburg, SC: Reprint Co., 1990); Charles Joyner, *Down by the Riverside: A South Carolina Slave Community* (Urbana: University of Illinois Press, 1984), 117–126.

[83] Stoney, *Plantations of the Carolina Low Country*.

[84] Ibid., 68 (second quotation), 72 (first quotation), 78 (third quotation), and passim. This mode of representation is consistent with what Jennifer Eichstedt and Stephen Small refer to as "symbolic annihilation" in their study of interpretation at plantation museums. See Jennifer L. Eichstedt and Stephen Small, *Representations of Slavery: Race and Ideology in Southern Plantation Museums* (Washington, DC: Smithsonian Institution Press, 2002), chap. 4.

Visual imagery augmented rhetorical practice. The photographs included in *Plantations* concentrate overwhelmingly on houses. Of the 125 that depict plantations, 51 show exterior views of planter's dwellings. Fifty-four show interiors, mainly ornate mantels, staircases, and door surrounds. Few photographs show other plantation elements. Three are of oak allées, two are of rice fields, and eight show outbuildings. Only two show slave houses, both of them views of the street of brick cabins at Boone Hall Plantation (perhaps the most substantially built structures of their kind in lowcountry). In sum, the images included in *Plantations* focus predominantly on big houses. That a mere four images show rice fields and slave cabins demonstrates the book's profoundly incomplete perspective.[85]

Although Stoney and the other contributors might have justified their approach on the grounds that they sought to study architecture, not entire plantations, the title of the book and their language demonstrates otherwise. They wrote about plantations and recognized plantations as their subject. They distinguished between plantations and plantation houses as needed and depicted each visually. What *Plantations* demonstrates is the extraordinary selective view of the past that dominated the era. At a time when most white Americans viewed black people as primitive, docile beings and recalled antebellum planters as noble aristocrats, representations of historical plantations focused on remains associated with the latter group. Planters' dwellings and associated spaces took precedence. Elegant architecture and refined landscapes suggested the power and authority of a ruling class that had presided over a society ordered according to "natural" hierarchies. Slave cabins and spaces of labor and production seemed thoroughly prosaic by comparison – uninteresting, unimportant, and undeserving of attention.[86]

Stoney, Simons, and Lapham, as the figures responsible for the vision expressed in *Plantations*, hardly qualified as impartial observers. Their ties to the old planter class made them anything but. Yet their view of

---

[85] Stoney, *Plantations of the Carolina Low Country*, 86–233.

[86] On historical memory of the Old South, see especially David W. Blight, *Race and Reunion: The Civil War in American Memory* (Cambridge: Belknap Press of Harvard University Press, 2001); David W. Blight, *Beyond the Battlefield: Race, Memory, and the American Civil War* (Amherst: University of Massachusetts Press, 2002); Alice Fahs and Joan Waugh, eds., *The Memory of the Civil War in American Culture* (Chapel Hill: University of North Carolina Press, 2004); Nina Silber, *The Romance of Reunion: Northerners and the South, 1865–1900* (Chapel Hill: University of North Carolina Press, 1993); Carol E. Janney, *Remembering the Civil War: Reunion and the Limits of Reconciliation* (Chapel Hill: University of North Carolina Press, 2013).

history differed little from that held by most Americans of the time. By the early twentieth century, most Americans viewed the Civil War as a glorious, heroic struggle that had forged a stronger nation, not a failed bid to deliver true equality. In this context, views of physical space matched the prevailing variety of historical amnesia. Veneration of the planter class steered attention toward big houses and the surroundings; marginalization of African Americans' role in the Civil War and emancipation obscured spaces and structures associated with their lives. In short, patterns of remembrance and attention to surviving remains shared an integral relationship, each buttressing the other and removing what they ignored from consideration.[87]

Above all, *Plantations* illustrates the deceptive allure of what scholars such as David Lowenthal have called "tangible pasts." Relics offer seemingly unmediated connections to distant eras. They tend to be prized for their evocative power and seen as more reliable guides to history than written accounts. The latter, after all, make selective use of sources and require interpretation; artifacts, by contrast, seemingly do not. Yet because social and economic conditions influence what survives from any given period, tangible remains tend to be profoundly misleading. Structures associated with elite actors are almost always overrepresented, while those associated with other social groups suffer the opposite fate. Consequently, tangible remains tend to be poor guides to history. No matter how powerful the sense of immediacy and authenticity may be, surviving edifices are generally more ornate, better built, and more visually impressive than historical norms.[88]

Plantations exemplifies these problems acutely. Although the exact reasons that Stoney and his collaborators chose the plantations discussed in the book are impossible to determine, accessibility, historical associations, and architectural qualities clearly played important roles, and other considerations may have had an influence. No matter what calculus they employed, their inventory aggrandized the lowcountry past. Plantations, in the authors' view, consisted mainly of stately houses and handsome landscapes; other features qualified as incidental. Emphasis on architecturally distinguished houses foregrounded material refinement

---

[87] Blight, *Race and Reunion*; Silber, *Romance of Reunion*; Janney, *Remembering the Civil War*.

[88] David Lowenthal, *The Past Is a Foreign Country* (Cambridge: Cambridge University Press, 1986), 238–249; David Lowenthal, "Past Time, Present Place: Landscape and Memory," *Geographical Review* 65 (1975): 1–36. See also Andrew Jones, *Memory and Material Culture* (Cambridge: Cambridge University Press, 2007).

while simultaneously obscuring everything else. Whether the authors recognized the biases inherent in their perspective is unclear, and whether any of their readers noticed is no more apparent. Yet even if some readers did, it apparently mattered little. *Plantations* achieved considerable success. Following the initial release in November 1938, copies of the book sold so rapidly that the publisher issued a second edition in January 1939. Subsequent printings followed in 1945, 1955, and 1964. Throughout the middle decades of the twentieth century, *Plantations* stood as the definitive study of lowcountry plantations. When people wanted to learn about plantations, they turned to its pages, to images of timeworn majesty and tales of virtuous families, handsome houses, and lost grandeur.[89]

Viewed in perspective, *Plantations* demonstrates the destabilizing force of sportsmen's and sportswomen's activities. Not until estate-making began in earnest did lowcountry people take steps to record the history of plantations. Restoration, rehabilitation, and associated transformations compelled lowcountry people to judge the significance of surviving remains. Although some forms of recognition preceded the peak of northerners' activities, the accounts that appeared during the 1930s demonstrate the reaction triggered by the estate-making boom. Sass's essay, Alice Ravenel Huger Smith's *A Carolina Rice Plantation of the Fifties*, and *Plantations* marked a committed effort to inscribe a particular vision of the past in memory. By the end of the decade, that vision had gained widespread acceptance.[90]

In 1981, David Lowenthal and Marcus Binney observed that efforts to "preserve" valued buildings and sites are motivated by interest in a past seen as (1) unlike the present, (2) crucial to the identity of a particular social group, and (3) threatened with destruction (in fact or in perception).[91] In the lowcountry, recognition of early plantations as vestiges of bygone eras developed by the 1880s if not sooner, and the region's white elite clearly saw such plantations as central to their individual and

---

[89] Samuel Gaillard Stoney, *Plantations of the Carolina Low Country* (Charleston: Carolina Art Association, 1939); Samuel Gaillard Stoney, *Plantations of the Carolina Low Country* (Charleston: Carolina Art Association, 1945); Samuel Gaillard Stoney, *Plantations of the Carolina Low Country* (Charleston: Carolina Art Association, 1955); Samuel Gaillard Stoney, *Plantations of the Carolina Low Country* (Charleston: Carolina Art Association, 1964).

[90] Sass, "The Low-Country"; Stoney, *Plantations of the Carolina Low Country*; Smith, Sass, and Smith, *A Carolina Rice Plantation of the Fifties.* See also "The Low Country."

[91] David Lowenthal and Marcus Binney, eds., *Our Past before Us: Why Do We Save It?* (London: T. Smith, 1981), 17–21.

collective identities. So long as rural areas remained dormant, however, only the slow-moving forces of weathering and decay posed a threat. Sportsmen's and sportswomen's remaking of old plantations overturned that balance. The estate-making boom of the 1920s caused lowcountry people to recognize that plantations throughout the region might be altered beyond recognition. Devotees of regional history rushed to catalog examples they viewed as important and assess their significance. The perspectives expressed in "The Low-Country," *Plantations*, and a host of other publications resulted. That plantations remade as sporting estates figured among those recognized as historically important is ironic but not surprising. As Lowenthal has observed, relics are always transformed to meet contemporary needs. Since lowcountry people saw most of the changes that northerners made as sympathetic, not destructive, they voiced support and approval.

CONCLUSION

Representations of northerners' estates played a crucial role in shaping beliefs about the lowcountry and its plantations. The visual and rhetorical idioms crafted by newspaper and magazine reporting informed popular views of new estates and their significance. The overwhelming emphasis on revival, restoration, and renewal obscured the newness of northerners' plantations, the innovations they embodied, and how dramatically they differed from earlier plantations. The consistent focus on plantation houses and immediately proximate spaces emphasized features associated with white elites while excluding those that reflected slavery and commercial enterprise. The resulting vision centered on stately mansions and omitted nearly everything else. Newspaper and magazine articles portrayed northerners' estates as handsome dwellings set amid lush surroundings. Although some historical plantations had possessed these elements, many did not, and all had other features that directly reflected labor and staple agriculture. Ignoring such elements produced a profoundly circumscribed view of what plantations had been historically. In effect, it removed from view African Americans and the primary spaces where they had lived, labored, and suffered.

Historical writing added strength and credibility to views expressed in newspapers and magazines while situating northerners' estates in a still-unfolding history. Not only did Sass's and Stoney's narratives view plantations in a similar fashion, but, by casting northerners' estates as the latest phase in a long line of plantation development, they marginalized differences across time. Sass's and Stoney's writings chronicled a history of landed estates that told little about the historical use and purpose of plantations and the vast majority of associated actors. Inattention to physical exertion, hardship, and suffering rendered a nostalgic portrait of aristocratic country seats. Even though lowcountry slavery had been among the harshest varieties on the North American mainland, historical writing of the 1930s ignored its rigors. Sass and Stoney depicted a virtuous planter class whose wealth flowed effortlessly from prosperous estates. In their hands, forced subjugation of black men, women, and children disappeared.

Views expressed in writings of the era gained credibility and influence partly because of the absence of countervailing perspectives. The rigidly exclusionary culture of the Jim Crow South left African Americans with few opportunities to challenge such accounts. Criticizing slaveholders' supposed benevolence, after all, amounted to an attack on the prevailing social order, and whites' vigilant defense of the color line made such behavior dangerous. Moreover, the challenges facing lowcountry blacks made contesting views of rich peoples' estates a low priority. In Charleston, African Americans concentrated on securing access to jobs, housing, schooling, police protection, and public services. In rural areas, maintaining self-sufficiency took precedence. Land-owning farmers and tenants had no interest in jeopardizing their autonomy. In this context, offering alternative views of the past seemed unimportant and arguably unwise. Practical concerns mattered more.[92]

The portrayals of the 1930s produced a body of literature that informed popular views of the lowcountry and its past for decades. Throughout the 1940s and 1950s, journalists referred to Martin's and Murray's sketches whenever newsworthy events occurred. Sales of large estates, construction of new houses, and deaths of long-term owners provided opportunities to recount plantation histories. Writers typically

---

[92] Fraser, *Charleston! Charleston!*, 359–364; Yuhl, *Golden Haze of Memory*, 15. See also Blain Roberts and Ethan J. Kytle, "Looking the Thing in the Face: Slavery, Race, and the Commemorative Landscape in Charleston, South Carolina, 1865–2010," *Journal of Southern History* 73, no. 3 (Aug. 2012): 639–684. On rural blacks, see T. J. Woofter, *Black Yeomanry: Life on St. Helena Island* (New York: Henry Holt and Co., 1930); Carolyn Baker Lewis, "The World around Hampton: Post-Bellum Life on a South Carolina Plantation," *Agricultural History* 58, no. 3 (July 1984): 456–476; Lawrence S. Rowland and Stephen R. Wise, *The History of Beaufort County, South Carolina*, vol. 3: *Bridging the Sea Islands' Past and Present, 1893–2006* (Columbia: University of South Carolina Press, 2015), 182–195, 218–221, 327–328.

used information from Martin's and Murray's sketches verbatim, without new research or even basic fact-checking. The thinly researched accounts of the 1930s thus became the basis for another round of reporting. Sheer repetition gave the information conveyed a guise of credibility.[93] At the same time, it dissuaded further inquiry. By suggesting a set of well-understood histories, later accounts suggested further research would prove fruitless.

Meanwhile, other writers drew from Martin's and Murray's articles and a host of other sources in producing books that emulated *Plantations of the Carolina Low Country* in form, style, and substance. In 1955, the publication of Alberta Morel Lachicotte's *Georgetown Rice Plantations* provided a similar compendium for plantations in Georgetown County. Lachicotte, a descendant of a rice-planting family, conducted more extensive research than Stoney and his cohort and wrote more detailed plantation histories, but her perspective on the past differed little. She also emphasized the grandeur of the pre–Civil War lowcountry and concentrated on planters' dwellings and their immediate surroundings. Like Stoney and his colleagues, Lachicotte saw plantations as vestiges of a genteel past.[94]

Remarkably, Lachicotte's book is not the most recent to espouse such a view. *Lowcountry Plantations Today*, a self-proclaimed successor to *Plantations of the Carolina Low Country*, presents a similar vision. Written by lowcountry author and naturalist William P. Baldwin and published in 2002, it tells the history of thirty-five plantations in the same manner – in brief sketches about elegant houses and planter families. Although the preface mentions the "dark cloud" of slavery and the hardships slaves endured, Baldwin fails to make slaves, labor, and human

[93] See, for example, James M. Lofton, Jr., "Do You Know Your South Carolina? Rice Hope," newspaper clipping cited as *News and Courier*, May 13, 1940, in Rice Hope (Georgetown Co.) Plantation file, South Carolina Vertical File Collection, Charleston County Public Library, Charleston, SC (hereafter SCVFC); "Rice Hope Plantation Opened," newspaper clipping cited as *News and Courier* March 21, 1954, SCVFC; "Rice Hope, of Historical Interest, Is One of Points on Georgetown Plantation Tours," newspaper clipping cited as *News and Courier*, March 28, 1954, SCVFC; "The Wedge, Santee Plantation, Sold to C. E. Woodward," newspaper clipping cited as *News and Courier*, Feb. 10, 1946, SCVFC; John M. Lorton, Jr., "Do You Know Your Lowcountry? Annandale," newspaper clipping cited as *News and Courier*, Sept. 30, 1940, SCVFC; "Do You Know Your South Carolina? Mulberry Plantation," *News and Courier*, Feb. 17, 1947, 10.
[94] Alberta Morel Lachicotte, *Georgetown Rice Plantations* (Columbia: The State Printing Co., 1955).

suffering part of the stories told in the ensuing pages. By casting plantations as handsome houses set amid stunning landscapes, Baldwin presents a romantic portrait of the antebellum lowcountry. As with *Plantations of the Carolina Low Country*, readers are left to wonder about the representativeness of the plantations depicted. The book makes no mention of the hundreds of others that historically existed. In sum, *Lowcountry Plantations Today* perpetuates the myth of genteel plantation past. By ignoring slavery, it renders plantations as magnificent country estates, nothing more and nothing less.[95]

Baldwin's book demonstrates the durability of the vision that took hold in the 1930s. As wealthy northerners created grand estates, narratives of a majestic past became closely tied to stately houses and elegant surroundings. Newspapers, magazines, and historical writings made a rose-tinted view of lowcountry history central to remembrance of plantations and understandings of new estates. As well-informed authorities recounted the history of the region, they lashed nostalgia for an imagined past to sumptuous retreats that bore little resemblance to the commodity-producing complexes of the colonial and antebellum eras. The harshness and depravations of lowcountry slavery went unrecognized, crowded out by veneration of the old planter class and the world it had created. Meanwhile, praise for new owners and their estates collapsed differences across time and space. As journalists, historians, and a handful of other commentators rendered judgments about the changes taking place, they credited sportsmen and sportswomen with reviving age-old traditions. In the process, they obscured the origins of northerners' "plantations" and their role in a resurgent lowcountry.

---

[95] William P. Baldwin, *Lowcountry Plantations Today* (Greensboro, NC: Legacy Publications, 2002).

7

Plantation Life

Varieties of Experience on the Remade Plantations of the Lowcountry

> Plantation life is enjoyed to the fullest by the new owners of the various tracts of land … It means different pleasures to the different owners.
> – Chlotilde R. Martin, October 1932[1]

In the late 1920s, the phrase "plantation life" became part of the low-country vernacular. First used by owners of large estates, it soon entered popular discourse. By the mid-1930s it became a mainstay of newspaper and magazine reporting. As the number of northern-owned plantations continued to grow, references to "plantation life" became common. By the end of the decade, the phrase routinely appeared in discussions of the lowcountry and its plantations.[2]

"Plantation life" served as more than a convenient shorthand for the activities that northerners came to the lowcountry to enjoy. More than mere catchphrase, it signaled continued development and maturity. By underscoring what some plantations had become, it made clear the

---

[1] Chlotilde R. Martin, "Low-Country Plantations Stir as Air Presages Coming of New Season," *News and Courier* (Charleston, SC), Oct. 16, 1932, A6.

[2] The earliest known use of the phrase is Benjamin R. Kittredge to S. Dana Kittredge, n.d. [ca. 1927], folder 4, box 15, Kittredge Family Papers, South Carolina Historical Society, Charleston, SC (hereafter Kittredge Family Papers). The earliest published instance is Martin, "Low-Country Plantations Stir as Air Presages Coming of New Season." For later mentions, see "Wedgefield Plantation," *Country Life* 75, no. 2 (Dec. 1938): 53; Edward Durell Stone, *The Evolution of an Architect* (New York: Horizon Press, 1962), 50. References to "plantation life" appear frequently in northerners' memoirs. See, for example, Gertrude Sanford Legendre, *The Time of My Life* (Charleston: Wyrick and Co., 1987), 70; Frances Cheston Train, *A Carolina Plantation Remembered: In Those Days* (Charleston: History Press, 2008), chap. 4.

breadth and depth of the changes under way. By highlighting the role of plantations in popular views of the lowcountry, "plantation life" underscored the influence of large estates. By denoting a realm of activity specific to grand "plantations," plantation life told of the purposes that many plantations now served.

Plantation life grew out of three intersecting trends. One was growing awareness of northerners' plantations and the lowcountry sporting scene. The wave of newspaper and magazine reporting that introduced northerners' estates to national audiences made them part of an imagined geography that anchored ideas about "the South," "the lowcountry," and "plantations." As media attention proliferated, northerners' estates and activities became widely known. In turn, Americans increasingly thought of plantations as grand estates where wealthy people spent their winters hunting, relaxing, and soaking up warm sunshine. No longer solely places of intensive agriculture and former sites of slavery, plantations now served purposes comparable to those of upper-class retreats elsewhere.

Second, plantation life demonstrated growing recognition of the lowcountry as a distinct region. Popular interest in northerners and their estates developed amid growing interest in the lowcountry and its past. From the late 1920s onward, an outpouring of fiction, visual art, and features in national magazines and newspapers focused attention on the region as never before. Lowcountry writers and artists cast the lowcountry as a land apart, a region with its own history, traditions, and people. In the most common formulation, it appeared as an especially exotic corner of a thoroughly exotic region. As the South at its most aristocratic, most historic, and least disturbed, the lowcountry appeared as a living vestige of the antebellum era. Emergent views of the region infused northerners' activities and estates with new meaning. As images of the lowcountry became etched in national consciousness, observers near and far viewed northerners and their estates against the backdrop of history. In turn, northerners increasingly appeared not simply as sporting enthusiasts but as successors to the rice-planting elite. Moreover, northerners' estates became more than old plantations put to new uses; increasingly, they appeared as places where age-old traditions lived on.

Beneath growing awareness and recognition lay continued progress and development. By the late 1920s, northerners' activities amounted to more than an amalgam of hunting, horseback riding, and assorted entertainments. In combination with convivial socializing and activities based on personal interests, these pursuits denoted a sporting lifestyle with a vital role in upper-class culture. By the 1920s, enthusiasm for field sports formed the basis of a peripatetic lifestyle that moved with the seasons. Commonly referred to as "country life," it emanated from the great estates and sporting clubs of Long Island. When the weather turned cold, the upper classes migrated southward, taking their preferred pastimes with them. Florida resorts, the hunting grounds of the Thomasville-Tallahassee area, the polo fields of Aiken and Camden, and the famed golf course at Pinehurst became seasonal playgrounds for wealthy, privileged people, forming an exclusive network of social and recreational venues along the southeastern seaboard.[3]

Plantation life merged the basic attributes of country life with the physical space of "plantations." It combined upper-class pastimes with settings where vestiges of earlier periods stood alongside new buildings and landscape features. It adapted elite behaviors for a region commonly recognized for an aristocratic heritage and for plantations with colonial and antebellum origins. By yoking the social and environmental conditions of old plantations to northerners' activities, plantation life fused upper-class rituals with places that northerners and others viewed as quintessentially southern. In turn, northerners merged the imagined southland of popular lore with lived experience. By making their estates

---

[3] Robert B. MacKay, Anthony K. Barber, and Carol A. Traynor, eds., *Long Island Country Houses and Their Architects, 1860–1940* (New York: Society for the Preservation of Long Island Antiquities in association with W. W. Norton and Co., 1997), 19–33; Clive Aslet, *The American Country House* (New Haven: Yale University Press, 1990), chaps. 11 and 14; Mark A. Hewitt, *The Architect and the American Country House, 1890–1940* (New Haven: Yale University Press, 1990), 10–14, 127. On Florida resorts, see Edward N. Akin, *Flagler, Rockefeller Partner and Florida Baron* (Kent, OH: Kent State University Press, 1988), chaps. 6–9; William B. Stronge, *The Sunshine Economy: An Economic History of Florida since the Civil War* (Gainesville: University Press of Florida, 2008), 81–85. On hunting in the Thomasville-Tallahassee area, see "From Thomasville to Tallahassee," *Country Life* 17, no. 4 (Feb. 1935): 11–14; Albert G. Way, *Conserving Southern Longleaf: Herbert Stoddard and the Rise of Ecological Land Management* (Athens: University of Georgia Press, 2011), chap. 1; Julia Brock, "Land, Labor, and Leisure: Northern Tourism in the Red Hills Region, 1890–1950" (PhD diss., University of California at Santa Barbara, 2012); William R. Brueckheimer, "The Quail Plantations of the Thomasville-Tallahassee-Albany Regions," in *Proceedings: Tall Timbers Ecology and Management Conference*, no. 16 (Tallahassee: Tall Timbers Research Station, 1982): 141–165; Clifton Paisley, *From Cotton to Quail: An Agricultural Chronicle of Leon County, Florida, 1860–1967* (Gainesville: University of Florida Press, 1968), vi, 74–98; Hank Margeson and Joseph Kitchens, *Quail Plantations of South Georgia and North Florida* (Athens: University of Georgia Press, 1991). On Aiken and Camden, South Carolina, see Mary Katherine Davis Cann, "The Morning After: South Carolina in the Jazz Age" (PhD diss., University of South Carolina, 1984), 427–428. On Pinehurst, see Richard Mandell, *Pinehurst: Home of American Golf* (Pinehurst, NC: T. Eliot Press, 2007).

stage sets for an array of performative, symbolically charged activities, northerners gave their activities greater significance.

Plantation life manifested itself in several ways. For northerners and their guests, it imparted meaning and vitality to the experience of a "plantation." By combining actual and imagined Souths, plantation life translated exoticism, intrigue, and regional stereotypes into almost palpable form. Although the imagined dimensions of plantation life depended on northerners' inclination to associate themselves and their experiences with popular imagery, in practice, many did so almost constantly. Northerners' ideological investments in "the South" and "the lowcountry" aligned their activities with popular views of the Carolina coast and the southern states as a whole, thereby imparting greater value to their activities than met the eye.

Plantation life also contrasted new social relations with the fabled hierarchies of the antebellum era. Northerners, lowcountry people, and a host of onlookers consistently saw estate owners and black laborers as occupying familiar roles. Most viewed owners as successors to the planters of the Old South and African Americans as primitive, usually docile laborers who willingly served their superiors in ways reminiscent of the supposedly benevolent race relations of the past. Social conditions on northerners' estates thus shaped the experiences of all involved while legitimating the racial hierarchies of Jim Crow.

Most significant, plantation life fueled the material development of northerners' estates. The advent of a lifestyle centered on performative display demanded further refinement and pretention. The creation of grander, more ambitious estates reflected northerners' desire to display wealth and status and the concomitant need for settings suited to recreational and social activities. Material form, use, and symbolism formed three sides of an interlocking whole, each informing the other and spurring on development of more refined and intricate spaces.

In sum, plantation life involved more than met the eye. Although some observers read it as conscious attempt to reclaim the fabled lifestyle of the old planter class, it grew out of social and cultural circumstances specific to northerners' estates and activities. As a realm of activity specific to grand plantations, plantation life merged allusions to a romantic past with the social practices of a contemporary elite. Lively, energetic, and characterized by equal parts rusticity and refinement, plantation life enthralled participants, awed onlookers, and drew frequent commentary. As a forum for carrying out rituals with a privileged role in upper-class culture, plantation life supplied wealthy, powerful people with rare opportunities. In an age of shifting class boundaries and new challenges from below, those proved as important as wealth itself.

PLANTATION LIFE AS LIVED EXPERIENCE

Recreational hunting formed the basis of plantation life. Just as northerners did at clubs and retreats, estate owners spent their days hunting ducks, quail, deer, doves, and other game. The same qualities that had attracted sportsmen to the lowcountry in the 1880s and 1890s remained just as important thirty and forty years later. Game populations declined over time, and the spectacular sort of game-taking that had prevailed early on quickly became a thing of the past. Still, the lowcountry offered generally favorable conditions. Sportsmen and sportswomen continued to see it as a prime destination. As conditions across the North and Midwest deteriorated, the lowcountry's status rose. Sportsmen and sportswomen held its gamelands in high regard.[4]

The style and manner of northerners' hunts changed over time. As heterosocial recreation became common, hunting became part of an array of leisure and sporting pursuits and less an undertaking unto itself. Although it remained the focus of northerners' activities, hunting became part of an ensemble that included boating, horseback riding, sightseeing, fishing, and convivial socializing. In turn, its significance changed. Less overtly focused on displays of masculinity, it became part of a repertoire of recreational pastimes with elite associations. Northerners continued hunting in exclusively male groups, especially at clubs and retreats. Moreover, at large estates, men sometimes took to the field under similar circumstances. In general, however, plantations became spaces of

---

[4] On declines in waterfowl populations, see Henry H. Carter, *Early History of the Santee Club* (Boston[?]: n.p., 1934), 11–15; Suzanne Cameron Linder, *Historical Atlas of the Rice Plantations of the ACE River Basin – 1860* (Columbia: South Carolina Archives and History Foundation, Ducks Unlimited, and the Nature Conservancy, 1995), 309–310; "Duck Season Finished," *News and Courier*, Feb. 1, 1930, 2; "Big Duck Hunting Season Expected," *News and Courier*, Nov. 14, 1933, 2; "Duck Season Ends Tomorrow," *News and Courier*, Dec. 18, 1935, 14. For a national perspective, see Paul A. Curtis, "Will We Have Good Duck Shooting Again?," *Country Life* 69, no. 1 (Nov. 1935): 34–35, 78–79. On declines on the once-prized gamelands of the Chesapeake, see C. John Sullivan, *Waterfowling on the Chesapeake, 1819–1936* (Baltimore: Johns Hopkins University Press, 2003), 71–82. On declines on the Outer Banks of North Carolina, see Zach Taylor, "Currituck's Grand Old Hunting Clubs," *Sports Afield* 180, no. 6 (Dec. 1978): 39. On the continuing significance of hunting at large estates for the upper classes, see Dixon Wecter, *The Saga of American Society: A Record of Social Aspiration, 1607–1937* (New York: C. Scribners' Sons, 1937), chap. 11.

heterosocial activity. Men and women participated in sporting and leisure pursuits on roughly equal footing.

In 1931, Willis E. Fertig described the hunting that he and his guests enjoyed at his Ponemah estate in Georgetown County. His summary offers a valuable glimpse of northerners' activities. Sportsmen and sportswomen took full advantage of the range of species available. Whereas sportsmen and sportswomen had initially hunted ducks and quail and showed limited interest in other species, by the late 1920s, they hunted a wide variety of game.

Fertig began by describing one of the most coveted varieties of hunting that northerners enjoyed. "Quail shooting," he explained, is done "on horseback with fast, wide ranging dogs." Hunters rode to locations where quail nested and then slowed to let their dogs work. "When the dogs find a covey," Fertig explained, "the gunners dismount to shoot on the covey rise." Speed and agility make quail immensely challenging. Quail fly erratically when flushed from cover, often scattering in multiple directions. Quail hunting demands quick reflexes, sharp aim, and skill. Although Fertig did not describe the experience of a quail hunt in detail, anyone who read his comments likely would have known its basic contours. By the early 1930s, lowcountry people recognized quail as one of northerners' preferred game birds.[5]

Fertig next spoke of turkey hunting, which he called "perhaps the most fascinating of all the sporting events obtainable." Fertig described turkeys as "wise and willy bird[s] with a keen sense of hearing and wonderful eyesight." He also emphasized their skill in evading sportsmen. "The turkey," Fertig explained, "is amply able to take care of himself under all conditions." Fertig characterized "the bagging of a turkey" as "a real event." As he put it, "there are few thrills to compare with that of bringing down a large gobbler, and thus securing his head for a trophy."[6]

Deer hunting also supplied plentiful opportunities for pleasure and excitement. Northerners generally shot driven deer. "Drivers" – men mounted on horseback – rode through brush and forest with trained hounds running alongside. By flushing deer from cover and sending them racing toward waiting hunters, driving eliminated the waiting that

---

[5] Chalmers S. Murray, "Ponemah, 'Happy Hunting Ground,'" *News and Courier*, Apr. 26, 1931, A6. On the cultural significance of quail hunting, see Way, *Conserving Southern Longleaf*, 39–41.

[6] Murray, "Ponemah, 'Happy Hunting Ground.'"

characterized the "still-hunt," the method that required hunters to wait for deer to come into view. In contrast to the patience that the still hunt demanded, driving created a fast-moving sequence of activity. When deer dashed into the open, the nearest hunter opened fire. Shooting at close range increased the chances of a kill, as did hunters' use of shotguns. Moreover, since deer ran past waiting hunters in full flight, driving ensured that hunters faced fast-moving targets. Fertig described driving as flushing "the game out of the swamps and thickets past the hunters." Hunters, he explained, "are placed on stands to shoot the fleet little animals as they run by at full speed."[7]

Other varieties of hunting rounded out northerners' activities. "The fox hunts," Fertig observed, "are usually brought off at night with packs of hounds which follow the trail until the fox is either killed or lost. As in the deer drives, friends and neighbors participate and are privileged to bring their hounds to add to the pack." Fertig also noted "there is raccoon and o'possum hunting at night with trained dogs." Moreover, he observed that "squirrel, rabbit, snipe and woodcock shooting furnish equally good sport." Northerners did not lack for options. Species diversity and large game populations supplied exceptional opportunities.[8]

Fertig emphasized the social dimensions that came with some varieties of hunting. "A deer drive," he observed, "is a sort of social affair in which friends and neighbors are invited to participate. Refreshments are served after the drive." Large parties proved common. Because of the number of hunters and drivers involved, gatherings of a dozen or more routinely occurred. In some cases, parties could number fifty people or more, not including guides and servants.[9]

Relaxed game laws contributed to northerners' pleasure and enjoyment. Fertig characterized the South Carolina hunting season as "long." It began in August with the opening of deer season. In September, "doves, rail and gallinule" became legal. In November, hunters turned their attention to "quail, ducks, geese, snipe, woodcock, squirrel and rabbit." Ducks and quail remained in season until March 1. All told, sportsmen and sportswomen had opportunities to hunt "continuously for six and

[7] Ibid.   [8] Ibid.

[9] Ibid. For a detailed description of a deer drive, see John J. Seibels to Joseph S. Frelinghuysen, Nov. 10, 1926, box 2, Joseph S. Frelinghuysen Papers, Archibald S. Alexander Library, Rutgers University, New Brunswick, NJ (hereafter Frelinghuysen Papers).

one-half months in each year." Moreover, Fertig noted that "when the shooting season is over there is fox hunting."[10]

Fertig's comments highlight important dimensions of northerners' activities. The advent of heterosocial hunting went hand in hand with new modes of behavior. Although men who traveled to the region early on did not limit themselves to duck and quail, the prevalence of these species, coupled with comparatively short stays in the region, effectively dissuaded interest in others. Hunting of deer, opossum, turkey, and other game took place, but sportsmen generally concentrated on more coveted species. As the years passed and northerners' activities evolved, owners of large estates developed a different orientation. By the late 1920s, the traditional emphasis on duck and quail coexisted with ample attention to other species.[11]

The turn toward more varied forms of hunting involved more than a desire to exploit the full range of recreational opportunities available. Rather than reflecting diminishing interest in species such as duck and quail or new favor for others, it represented an ad hoc response to needs that developed over time. Hosting visitors for stays of varying duration made more varied forms of recreation essential. Hunting quickly evolved into a pastime that accommodated skilled sportsmen and sportswomen, less accomplished enthusiasts, and relative novices. Owners of large estates turned to lesser species in order to provide options for everyone present. Greater inclusivity turned hunting into an eminently social pastime that accommodated varying levels of skill, interest, and enthusiasm.

Logs kept by owners of large estates illustrate the shift. At Boone Hall Plantation north of Charleston, for example, Thomas and Alexandra Stone hunted multiple varieties of game with a continually changing cast of companions. At times they took to the field with other plantation owners and a few lowcountry natives. In other instances, they hunted with guests from afar. During the fall and winter of 1936–1937, the Stones hunted several species of wildlife at Boone Hall and in a few other locations. Records of their outings demonstrate varied experiences and frequent participation by women.

In the fall, the Stones hunted deer. Although busy with construction of a new house and an ambitious campaign of improvements, the couple still

---

[10] Murray, "Ponemah, 'Happy Hunting Ground.'"

[11] Some sportsmen focused on particular species. Jacquelin S. Holliday, for example, concentrated on quail hunting. See Chalmers S. Murray, "Holliday Winters in Peedee," *News and Courier*, Aug. 30, 1931, B11.

found time for recreation. In January, Thomas Stone noted they had killed three bucks during the season. "I shot one, which was my first," he recalled. In keeping with an age-old ritual, one of Stone's companions smeared his face with blood from the freshly killed animal to commemorate the occasion. As Stone recounted, "I was well bloodied by Ephie Seabrook, who did his job with obvious enjoyment."[12]

In January, unseasonably warm temperatures put a temporary halt to the Stones' outings. As the month drew to a close, Stone noted in his diary that he had done "practically no shooting or hunting since Christmas." As he explained, "It is too hot for the dogs, and the birds are away in the deep woods." Since finding birds proved "practically impossible," the Stones occupied themselves with other activities.[13]

In early February, conditions improved. The Stones immediately turned their attention to quail. While hunting on Parker's Island in the Wando River, they found a "large covey" that Alexandra and a friend, Doug Deeks, had previously located. The birds "flushed wild," which left the Stones and their companions unable to get a good shot. "Before we could get anywhere near them," Stone recounted, they "went so deep in the woods that we did not get any single shooting." He attributed the birds' behavior to a pattern seen throughout the lowcountry that spring. "George Buist tells me that almost everybody has been having just this experience all year."[14]

In contrast to their quail shoots, the Stones had "great success" with a turkey hunt, even though they did not kill a single turkey. "We only killed two coons," Stone noted. "I had an enormous flock of between twenty-five and thirty [turkeys] come to my blind," he recalled, but, "before I could make up my mind how to get at them with the gun, something frightened them." The whole flock "took to the air with the rapidity of a covey of quail." The Sharps, a couple who accompanied the Stones on the outing, "each had a long shot at the same bird, and were both persuaded that it was wounded," but the party never found the animal. "We had to give up after a long hunt for it through the woods," Stone noted. Fortunately, happenstance intervened. A quick turn of events ensured that the group did not return empty-handed. "Cecil said that he thought he saw a

[12] Thomas A. Stone Journal, Jan. 26, 1937, in Boone Hall Scrapbook 2, South Carolina Historical Society, Charleston, SC (hereafter Stone Journal). For another mention of the same ritual, see Neal Cox, *Neal Cox of Arcadia Plantation: Memories of a Renaissance Man* (Georgetown, SC: Alice Cox Harrelson, 2003), 35–39.
[13] Stone Journal, Jan. 26, 1937.   [14] Ibid., Feb. 9, 1937.

possum's head sticking out of what looked like a squirrel's nest high up in a bare tree," Stone recalled. "I fired at it with both barrels in rapid succession, and two enormous coons came tumbling down. One of them was running when it hit the ground," he noted. It ran "straight for Ruth Sharp, who shot and killed it." The other raccoon, Stone noted, "was already dead from my shot."[15]

The Stones hunted frequently throughout the spring. A few days after the turkey shoot, Thomas Stone mentioned that he would again "go for turkeys on Sunday morning." Several days later, he wrote about having done "a little quail shooting," with "three birds to show for it." The following week, he and several companions "tried the turkeys." They took to "the Boone Hall blinds, but without success." Although "Kinsey saw the flock of 30 near the blind in which he and Bill and Alex were," the birds stayed too far away for a good shot.[16]

Still later in February, Stone mentioned hunting as one of several activities that he and Alexandra engaged in with guests. "We had a fine week-end with the Wrongs, the Achesons, and the Wilkinsons," he recounted. Heavy rains kept the party indoors for the better part of two days, but the group went outside as soon as the weather cleared. As the festivities drew to a close, Stone noted that "everyone has done a little bit of riding and a little bit of shooting and a little bit of everything."[17]

The Stones continued to hunt for the remainder of the season. On March 1, the last day of quail season, the Stones again hunted with the Sharps. "We had a wonderful day," Stone observed. "Found six covey and killed nine birds."[18]

The Stones' activities fit within a well-established pattern. Although practices varied, by the 1930s, owners of large estates tended to hunt in a similar fashion. Records kept by other northerners reveal comparable behavior. Benjamin R. Kittredge, Jr., for example, hunted frequently at the Oakland Club, with the Legendres at Medway Plantation, and with an ongoing parade of guests. Throughout the second half of the 1930s, Kittredge routinely hunted doves at Oakland (sometimes bagging as many as eighty-eight birds), enjoyed outings with the Legendres, and shot game at his Dean Hall Plantation.[19] At Friendfield Plantation near Georgetown, Radcliffe Cheston, Jr., and his family spent their days

---

[15] Ibid., Feb. 9, 1937. [16] Ibid., Feb. 23, 1937. [17] Ibid., Feb. 23, 1937.
[18] Ibid., March 6, 1937.
[19] Benjamin R. Kittredge diaries, 1935–1940, folders 1 and 2, box 25, Kittredge Family Papers.

"hunting quail, ducks, deer, and turkey." Fox hunts often started before dawn, and hunters also stalked coon and wild boar. "Everyone loved to hunt," recalled one of Cheston's daughters.[20] Bernard Baruch and his guests maintained a similar routine at Hobcaw Barony.[21] Across the lowcountry, northerners seized opportunities for "good sport."

The style and manner of hunting at large estates differed modestly from hunting clubs and retreats. Different social circumstances and somewhat different practices produced subtle variations, not dramatic contrasts. In each case, sportsmen and sportswomen hunted in a manner rooted in longstanding traditions of elite recreation. Participants displayed virtue, self-control, and skill before audiences of peers. They also demonstrated accomplishment, displayed wealth and status, and socialized in exclusive groups. The difference came with the presence or absence of other forms of recreation and heterosocial circumstances. At bottom, hunting at clubs and retreats privileged masculine display. Their gentlemanly ethos came directly from the elite sporting traditions of the nineteenth century. By contrast, the heterosociality of northerners' estates allowed for greater variety. Hunting remained an exercise in virtue and self-control, but it accommodated moments of light-hearted frivolity and casual banter. Not nearly as much a test of personal character and will, it offered looser, less demanding conditions and opportunities for revelry and relaxation of a kind that had little place at clubs and retreats.

The lowcountry's popularity with elite hunters derived partly from the survival of practices that had died out elsewhere. Comparatively lax game laws allowed for varieties of hunting that had vanished from other locales. Deer driving offers a prime example. The practice of driving game had English roots. Nobles preferred having dogs or men chase animals toward waiting hunters. Nineteenth-century Americans recognized driving as an aristocratic practice. Whereas the still-hunt originated with frontiersmen and Indians, driving game came from European elites. The former embodied egalitarian ideals; the latter demonstrated elite authority.[22]

American hunters debated the relative merits of each type of hunting. Although some hunters regarded driving as unsporting, others considered it challenging. Some considered driving "manly" and derided the

---

[20] Train, *A Carolina Plantation Remembered*, 102.

[21] Lee Brockington, *Plantation between the Waters: A Brief History of Hobcaw Barony* (Charleston: History Press, 2006).

[22] Daniel Justin Herman, *Hunting and the American Imagination* (Washington, DC: Smithsonian Institution Press, 2001), 155–157.

still-hunt as "murderous." Others took the opposite view. Over time, the still-hunt won out. Declining game populations proved more influential than questions about sport and manhood. Since driving tended to destroy large numbers of animals, most northern and Midwestern states outlawed the practice. By the early 1900s, driving had become rare above the Mason-Dixon Line. Southern states, however, continued to allow it. As a result, southern gamelands became havens for men and women who preferred driving, whether because of its aristocratic heritage, attendant thrills, or social pleasures.[23]

The continuing popularity of driving hints at reasons for the low-country's appeal beyond the obvious qualities of inexpensive land, plentiful wildlife, and mild weather. Although these factors mattered, they formed part of a more complex equation. Emergent ideas about the low-country played a crucial role in shaping northerners' experiences. Old plantations made a seemingly majestic past an abiding presence. African American workers showed the persistence of racial hierarchies and the supposedly benevolent race relations of the antebellum era. "Gullah," as a language and set of behaviors associated with dark-complected blacks living in meager and isolated circumstances, underscored the exoticism of the lowcountry and social and cultural conditions rooted in slavery. The lush landscape and coastal scenery echoed portrayals of the South as an unspoiled paradise. In sum, sensory, social, and symbolic qualities accentuated a wide range of real and perceived associations. Coupled with the survival of practices such as driving, the lowcountry affirmed beliefs held by people who saw themselves as members of a social and cultural elite. Recreation thus became an exercise in elite authority and a means of self-definition. Whether any onlookers recognized what it meant to estate owners and their guests mattered little. As with other upper-class activities, recognition among peers mattered most.

VARIETIES OF RECREATION

By the late 1920s, northerners employed an expansive definition of recreation. Although hunting and horseback riding remained favored pursuits, northerners added other activities at a brisk pace. Fishing, boating, carriage rides, and sightseeing became common. These pastimes supplied pleasure, enjoyment, and variety. They added to the appeal of

---

[23] Ibid., 156–157.

the lowcountry, diversified northerners' activities, and provided options for entertaining guests.

Fishing, for example, became a favorite pursuit of some plantation owners. At Medway Plantation, Gertrude and Sidney Legendre often spent afternoons fishing on Back River.[24] Fertig enjoyed "exceptional fishing" at Ponemah, where he caught "bass, bream, and red breast."[25] At Friendfield Plantation, Frances Cheston and her family fished for catfish on the Sampit River.[26] Baruch and his guests routinely fished at Hobcaw Barony.[27] Although hunting provided plantation life with its main attraction, fishing followed close behind.

Boating for pleasure and for transportation also proved popular. Because most estates lay along major rivers, many owners preferred traveling by water rather than by automobile. Poor-quality roads and the pleasure of viewing the lowcountry landscape from coastal rivers and streams provided added incentives. Most plantation owners kept at least one boat at the ready, and many had several. Jacquelin and Florence Holliday, for example, maintained a sizable fleet at Nightingale Hall that included a yacht, a speed boat, a run-about, several rowboats, and eight duck boats.[28] George D. B. Bonbright kept several boats at his Pimlico Plantation, among them "a sea plane, a sea sled, and a Cris-Craft" that he used "for his pleasure and that of his guests."[29] Jesse Metcalf and his wife sped along the Pee Dee River on a "luxurious speed boat capable of making forty-five miles an hour." The Metcalfs used the vessel to reach duck-hunting domains, to visit friends and neighbors, and for occasional trips to Georgetown.[30]

[24] See, for example, Sidney J. Legendre, "Diary of Life at Medway Plantation, Mt. Holly, South Carolina," private collection (copy in author's possession), May 17, 25, and 28, 1937, and May 23, 1939.

[25] Murray, "Ponemah, 'Happy Hunting Ground.'"

[26] Train, *A Carolina Plantation Remembered*, 103.

[27] Brockington, *Plantation between the Waters*, 53, 94, 106. Promotional literature touting the quality of lowcountry gamelands identified fishing as a pastime that could be enjoyed year-round. See N. L. Willett, *Game Preserves of Beaufort, Colleton, and Jasper Counties, South Carolina* (Beaufort, SC: Charleston and Western Carolina Railway Co., 1927), 4, 10–13. Thomas A. Stone also mentioned fishing on the Wando River in his journal. See Stone Journal, May 17, 1937.

[28] Murray, "Holliday Winters in Peedee."

[29] Chlotilde R. Martin, "New Yorker Reclaims Pimlico," *News and Courier*, May 24, 1931, B2.

[30] Chalmers S. Murray, "Metcalf Builds on Peedee Bluff," *News and Courier*, March 22, 1931, B14.

In some cases, boating served recreational purposes. In May 1937, after the quail and duck seasons had ended, the Stones organized a trip with the Legendres and several others. The Legendres arrived at Boone Hall shortly before noon on May 29 and boarded the Stones' boat, the "Wampancheone." The party sailed up the Cooper to Medway, where they spent the night. From there they traveled farther up the river to visit the Kittredges at Dean Hall and then turned back. The highlight of the trip, at least in Thomas Stone's view, came on the return leg, when the group spotted an alligator in the water estimated at "at least eight feet long." Stone shot the animal "in the nose" with a .22-caliber rifle. "It went to the bottom with a terrific splashing and didn't come up again," he later recounted.[31]

Some northerners made waterborne transportation their preferred means of traveling to and from the lowcountry. Franklyn Hutton maintained a boat landing on the Edisto River specifically for his personal yacht. According to the Charleston *News and Courier*, Hutton frequently brought the vessel to South Carolina.[32] Other northerners traveled to the lowcountry in similar fashion.[33] Thomas Yawkey, for example, routinely brought his yacht to his South Island retreat. Like "several other winter colonists," he used it for transporting guests to and from Georgetown.[34] Meanwhile, some northerners opted for "air yachts" instead of seagoing vessels. In April 1935, the *News and Courier* reported that Walker P. Inman, "millionaire sportsman and half-brother of Doris Duke Cromwell," traveled to the lowcountry in a $55,000 airplane. The two-engine propeller-driven Lockeed Electra, one of three of its kind, cruised at 200 miles an hour and weighed 7,000 pounds. Inman piloted it "with the aid of a robot" (presumably an autopilot device). Other winter colonists also flew to the lowcountry. Colonel Robert L. Montgomery, owner of Mansfield Plantation, reputedly kept "several of his planes"

---

[31] Legendre, "Diary of Life at Medway Plantation," May 28, 1937; Stone Journal, May 25, 1937.

[32] "Do You Know Your Lowcountry? Prospect Hill."

[33] See, for example, "Vanderbilt and His Sea, Air Yachts Arrive at Charleston," *News and Courier*, Oct. 14. 1934, A8; Chalmers S. Murray, "Attracted by Climate Northerners Become Land Barons of the Carolina Coast," *News and Courier*, Dec. 15, 1929, A3.

[34] "South Island, Where Cleveland Hunted," *News and Courier*, Feb. 8, 1931, B9. On other estate owners' use of yachts, see Lawrence S. Rowland and Stephen R. Wise, *The History of Beaufort County, South Carolina*, vol. 3: *Bridging the Sea Islands' Past and Present, 1893–2006* (Columbia: University of South Carolina Press, 2015), 287–288.

at the Georgetown airfield, and Dr. James Cowan Greenway and his wife also traveled by air.[35]

Sophisticated flying machines and seagoing vessels set northerners apart. Not only did these vehicles display material wealth, they showed northerners to be on cutting edge of transportation technology. Although northerners favored the lowcountry in part because it offered opportunities to escape the "modern" world, estate owners saw no reason to forgo the latest advancements during their winter sojourns. Even while enjoying the pleasures of a region steeped in the ambiance of an earlier time, northerners insisted on using current technology. In this fashion they brought two worlds into close association. If anyone found the juxtaposition curious, it went unmentioned. For owners and onlookers alike, the mingling of new machinery and old estates seemed just one part of the scene that developed with the rise of plantations dedicated to sport and leisure.

VISITING, RECREATION, AND PERFORMANCE: SOCIAL DIMENSIONS OF PLANTATION LIFE

As the number of northern-owned estates grew, plantation life took on new dimensions. The social scene of the late 1920s and 1930s announced the arrival of a status-driven realm of activity. Plantations ceased to be spaces of quiet relaxation and refuge. Instead, they became settings for vigorous social intercourse. In turn, northerners' activities took on qualities reminiscent of practices in better-established locales. An expanding range of social and recreational activities placed new emphasis on polite behavior and ritualized displays. Although plantation life never came close to approximating the status-seeking and pretention that characterized upper-class culture in the North, it nonetheless became a carefully choreographed realm of activity, replete with its own protocols, expectations, and boundaries. Plantation life differentiated northerners' estates from other holdings and announced a new stage in the development of grand "plantations."

The basic reason for the shift came from growing numbers of visitors. By the late 1920s, estate owners routinely hosted guests for stays ranging

---

[35] "New 'Bird' to Winter in South Carolina Lowcountry," *News and Courier*, Apr. 28, 1935, C3; "Many Motorists Use Coast Route," *News and Courier*, Apr. 15, 1936, D3; "Ol' Maussa Greenway Gives Party for Plantation Hands," *News and Courier*, Jan. 10, 1935, 3.

FIGURE 7.1 Cocktail party at Gippy Plantation, 1936. At the peak of northerners' activities, social gatherings typical of upper-class havens in the North became common.

From the Grimball-Gaud Papers, South Carolina Historical Society, Charleston, SC.

from a few days to several weeks. Socializing became a major focus. Lunches, dinner parties, and the simple pastime of visiting became routine. For some owners, sightseeing, fishing, and horseback riding became nearly as common as hunting. As recreation became increasingly social, estate owners found themselves saddled with hosting and entertaining. Whereas northerners had previously enjoyed relatively quiet circumstances, by the late 1920s, plantation life became a constant blur of activity. For many participants, it became an ongoing parade of parties, teas, hunts, and gatherings that mirrored social life in more established locales (see Figure 7.1).

Correspondence, journals, and newspaper accounts bring these developments into focus. In March 1940, Sidney Legendre observed "the spring season is open season on plantation owners." With these words he identified a plight that befell many estate owners.[36] Although northerners sought seclusion and quietude, plantation life left many longing for both. Continual flows of visitors, social gatherings, and recreational outings left owners struggling to keep up with the ongoing rush of activity and yearning for moments alone.

Thomas Yawkey, for example, "entertain[ed] extensively during the winter months." He and his wife insisted that friends and family share in "the pleasures of field and stream" at their South Island estate. As owner of the Boston Red Sox, Yawkey routinely brought members of his baseball team to the lowcountry to hunt and fish. Stars such as Ty Cobb, Tris Speaker, and Eddie Collins all visited the Yawkeys' retreat. One guest, pioneering aviator Lieutenant Pete Quesada, opined that he would like nothing better than the chance to spend "several months" at South Island. Quesada viewed the prospect of "lying around in the sunshine, taking an occasional hike through the scented woods, and once in a while trying his luck with deer and ducks" as tremendously appealing.[37] Senator Joseph S. Frelinghuysen sometimes hosted so many guests at his Rice Hope Plantation that he had to use an older house as a "dormitory of a sort" to accommodate "overflow" from the main residence.[38]

At Medway Plantation, Gertrude and Sidney Legendre hosted a steady stream of visitors. Throughout the fall and winter months, the Legendres entertained guests from afar, other plantation owners, and a handful of lowcountry people. Some came when invited; others dropped in unexpectedly. In some cases the Legendres hosted large gatherings, but most of the time they opted for smaller groups. The Legendres also participated in outings organized by other plantation owners. In all, the Legendres' version of plantation life involved extensive socializing.

Rarely did more than a few days pass without an unannounced visit. One day in early May 1937, Sidney Legendre returned from a short trip to Summerville to find "the Stones at the gate of the place." "They had come over for tea," he noted. Several days later, he and Gertrude hosted an evening gathering. As Sidney recorded in his diary, "George Buist, the Stones, and the Sharps came to dinner last night." The following day,

[36] Legendre, "Diary of Life at Medway Plantation," March 29, 1940.
[37] "South Island, Where Cleveland Hunted," *News and Courier*, Feb. 8, 1931, B9.
[38] "Frelinghuysen Builds at Rice Hope," *News and Courier*, Aug. 9, 1931, B12.

"Henry Lowndes and his wife came out to the plantation in the afternoon." The Legendres talked with their guests about the lowcountry and how Mr. Lowndes "would rather live in it than any place in the world." On the morning of March 17, "the Morowitzes came over from Fenwick Hall with two guests, Mr. George Rublee and Mr. Dow." The group talked for a time before the Morowitzes and their companions departed. The Legendres immediately got ready to go fishing, but, only minutes later, "the Simmons Warrings came calling." Later in the day, Sidney Legendre recounted the discussion that ensued. Musing about his talent for spur-of-the-moment chitchat, Legendre observed, "conversation has become automatic with me." "I rarely know what I am saying or what the other person is saying," he noted. When "you are continually seeing people," Legendre recounted, "you do not have any time to think yourself unless you do it while you are with them." Legendre had apparently trained himself to feign attention while thinking about other matters. "I listen to them criticize the Government [while] thinking where am I going to put the new corn crib, or when we should begin clearing the new field for the next year's crop," he observed.[39]

At times, the flow of guests became bothersome. In one remarkable entry, Legendre recalled an onslaught of visitors that left him exhausted and longing for solitude. "We had an incessant, never-ending stream of guests for the months of March and April. People in the house, people coming for three meals a day," he wrote. The whirl of activity prevented Legendre from working on a number of projects and putting crops in the ground. Speaking of his efforts to catch up after the guests' departure, Legendre recounted, "it was the maddest race against time that I have ever gone through." The experience left him "so exhausted" that he took "dinner in bed" two nights in a row.[40]

The following spring, the Legendres again hosted a large number of visitors. By the end of March, the couple felt they had seen everyone who "has a car and is tired of the big cities, or who does not want to motor to the north from Florida without breaking up the trip." Privately, Sidney complained vehemently about the situation. "They descend like locusts, eating all the thoughts out of your mind [and] leaving only the husk," he wrote. Although he noted that "people are always amusing, polite, generally interesting ... and flattering in their admiration of [Medway]," Legendre had grown weary of being a gracious host. He wondered if

[39] Legendre, "Diary of Life at Medway Plantation," May 10–13 and 17, 1937.
[40] Legendre, "Diary of Life at Medway Plantation," May 9, 1939.

others found social activity as taxing as he did. "Perhaps one is born sociable as you are born musical," Legendre mused. If so, he concluded that it would easily be possible to "strive all your life and never attain the true feeling and appreciation of the genius born to his trade."[41]

Although Legendre's comments may seem extreme, other estate owners had similar experiences. In the spring of 1937, Thomas and Alexandra Stone found themselves beset with more visitors than they would have liked. In mid-February, Thomas Stone wrote in his journal, "At this season of the year Alex dreads her morning mail." He continued, "Practically the whole Eastern Seaboard moves South, and they all 'would like so much to stop over and spend a few days with us.'" The Stones had already hosted a large number of visitors and expected more in the coming weeks. As Stone noted, "We have a big week-end for Washington's Birthday ... Bill and Fordy arrive Tuesday morning next, for the week; the Stone family arrives about February 26th or 27th ... the Charlie Stones are coming sometime, the Beers are coming sometime ... the Wolvens want to come – and so on, ad infinitum." Looking ahead to the next round of guests, Stone mused, "Next week and the week-end ... will be a very good test to see how elastic the house is."[42]

Several weeks later, as the season drew to a close, the Stones looked forward to having Boone Hall to themselves. On March 17, Thomas Stone wrote, "After this weekend ... it looks as if the stream of guests will stop for a while – thank the Lord. Alex and I are getting pretty fed up with the people."[43]

The experiences of the Legendres and the Stones underscore profound changes in plantation life. Although northerners prized peace and solitude, developments in the 1920s made opportunities to enjoy either scarce. The estate-making boom, the lowcountry's growing popularity, improved transportation connections, and more cohesive social networks eroded the isolation and relative quietude of earlier years. In turn, owners' experiences changed. Although plantations continued to allow for relaxation and rejuvenation, they became intensely social domains. In the process, owners' priorities shifted. Although northerners' estates had always offered possibilities for performative display, the importance of those performances and their frequency grew over time. By the early 1930s, socializing in a manner consistent with the prevailing norms of upper-class culture became central to plantation life.

---

[41] Ibid., March 29, 1940. [42] Stone Journal, Feb. 12, 1937.
[43] Ibid., March 17, 1937.

IDEAS

Sportsmen and sportswomen drew meaning from beliefs about the lowcountry, its past, and its plantations. By viewing the lowcountry in relation to other parts of the South and other upper-class recreational destinations, they particularized their experiences. In turn, northerners gave meaning to their "plantations" and distinguished themselves from other elites. In part, northerners' views incorporated common beliefs about "the South" and qualities that differentiated southerners from other Americans. At the same time, estate owners also drew upon knowledge of other locales and familiarity with the lowcountry, their plantations, and their favored activities. Northerners' views legitimated their choices and behaviors. They affirmed investments in lowcountry estates and the pastimes that northerners came to the region to enjoy. In this fashion, plantation life developed an ideological basis. Although the views of journalists, lowcountry people, and other observers shaped popular knowledge of northerners' estates and activities, northerners also influenced the perspectives that developed.

In formulating understandings of the lowcountry, northerners took practical and imaginative considerations into account. The first category encompassed factors such as climate, ease of access from northern cities, the availability and cost of land, and the type, variety, and quality of hunting available. These characteristics formed the basis of distinctions with Florida and other parts of the South. The second category encompassed less tangible beliefs about the lowcountry and the South. It included ideas about history, culture, social hierarchies, and qualities northerners saw as specific to the lowcountry. These categories, rather than being mutually exclusive, overlapped. Prevalent beliefs about the South found such widespread acceptance in 1930s-era America that they held the status of near facts. So did emergent beliefs about the lowcountry. As promotional literature and newspapers and magazines increasingly depicted the lowcountry as a region with a distinctive history, traditions, and society, a relatively narrow set of characteristics became generally accepted, almost to the extent of being seen as innate.

Climate figured prominently among the reasons northerners prized the lowcountry. Upon arriving in Georgetown one fall, Bernard Baruch declared that "this section of the country has the best climate in the world." He considered it preferable to the "extreme south" – a thinly veiled reference to Florida – where he viewed temperatures as often "too warm to engage vigorously in field sports."[44] Other northerners concurred. In a letter to an acquaintance, Senator Joseph S. Frelinghuysen touted the mild temperatures that prevailed in the lowcountry throughout the fall and winter months. "South Carolina affords a better climate than Florida for the simple reason that it is not enervating," he wrote. Frelinghuysen described it as "the October climate all winter." As he told his correspondent, "You get a little touch of frost now and then" but rarely anything harsher. Overall, he found conditions in the lowcountry "very stimulating." "With the pines and sea air," Frelinghuysen added, "it [is] an ideal climate in the winter time."[45]

Northerners also saw the lowcountry as possessing qualities that made their stays especially pleasurable. In addition to the attraction of unspoiled coastal scenery, northerners saw less noticeable characteristics as contributing to their enjoyment. According to one account, many regarded the air of the Carolina lowlands as having a "wine-like quality" not found farther south.[46] Gertrude Legendre once longed for the "cool, wet, aromatic Carolina air" of her Medway Plantation while traveling overseas.[47] For her husband, Sidney, nothing could match the pleasure of an early-morning horseback ride at Medway in "morning air ... filled with the perfume of the pines, and honey suckle."[48] Statements such as these indicate the importance that northerners ascribed to sensory experiences. Such experiences gave meaning to northerners' visits and added to the differences they saw between the lowcountry and other locales.

Northerners also connected their experiences to familiar tropes about the South. Just as journalists cast the lowcountry as a land of unspoiled charm and seclusion, northerners viewed their plantations in similar fashion.[49] John A. Miller, for example, considered his Estherville Plantation the "one place where he [could] relax and forget about the cares of the business world."[50] Bernard Baruch saw Hobcaw Barony as a site of refugee and rejuvenation that offered opportunities for what he called

[44] Murray, "Attracted by Climate."

[45] Joseph S. Frelinghuysen to George D. B. Bonbright, May 8, 1935, box 1, Frelinghuysen Papers. For similar views, see "Mr. Harmon on Charleston," *News and Courier*, Sept. 22, 1925, 4; "Ponemah, 'Happy Hunting Ground'"; "Northerners Lease 12,000-Acres from Rhem Family," *News and Courier*, Jan. 30, 1930, 7.

[46] Murray, "Attracted by Climate."   [47] Legendre, *Time of My Life*, 145.

[48] Legendre, "Diary of Life at Medway Plantation," May 17, 1937.

[49] See, for example, E. T. H. Schaffer, "Sea Island Lure," *American Motorist* 12, no. 2 (Feb. 1929): 12–13, 34–37; James C. Derieux, "The Renaissance of the Plantation," *Country Life* 41, no. 3 (Jan. 1932): 34–39.

[50] "Esterville, Transformed Coastal Estate," *News and Courier*, Feb. 15, 1931, B7.

"detached contemplation." "In this hectic age of distraction," Baruch observed, "all of us need to pause every now and then" and step outside "the rush of the world." His most restful and restorative pauses came at Hobcaw, which he described as a "veritable Shangri-La." With "the finest duck hunting in the United States ... four rivers and a bay abounding in fish; vast stretches of almost primeval forest, and – no telephone," Hobcaw supplied ample opportunities for renewal.[51]

Idealization of physical spaces and places figured prominently in northerners' views of the lowcountry. Sidney Legendre saw his Medway Plantation in ways that suggest deep attachment. Walking across the grounds of the main house late one evening, Legendre found himself in "fairy land." Recounting the moment later, he wrote, "The giant oaks threw the moon on to the ground in a series of intricate patterns, and the fire flys lit up the lawn with their sparking light. The old house sat there and dreamed of its three hundred years of existence, and the people that fought, loved and died inside its walls." On another occasion, Legendre wrote of arriving at the plantation after dark and being awed by the majesty of the house and grounds. With the house "flooded in moonlight," he considered it a "crime" that swarms of mosquitoes forced him and Gertrude "to remain in doors after sunset."[52]

Although Legendre found the house at Medway especially enchanting, other dwellings also captured his attention. In one instance, Legendre expressed strong sentiments about a house at Pine Grove Plantation, which abutted Medway. He wrote:

When a house is over one hundred and fifty years it begins to take on character and like an old person broods over the days that used to be. When you walk its halls it tries to tell you what has passed through them, of the lives, tragedies and happinesses that have taken place within its encircling arms. Every time I enter the old place it seems to beg me to save its life, to prevent its solid old timbers from decaying to bring to its silent walls the sound of human voices and laughter and tears.[53]

Legendre's vision of old houses as repositories of human history is suggestive of the thinking that inspired northerners' actions. Viewing history as an embodied presence formed a necessary precursor to restoration. Seeing material objects as vestiges of another time created possibilities for sympathetic, respectful treatment. Although it is difficult to say how

[51] Bernard, *Baruch: My Own Story*, 267–268.
[52] Legendre, "Diary of Life at Medway Plantation," Feb. 24, 1937.
[53] Ibid., Feb. 21, 1941.

many northerners saw old houses the same way, their collective actions indicate that many did. Otherwise, restoration of select structures and purposeful retention of distressed elements is difficult to explain. In Legendre's case, his comments illuminate sentiments that he and Gertrude expressed in rehabilitating the house at Medway. The Legendres saw themselves as caring for a valued artifact and made choices based on that understanding. Retaining the weathered appearance of the exterior seemed, in their view, an appropriate method of highlighting the building's age and longevity.

Of course, at bottom, hunting supplied the main reason for the lowcountry's appeal. Willis E. Fertig made his estimation of the lowcountry's gamelands clear when he declared, "Not even in England and Scotland, where outdoor sport is almost a profession, can game preserves be found where sport is enjoyed every month of the year as it is in Georgetown county."[54] Other evidence indicates that Fertig did not exaggerate. Northerners continually extolled the virtues of the lowcountry as a sporting destination, and at least twenty estate owners had sufficient experience to make informed judgments. At least eight had hunted grouse in the Scottish Highlands, an exclusive pursuit available only to exceptionally wealthy, well-connected hunters,[55] and others had hunted in exotic destinations around the globe. Dr. A. W. Elting, for example, had hunted in Africa on at least one occasion and Robert R. M. Carpenter had hunted extensively in Africa and South America. Franklyn Hutton spent portions of his summers hunting in Czechoslovakia, which offered conditions similar to the Scottish Highlands. The Legendres, of course, had hunted in Africa, the Middle East, Southeast Asia, British Columbia, and Alaska.[56]

---

[54] Murray, "Ponemah, 'Happy Hunting Ground.'"

[55] Owners of large estates who hunted grouse include Herbert L. Pratt, James Paul Mills, Herbert Pulitzer, Bernard M. Baruch, Robert Goelet, Payne Whitney, Harry Payne Whitney, and Robert H. McCurdy. See "Americans Facing toward Homeland," *New York Times*, Sept. 8, 1907, C2; "Whitney's $50,000 Shoot," *New York Times*, Aug. 17, 1909, 1; "Mackay and Baruch to Shoot in Scotland," *New York Times*, July 23, 1923, 13; "Americans Invade Scotland to Shoot," *New York Times*, July 5, 1925, E3; "Grouse Abundant on Scottish Moors," *New York Times*, Aug. 12, 1934, E3; "Americans Sail to Shoot Grouse," *New York Times*, Aug. 6, 1939, D1. Horatio S. Shonnard, owner of Harrietta Plantation near Georgetown, also maintained a shooting lodge in Scotland for a time. See "Horatio S. Shonnard, Once a Stockbroker," *New York Times*, Oct. 12, 1946, 14.

[56] The Legendres' exploits are detailed in "All Animals, Even Fiercest, Run from Man, Berkeley Big Game Hunter Says," *News and Courier*, Feb. 2, 1936, C2; Legendre, *Time of My Life*; Sidney Jennings Legendre, *Land of the White Parasol and the Million Elephants* (New York: Dodd, Mead, and Co., 1936); Sidney Jennings Legendre,

In short, some estate owners knew the difference between average gamelands and truly exceptional shooting. When these men and women championed the lowcountry, they spoke from experience.

RACE

For owners and their guests, plantation life involved immersion in settings where dark-skinned peoples outnumbered them by as many as ten to one. Lowcountry blacks' seemingly primitive behaviors, speech, and appearance made them the apparent equals of exotic peoples in faraway locales. "Gullah Negroes," with their strange dialect, beliefs, and behaviors, continually awed, astounded, and intrigued. Commonly viewed as a living legacy of slavery, lowcountry blacks seemed a direct result of the stagnation that had set in after the Civil War. The same conditions that had created a timeworn landscape also resulted in an African American population that seemed barely removed from slavery. Although northerners' interaction with African Americans took many forms, estate owners generally saw blacks as laborers, as social inferiors, and characteristic of "the South." These categories gave meaning to northerners' experiences and showed acceptance of prevailing beliefs about race and social hierarchies.[57]

Plantation superintendents experienced plantation life on different terms. For these men, plantation life involved managing a large estate with a predominantly black workforce. Plantation superintendents fell

Okovango, Desert River (New York: J. Messner, 1939). On Elting's African safari, see Chlotilde R. Martin, "Pine Island, Beaufort County, Offers Seclusion and Sport," *News and Courier*, Dec. 20, 1931. On Carpenter's adventures, see "R. R. M. Carpenter, Industrialist, Dies," *New York Times*, June 12, 1949, 76. Franklyn L. Hutton reputedly went abroad each summer, with "hunting in Czechoslovakia being a chief diversion." See J. V. N., Jr., "Do You Know Your Lowcountry? Prospect Hill," *News and Courier*, May 9, 1938, 10. The Legendres may also have hunted in Czechoslovakia. See Legendre, "Diary of Life at Medway Plantation," May 17, 1937.

[57] For references to "Gullah Negroes," see N. L. Willett, *Game Preserves and Game of Beaufort, Colleton and Jasper Counties* (Beaufort, SC: Charleston and Western Carolina Railway Co., 1927), 16; Duncan Clinch Heyward, *Seed from Madagascar* (1937; reprint, Columbia: University of South Carolina Press, 1993), 163. On lowcountry blacks as a living legacy of slavery, see E. T. H. Shaffer, "The Ashley River and Its Gardens," *National Geographic* 49, no. 5 (May 1926): 530–531; Julia Wood Peterkin, *Roll, Jordan, Roll* (New York: Robert B. Ballou, 1933), 17. On northerners' views of southern blacks in general, see Nina Silber, *The Romance of Reunion: Northerners and the South, 1865–1900* (Chapel Hill: University of North Carolina Press, 1993), 78–81; Rebecca Cawood McIntyre, *Souvenirs of the Old South: Northern Tourism and Southern Mythology* (Gainesville: University Press of Florida, 2011), chap. 4.

into two categories. Northerners sometimes hired native southerners to manage their estates. These men often had ties to the estates they managed and relationships with laborers and local merchants. All knew the social and cultural norms of the region, which facilitated interaction with workers, neighboring landowners, and suppliers.[58] In other instances, northerners hired northern men as managers. Members of this group held prejudices characteristic of working-class whites in northern cities. Popular stereotypes informed their views of African Americans and southern whites. Northern superintendents recognized themselves and their employers as belonging to a different culture, with markedly different beliefs, values, and behaviors than southerners.[59]

African Americans experienced plantation life differently. For black plantation laborers and tenants, plantation life amounted to a continually unfolding series of interactions with white people who spoke and acted differently from any others they knew. African Americans recognized northerners as different from white southerners. Having traditionally worked for members of the planter class, African Americans saw northerners as similar to, yet different from, their usual employers. Exactly how lowcountry blacks regarded northerners and what differences they recognized is unclear. Limited sources make it difficult to know African Americans' views with certainty. The historical record offers suggestive glimpses, however, all of which are best understood in relation to lowcountry blacks' general approach to wage labor. As scholars such as John Scott Strickland have shown, lowcountry blacks resisted arrangements that limited their autonomy. Preferring self-sufficiency to monetary gain, African Americans generally farmed for their own consumption and for market, hunted and fished for the same purposes, and worked for wages to supplement household production. This approach allowed African Americans to take advantage of wage labor without becoming dependent on it. The rise of northern-owned estates greatly expanded opportunities for wage-earning in many parts of the lowcountry. Some blacks

[58] At Friendfield Plantation, Radcliffe Cheston, Jr., employed Pat McClary, the son of the former owner, as his superintendent. See Train, *A Carolina Plantation Remembered*, 50–51. Bernard M. Baruch also employed local men. See Brockington, *A Plantation between the Waters*, 54–55. James Derieux also noted that some northerners employed former owners as managers. See Derieux, "The Renaissance of the Plantation," *Country Life* 61, no. 3 (Jan. 1932): 36.

[59] Joseph Frelinghuysen, for example, employed a northerner named Herman Hansen as his superintendent at Rice Hope Plantation. See correspondence in box 6, Frelinghuysen Papers.

seized options for full-time employment, but most chose part-time labor. Regardless, northerners' demand for labor benefited African Americans, for it offered more reliable employment and better working conditions than virtually any of the alternatives.[60]

Owners of large estates and their guests saw plantation life as an exotic realm where uncommon experiences became the norm. Just as owning a plantation supplied tangible links to a select part of the American past, the experience of a plantation involved immersion in a distinctively southern milieu. African Americans' presence in the physical space of plantations supplied plantation life with characteristics that northerners viewed as distinctive. Northerners saw plantations as quintessentially southern spaces where racial hierarchies cast themselves in sharp relief. Moreover, since plantations had once been landed estates of a kind, northerners also viewed them as connoting status and authority. The history of plantations, as Americans of the era understood it, supplied ties to a quasi-feudal society. Together, these beliefs shaped plantation life. By delineating the relationship between lived experience and the imagination, they made plantation life considerably more than a group of leisure pursuits carried out with agreeable company.

Thus, whereas plantation life centered on labor for the majority of people present, northerners saw it as a realm where characteristically southern experiences became common and popular stereotypes entered into lived experience. Plantation life tied imagined realms to places called plantations and made them settings for experiences characteristic of "the South," "the lowcountry," and traditions, beliefs, and cultural practices associated with each. Although popular discourses proved crucial, the practical requirements of northerners' estates did also. Black labor not only sustained northerners' estates and activities; northerners considered white supervision of black workers typical of "the South," plantations especially. In this sense, the practical and imagined dimensions of northerners' experiences reinforced one another.

In general, African American labor organized itself into three categories. One encompassed domestic service. Virtually all plantation owners

[60] John Scott Strickland, "Traditional Culture and Moral Economy: Social and Economic Change in the South Carolina Low Country, 1865–1910," in *The Countryside in the Age of Capitalist Transformation: Essays in the Social History of Rural America*, ed. Steven Hahn and Jonathan Prude (Chapel Hill: University of North Carolina Press, 1985), 141–178; Scott E. Giltner, *Hunting and Fishing in the New South: Black Labor and White Leisure after the Civil War* (Baltimore: Johns Hopkins University Press, 2008), 130–132.

employed at least one African American housekeeper and cook. At large estates, two or more proved common. Men and women working in this capacity prepared food, kept house, did laundry, cleaned, assisted with personal needs, and attended to miscellaneous chores. Women predominated. Although some northerners employed black men as butlers, women performed most domestic labor.[61]

A second category centered on general farm labor. African Americans performed tasks ranging from cultivation of row crops to upkeep of buildings and grounds, care of animals, maintenance of water-control systems, planting feed crops, forest maintenance, and a host of other activities. This variety of labor required few skills, paid poorly, and often subjected workers to difficult conditions. It encompassed the largest number of jobs on northern-owned estates.[62]

A third category of labor pertained to recreation, hunting especially. Just as at hunting clubs and retreats, African Americans performed valuable service as guides and assistants on northerners' estates. Estate owners, like other northerners, prized blacks' knowledge of the natural world and their ability as hunting guides. Northerners depended on black guides to lead them to good shooting locations and for information about animal behavior. Some African Americans also possessed specialized skills that benefited northerners – the ability to drive game, for example. General laborers drove wagons, ported equipment, cleaned game, and tended to sundry chores. In sum, northerners depended on black workers in multiple ways. Without black labor, sportsmen's and sportswomen's stylized brand of hunting did not occur.[63]

By placing lowcountry blacks and upper-class northerners in close contact, plantation life brought together two groups of people who could hardly have been more different. On the one hand, northern plantation owners and their guests led privileged lives. Wealthy, well educated, and socially and politically connected, they belonged to a class of people who

---

[61] This composite portrait is drawn from the following sources: Legendre, "Diary of Life at Medway Plantation"; Brockington, *Plantation between the Waters*, 54–55; Train, *A Carolina Plantation Remembered*, chap. 4; Clare Boothe to "My gentle little Paolo" [Paul Gallico], Feb. 31, 1933[?], folder 4, box 795, and Clare Boothe Luce to Mike, [Jan. 1935], folder 5, box 792, both in Clare Boothe Luce Papers, Library of Congress, Washington, DC; Clare Boothe Luce, "The Victorious South," *Vogue*, June 1, 1937, 79–81, 120–122; Agreement between Joseph S. Frelinghuysen and Charles E. Bedeaux, Nov. 23, 1933, 4, box 1, Frelinghuysen Papers.

[62] Giltner, *Hunting and Fishing in the New South*, 129–131.   [63] Ibid., 86–93, 113–134.

viewed, acted, and styled themselves as an elite. Overwhelmingly Anglo-Saxon and Protestant, they held conservative political views. Although nominally egalitarian, northerners stood firmly convinced of their social and racial superiority. For them, travel to the lowcountry represented immersion in a locale as exotic as any on North American shores and comparable to others farther afield. As a region with an aristocratic heritage, stunning scenery, and a massive population of African Americans, the lowcountry possessed qualities that made it "the South" yet unique. Ultimately, the lowcountry appeared as a land unto itself. Sportsmen and sportswomen viewed the region as the South at its most aristocratic, most exotic, and least disturbed. For many, the region satisfied longings for uncommon encounters and experiences possessed of exceptional authenticity.

By contrast, lowcountry blacks lived dramatically different lives. Impoverished, disenfranchised, and barred from social and economic advancement by the rigid caste system of the postbellum South, they lived in circumstances as meager as those known to any Americans of the time. Many continued to live in houses built for slaves; others occupied crude dwellings of more recent origin. Few knew modern comforts and conveniences. Amenities such as indoor plumbing and electricity had no place in the rural lowcountry of the 1930s. Not until the rural electrification crusade of the late 1930s did that begin to change but, even then, African Americans lagged behind. Rural whites generally received electricity, indoor plumbing, and other amenities sooner than rural blacks.[64]

Agriculture sustained the lives of lowcountry blacks. Although African Americans achieved remarkably high levels of land ownership in some areas, that alone did not ensure social advancement. As Eric Foner has observed, the post-emancipation history of the lowcountry is an object lesson in the ambiguity of historical outcomes. Although lowcountry blacks came closer to securing economic and political power than in other parts of the South, by the beginning of the twentieth century, disenfranchisement, economic coercion, and outright violence erased most of the gains made in the immediate postwar period. Amid these circumstances, unusually high levels of land ownership and deep bonds of community helped African Americans weather Jim Crow but did nothing to challenge

---

[64] Walter B. Edgar, *History of Santee Cooper, 1934–1984* (Columbia: R. L. Bryan Co., 1984); T. Robert Hart, Jr., "The Santee-Cooper Landscape: Culture and Environment in the South Carolina Lowcountry" (PhD diss., University of Alabama, 2004), 35, 94–95, 106–140.

the institution. Although many lowcountry blacks attained greater autonomy and self-sufficiency than their counterparts elsewhere, they still faced extraordinary hardships. Most relied on a combination of farming, hunting and fishing, and intermittent wage labor. Poverty, social and economic marginalization, and severely constrained opportunities characterized the lives of lowcountry blacks, with debilitating consequences for the region and the cause of black freedom in general.[65]

Culturally, lowcountry blacks differed from other African Americans. The Gullah language and folklore, expressive religious practices, and exceptional solidarity consistently drew notice from visitors. Social networks rooted in antebellum plantation communities fostered the sense of a people barely removed from slavery. During the 1920s and 1930s, investigations by sociologists and anthropologists highlighted conditions that explained the persistence of folk traditions and the survival of apparently "African" behaviors and beliefs. Researchers quickly discovered social structures descended from antebellum plantations and sustained by fierce commitments to self-sufficiency. Other scholars traced the African roots of religious and cultural practices, and later studies determined that the Gullah people of the lowcountry had low levels of admixture with Native Americans and whites. In sum, when people of the 1920s and 1930s spoke of lowcountry blacks' "African" qualities and portrayed them as different from other African Americans, their judgments had a basis in fact.[66]

---

[65] Eric Foner, *Nothing but Freedom: Emancipation and Its Legacy* (Baton Rouge: Louisiana State University Press, 1983), 110. On black landownership, see Bureau of the Census, *Fifteenth Census of the United States: 1930, Agriculture*, vol. II (Washington, DC: Government Printing Office, 1932), 163–164. On the lives of rural blacks, see T. J. Woofter, Jr., *Black Yeomanry: Life on St. Helena Island* (New York: Henry Holt and Co., 1930); Clyde Vernon Kiser, *Sea Island to City: A Study of St. Helena Islanders in Harlem and Other Urban Centers* (New York: Columbia University Press, 1932), chaps. 2–4; Rowland and Wise, *Bridging the Sea Islands' Past and Present*, 178–195, 207–210, 217, 218–220, 327–328; Carolyn Baker Lewis, "The World around Hampton: Post-Bellum Life on a South Carolina Plantation," *Agricultural History* 58, no. 3 (July 1984): 456–476.

[66] Woofter, *Black Yeomanry*; William S. Pollitzer, *The Gullah People and Their African Heritage* (Athens: University of Georgia Press, 1999); Margaret Washington Creel, *A Peculiar People: Slave Religion and Community-Culture among the Gullahs* (New York: New York University Press, 1988); Patricia Guthrie, *Catching Sense: African American Communities on a South Carolina Sea Island* (Westport, CT: Bergin and Garvey), 1996. See also Stephen R. Wise and Lawrence S. Rowland, *The History of Beaufort County, South Carolina*, vol. 2: *Rebellion, Reconstruction, and Redemption, 1861–1893* (Columbia: University of South Carolina Press, 2015), chaps. 8 and 11; Rowland and Wise, *Bridging the Sea Islands' Past and Present*, chap. 8.

Characterizing relations between northerners and African Americans, even in the most general terms, is difficult. Northerners held differing views of lowcountry blacks and individual owners often exhibited contradictory behaviors and beliefs. Like white southerners, northerners' relationships with African Americans varied greatly. Some amounted to simple relations of authority; others involved great complexity. In many cases, northerners barely knew the black men and women who lived and worked on their lands. Frequent references to "the Negroes" evinced a tendency to see all African Americans as more or less the same. Moreover, use of superintendents for plantation management placed intermediaries between owners and blacks. Northerners who visited their plantations for short stays and relied heavily on superintendents had detached relations with laborers. Northerners who spent more time in the lowcountry and participated directly in plantation operations had closer contact with at least some workers. In general, northerners had close interaction with house servants and hunting guides and more distant relations with general laborers.[67]

Regardless of how much contact northerners had with African Americans, northerners embraced the pervasive racism of the age. Northerners viewed lowcountry blacks as racially inferior – less intelligent, predisposed toward indolence and sloth, and prone to acting on base instincts. Whether or not northerners believed blacks to have regressed since slavery, as many white southerners did, is an open question. Certainly, most northerners believed emancipation had been necessary for the moral and economic progress of the nation, but had it benefited blacks as a race? No doubt many thought it had not. Regardless of northerners' views on the subject, most saw lowcountry blacks as primitive beings with unrealized potential but not the capacity for social equality. Northerners recognized that some African Americans might attain education and rise in society but believed that as a race they belonged with the laboring masses.[68]

Beyond these beliefs, northerners saw lowcountry blacks through several overlapping lenses. First, they appeared as products of history – a people who had once labored on large estates as human chattel, in a manner akin to feudal serfs. In this sense, African Americans

---

[67] On northerners' interactions with African Americans, see Legendre, "Diary of Life at Medway Plantation," May 23, 1939; Feb. 26, 1940; Apr. 17, 1941.

[68] For examples of estate owners' racialized thought, see, for example, ibid., May 17, 1937; May 23, 1939; Feb. 21, 1941; Apr. 17, 1941.

represented a tangible link to the past and a manifestation of the seemingly intractable "race problem" of the era.[69] Second, northerners saw African Americans as part of the South's picturesque charm. As part of a seemingly endless panorama of scenic beauty, local color, and "otherness," African Americans figured among the qualities that differentiated the South from the rest of the nation and the lowcountry from other parts of the South.[70]

Third, northerners recognized lowcountry blacks as more docile and deferential than other African Americans, especially those in the North. In the eyes of many observers, they seemed accepting of their subordinate status and uninterested in challenging the color line. Indeed, some northerners explicitly identified stable racial hierarchies as a reason for the lowcountry's appeal. Lowcountry boosters routinely touted the region's predominantly native-born population and stable race relations. Like industrial promoters, they saw these attributes as selling points. Many winter colonists agreed. The absence of class conflict, stable racial hierarchies, and a plentiful supply of compliant labor enhanced the lowcountry's appeal. Although estate owners saw all lowcountry blacks as deferential, they considered those living in rural areas the most acquiescent of the lot. Like lowcountry whites, northerners distinguished between "city negroes" and "plantation negroes." In Charleston, it seemed, blacks displayed some of the "uppity" tendencies that characterized African Americans in northern cities. In rural areas, blacks showed deference and respect. Moreover, large numbers of aged African Americans – "old-time darkies," in popular parlance – seemed a throwback to the antebellum era. In the rural lowcountry, then, a social order closer to what northerners and white southerners saw as desirable prevailed.[71]

In sum, northerners came to the lowcountry with well-established views on race and had no intention of trying to change the status quo.

---

[69] See, for example, Peterkin, *Roll, Jordan, Roll.*

[70] See, for example, ibid.; Stoney and Matthews, *Black Genesis,* ix–xxv.

[71] See, for example, the remembrances of Willie Washington that appear in Virginia Beach, *Medway* (Charleston: Wyrick and Co., 1999), 37. A similar view is expressed in Lewis, "The World around Hampton," 475. Sociologists studying rural African American communities in the late 1920s immediately recognized imbalances in sex and age distributions caused by outmigration of working-age men and women to urban areas. As T. J. Woofter observed, "old people and women form a disproportion number of those left behind." See Woofter, *Black Yeomanry,* 90–91. On race and class relations in general, see Yuhl, *A Golden Haze of Memory: The Making of Historic Charleston* (Chapel Hill: University of North Carolina Press, 2005), 134–135, 178–179.

The ready availability of African American labor and blacks' marginalized status figured among the reasons for the region's appeal. Northerners sometimes showed kindness and generosity toward black workers and some took an interest in their health and well-being, but reforming the social order of the Jim Crow South had no place in their agenda. Northerners' aims lay elsewhere.[72]

On the question of how African Americans viewed northern plantation owners and their guests, the picture is less clear. Tradition holds that rural blacks welcomed northerners for two reasons. One was economic opportunity. Northern-owned estates created a significant demand for labor. Construction of new houses and outbuildings created an initial wave of jobs, and upkeep and operation of northerners' estates established ongoing demand. In practical terms, northerners offered opportunities for black workers to earn steady wages year-round. In some areas, northerners' estates marked the first time African Americans had opportunities for consistent employment. Agriculture and timber camps, the main alternatives, varied seasonally and with the fall and rise of business cycles. Northerners' estates thus offered rural blacks valuable wage-earning opportunities.[73]

African Americans are also reputed to have favored northerners because they showed greater racial tolerance. Northerners may have treated African Americans somewhat better than white southerners, but the historical record shows northerners held thoroughly racist beliefs. Northerners' racism differed little if at all from that of white southerners. Northerners' paternalism may have assumed different form, but the contrasts appear to have been superficial. Although popular memory associates northerners' estates with improved race relations, the historical record paints a different portrait.[74]

Plantation life created three principal modes of black-white interaction. The first centered on northerners' role as landowners and employers and African Americans' status as residents and employees. In general,

---

[72] Examples of paternalistic benevolence abound. At Litchfield Plantation, Dr. Henry Norris built an infirmary for blacks living on his lands and in neighboring areas. See Chalmers S. Murray, "June Finds Norrises at Leitchfield," *News and Courier*, Aug. 16, 1931, A6. Bernard M. Baruch provided several forms of assistance to African Americans who lived on his lands. See Brockington, *Plantation between the Waters*, 55–56. At Arcadia Plantation, Isaac Emerson built a school and church for resident families. See Cox, *Neal Cox of Arcadia Plantation*, 87–88. On northerners' general attitudes toward blacks, see Rogers, *History of Georgetown County*, 489.

[73] Personal communication with Alan Jenkins, May 10, 2007.   [74] Ibid.

interactions of this sort approximated relations characteristic of farm labor in settings across the South. Although some differences existed, the general pattern exhibited the routine, monotony, and conflicts that typified low-skill labor in rural settings. That many blacks employed by estate owners also lived as tenants on their estates created opportunities for coercion and exploitation.

A second variety of interaction encompassed African Americans' service as domestics, general laborers, and hunting guides. These interactions also amounted to employer-employee relationships of a kind, but they differed in the degree of trust and personal intimacy involved. Northerners developed close associations with laborers who served them in specialized roles. These interactions produced a small group of African Americans who knew estate owners and their regular hunting companions more closely than most.[75]

A third mode of black-white interaction centered on performance. At plantations across the lowcountry, African Americans sang, danced, worshiped, and played in ways that entertained northern owners and their guests. In carrying out such activities, African Americans fulfilled northerners' expectations and elicited praise and approval. How African Americans viewed such activities is unclear. Since they could have refused to perform, the fact that such occasions happened at all is significant. How much blacks valued whites' attention is uncertain, however. They may have regarded it as welcome, as a minor irritant, or thoroughly patronizing. Still, in the context of contemporary race relations, approval and acknowledgment proved significant, for it showed white appreciation of black culture. As instances of white attention to black folk practices, performances on northerners' estates marked a break from prevailing norms.

Consider, for example, the ritual that took place at Medway Plantation on Saturday evenings. At dusk or soon thereafter, several African Americans would gather in the yard on the east side of the main house, close by the front door. There, the Legendres and any guests present would sit by a small bonfire, talking and enjoying the evening air. The African Americans would arrange themselves in a row and begin signing. One would lead: "Who buil' duh Aa'k?" In deep, rich voices, the other would respond: "Norah, Norah Lawd." As the song progressed, the singers would begin to sway and move, clap their hands, and harmonize.

---

[75] For examples of interaction with laborers, see, for example, Legendre, "Diary of Life at Medway Plantation," Feb. 5 and 26, 1940, and Feb. 21 and 24, 1941. See also Giltner, *Hunting and Fishing in the New South*, 130–134.

As the depth and power of their voices soared, the beauty of African Americans' folk traditions became apparent. Increases in tempo produced a commensurate quickening of vocal expression and bodily movements. Some songs culminated in dramatic crescendos. Others tapered off into the night air. In all likelihood, the Legendres and their guests clapped, smiled, and perhaps even hummed along as they took in the performance.[76]

The peak of northerners' activities in the lowcountry coincided with growing interest in spirituals and other forms of African American culture. Concerns about the potential for the spirituals tradition to be lost for good prompted whites to begin recording and transcribing African American songs. In the lowcountry, the Society for the Preservation of Negro Spirituals (SPNS) led the charge. Founded in Charleston in 1922, this group dedicated itself to transcribing, recording, and reproducing spirituals "with affectionate fidelity."[77]

A typical SPNS performance involved between fifteen and twenty members of the group dressed in costume singing Negro spirituals in Gullah. Performers sang, shouted in the tradition of lowcountry blacks, and commented on the origin and meaning of spirituals and the character of the society that had produced them. In the view of SPNS members, spirituals captured the religious devotion of lowcountry blacks without expressing any sort of resistance to slavery or longing for freedom. SPNS members, like other lowcountry elites, held a highly circumscribed view of the past, and their idealized conception of regional history found expression in their performances.[78]

Spirituals, in the eyes of the SNPS, represented a vestige of a lost era, a living link to the past. As a tradition carried on by "plantation negroes," spirituals represented a folk practice from a near-vanished era of racial harmony and social stability. For their performances, SPNS members wore clothes typical of antebellum slaveowners. Thus, they assumed the persona of masters, not slaves, to perform and interpret a music created by slaves. Symbolically, this presented white elites as the most knowledgeable authorities on African American culture. Through their costumes, the SPNS cast their ancestors as better able to understand and explain the culture of the people they had owned as property than those people could

---

[76] "Medway Plantation," *Town and Country* 103, no. 4318 (March 1949): 76–79; Yuhl, *Golden Haze of Memory*, 143. Other scholars have also noted African Americans' role in performing for white hunters as part of the latter's notion of a "complete sporting experience." See, for example, Giltner, *Hunting and Fishing in the New South*, 94.

[77] Yuhl, *Golden Haze of Memory*, 127–134.   [78] Ibid., 142–146, 154–156.

for themselves. By extension, this practice also presented SPNS members as best able to tell the spirituals story. As descendants of a cultured, educated class who had made the lowcountry a great and powerful region, they could best explain the role of spirituals in regional history and culture.[79]

In the hands of the SPNS, spirituals became a powerful means of promoting a particular vision of the past. As Stephanie E. Yuhl has observed, the SPNS's efforts amounted to a campaign to preserve a portrait of a particular type of lowcountry African American – the "old-time darkey" symbolic of the social order of the antebellum South. Each SPNS concert became a "history" lesson that expressed a nostalgia-laden view of the past. The popularity of SPNS performances and, more generally, the interest in spirituals that prevailed throughout the 1930s illustrates popular fascination with authentic vestiges of the antebellum South.[80]

Northerners such as the Legendres likely viewed spirituals in the same way as did lowcountry whites. Northerners generally recognized local whites as authorities on the history, folklore, and traditions of the region, even when it came to the practices and beliefs of African Americans. No record of the Legendres attending a SPNS performance exists, but it is possible that they did. Other northern plantation owners attended SPNS concerts and supported the organization by purchasing memberships and copies of *The Carolina Low-Country*.[81] Moreover, spirituals received considerable attention in the northern press. Accounts in leading newspapers and magazines presented spirituals in much the same way as the SPNS did. An article published in *Country Life* in 1935, for example, characterized spirituals as an "exotic music" that survived from the era of slavery. According to the author, it represented "a heritage we must not lose."[82]

Performances of the kind that occurred on Saturday nights at Medway took place throughout the lowcountry. In 1935, for example, the *News and Courier* reported on a performance at Cypress Gardens, an aquatic

---

[79] Ibid., 130–131, 142–145. For an earlier example of southern whites acting as experts on the behavior of southern blacks, see Silber, *Romance of Reunion*, 139–141.

[80] Yuhl, *Golden Haze of Memory*, 146–156.

[81] The names of several plantation owners and other northerners appear in a sales ledger of copies of *The Carolina Low-Country*. See Account Book, 1931–1933, folder 14, box 715, Records of the Society for the Preservation of Spirituals, South Carolina Historical Society, Charleston, SC (hereafter SCHS).

[82] Lydia Parrish, "A Heritage We Must Not Lose," *Country Life* 69, no. 2 (Dec. 1935): 50–55, 62.

park that Benjamin Kittredge had created at Dean Hall Plantation. According to the newspaper, "negro plantation hands" sang in the moonlight for a crowd gathered around a small bonfire. Attendees noted that "the negroes' voices had a mysterious African aspect."[83] Similar performances occurred elsewhere. At the Oaks Plantation in Goose Creek, "the Negro farm-hands" sometimes gathered outside the main house on moonlit nights to regale "the 'company' with 'shouts' and 'singing games' and 'buck and wings.'" According to one visitor, these episodes took place "quite in ante-bellum fashion."[84]

Other ritualized practices provided similar forms of interaction. At Springbank Plantation in Williamsburg County, retired insurance magnate Howard S. Hadden and his wife staged Christmas celebrations for their black laborers and others living in the surrounding area. As the *News and Courier* reported, "Every Christmas Mr. and Mrs. Hadden delight in playing Santa Claus to the negroes." It described the scene at the 1937 celebration:

One old "uncle" is dressed up as Saint Nick. He stands pompously before the tall columns of "Springbank" and claps his hands as a gift is handed to each member of his race as they file past. The line this Christmas was just 260 darkies long. And not a one left empty handed. From the veranda and the lawn one hundred fifteen white guests looked on while the negroes sang spirituals and danced.[85]

The Haddens' version of plantation life replicated a well-known ritual of the antebellum era. At plantations across the South, slaveowners had observed Christmas by giving gifts to slaves and allowing a day of rest. In casting themselves as Mr. and Mrs. Claus, the Haddens emphasized their power and authority. Gift-giving, in this sense, represented a display of wealth and status. In addition to emphasizing the Haddens' role as landowners and employers, it highlighted the apparent subservience and deference of hundreds of African Americans.[86]

---

[83] "Plantation Singers by Moonlight at Cypress Gardens," *News and Courier*, Apr. 19, 1935, A14.

[84] "The Oaks, a Restored Mansion of the South," *Country Life* 29, no. 2 (Dec. 1915): 54. In 1949, *Town and Country* magazine billed the practice of signing spirituals for plantation owners as "an old Low Country tradition." See "Medway Plantation," 78.

[85] Laura C. Hemingway, "Williamsburg Boasts Many Game Preserves for Hunt," *News and Courier*, Jan. 11, 1937, 3.

[86] For descriptions of Christmas on lowcountry plantations, see David Doar, *Rice and Rice Planting in the South Carolina Low Country* (Charleston: Charleston Museum, 1936), 33; Charles Joyner, *Down by the Riverside: A South Carolina Slave Community* (Urbana: University of Illinois Press, 1984), 101–102, 127, 134–137.

In other cases, holiday celebrations assumed somewhat different contours. In January 1935, the *News and Courier* reported on a New Year's celebration at Bleak Hall Plantation. According to the newspaper, when Dr. James Cowan Greenway and his wife arrived at Bleak Hall soon after Christmas, "the negroes" immediately "appeared in a body at the maser's [*sic*] cottage." Each held "a chicken for the doctor and his wife." Impressed by the thoughtfulness of their "people," the Greenways decided to throw a picnic. The event took place several days later, with "everyone on the plantation" attending. The Greenways served cake and ice cream and gave each of the women five dollars. "One old man," the newspaper reported, enjoyed himself so much that "he ate seven ice cream cones in spite of the rather chilly weather."[87]

In the spring of 1937, the Stones orchestrated similar festivities at Boone Hall Plantation. On May 17, 1937, Thomas Stone noted in his journal, "We are having a party for all the colored people on the plantation – and, I expect when the word gets around, for all the colored people in the Parish." He and Alexandra, he wrote, planned to serve "mulatto rice, beer, and pop." For entertainment, the Stones hired the Jenkins Orphanage Band, Charleston's best-known African American musical troupe. Two days later, Stone reported the party to have been "a great success." About 150 people attended – a figure well over the 50 he had expected. "Everyone seemed to enjoy themselves enormously," he noted. Singing and dancing took place, and "the mulatto rice, cake, ice cream, beer, soft drinks and candy disappeared like the dew on a hot summer morning." A lightwood fire and a full moon supplied a magical combination of light. According to Stone, the "whole scene ... was most entrancing." He added that "absolute perfection was missed only because of the rank smell from about 100 2½ cent cigars being smoked furiously at the same time."[88]

Although African Americans supplied plantation life with the better part of its racialized character, northerners also recognized lowcountry people as profoundly different. Some even saw them as a distinct race. Sidney Legendre, for example, expressed such a view on at least one occasion. Shocked by the indolence of lowcountry whites, he confided

[87] "Ol' Maussa Greenway Gives Party for Plantation Hands."

[88] Thomas A. Stone Journal, May 17 and 19, 1937, Boone Hall Scrapbook Number 3, SCHS. A somewhat similar birthday celebration took place at Prospect Hill Plantation in January 1934, when Franklyn L. Hutton and his wife celebrated their son's twenty-first birthday. See "F. L. Huttons Give 21st Birthday Party for Woolworth Donahue," *News and Courier*, Jan. 11, 1934, 5.

in his diary, "It is incredible to think that there is a strange low country race without ambition or hope who only live in the past and talk about how wonderful their ancestors were, and do nothing to emulate them." On another occasion, he complained about the characteristically "southern" behaviors that his superintendent, a northern man, had developed. "There is something quietly mysterious about the degeneration of the caucasion [*sic*] race in the south," Legendre wrote. "From an industrious energetic man you gradually see northern overseers drift down the ladder to where they have forgotten to shave and have the listless attitude of the South." Witnessing such a change, he added, "is frightening."[89]

Other northerners echoed Legendre's comments. In a 1937 article, Clare Boothe Luce highlighted typically "southern" behaviors that she viewed as endemic to the lowcountry. Recounting her experiences in the region, she explained, "the hustle, bustle, frenzy, the ban and push, and the scrabble and scurry – the normal, accepted, and miraculously efficient tempo in which things are done 'up North' – just simply does not operate in the South." Southerners, she insisted, are "a race of people who do things slowly." According to Luce, they would never "be bent to the North's frenetic dynamics."[90]

As Luce's and Legendre's comments make clear, northerners recognized the lowcountry as a thoroughly different place. Travel to the region involved a journey in time and space and immersion in a different culture. Exoticism, primitiveness, and vestiges of a slaveholding past commingled with a semitropical climate, lush foliage, and a timeworn landscape to create a sense of place like no other. Racial differences made themselves conspicuously apparent. Northerners' interactions with lowcountry people likely heightened their awareness of such differences. Northerners' contact with whites mainly involved educated elites. Interaction with African Americans limited itself to plantation laborers and other rural blacks. Social and racial categories thus ran along parallel paths. Virtually all African Americans whom northerners encountered belonged to the rural poor; most whites belonged to the lowcountry elite.

In many ways, plantation life replicated patterns of authority common through the South. Owners and superintendents supervised black workers in a manner that differed little from sharecropping plantations, with roughly comparable social arrangements. Tellingly, the atmosphere of

[89] Legendre, "Diary of Life at Medway Plantation," May 23, 1939.
[90] Luce, "The Victorious South," 121–122 (first through third and sixth quotes on 121; fourth and fifth on 122).

northerners' estates struck some visitors as characteristic of plantations operated as agricultural enterprises. While visiting Boone Hall Plantation for the first time in the spring of 1937, Thomas Stone's mother remarked, "One great charm about Boone Hall is its great activity. There seems to be so much doing every place you go. About a hundred negroes are employed just now, so you are constantly seeing someone raking, cleaning, picking nuts, planting, or doing something else."[91]

Northerners' version of plantation life mirrored earlier iterations in countless ways. White and black people continued to share the same space, to live, work, and play alongside one another, to interact daily and yet occupy different worlds. Owners found managing and maintaining a large estate endlessly challenging, sometimes to the point that they questioned the rewards. Like antebellum planters, northerners complained frequently about laborers, fretted about weather and pests, and worried that their efforts would fall short. No matter how much time and energy they expended, the results always seemed disappointing. Meanwhile, laborers relied on well-honed methods of accommodation and resistance in negotiating the directives of owners and superintendents. Although labor on northerners' estates offered advantages over other options, it also provided limited rewards. Lowcountry blacks took advantage of it while continuing activities that provided autonomy and maintained economic independence. For them, northerners' estates represented a short-term opportunity rather than a permanent solution to poverty and disenfranchisement.

If similarities between northerners' estates and earlier plantations abounded, so, too, did the differences. All flowed from the purpose of former. Compared with plantations devoted to commercial agriculture, northerners' estates marshaled their resources for leisure and recreation, for the pleasure of owners and guests. The word "plantation" remained the same, but its meaning had changed profoundly. Dramatic transformations had turned it into something new. In the process, earlier meanings had become obscured and new forms of symbolism had developed. A hundred years earlier, plantations had supplied wealth and power through productive capacity. Control of land, labor, and marketable

[91] Flora M. C. Stone to My Dear Ones, March 5, 1937, Boone Hall Scrapbook Number 2, SCHS. Stone's comments are strikingly similar to those of the eminent historian Ullrich B. Phillips, who characterized social conditions at a US Army training installation in Georgia during World War I as having "a plantation atmosphere." He attributed it a combination of white authority and black subordination. See Ullrich B. Phillips, *American Negro Slavery* (New York: D. Appleton and Co., 1918), viii–ix.

commodities had given planters status and authority. By contrast, sporting estates consumed wealth. They displayed authority through physical appearance and the activities they made possible, all on the basis of wealth earned elsewhere. No matter what the apparent similarities, the differences could hardly have been greater. From spaces of production to realms of consumption, plantations throughout the lowcountry had changed dramatically. Northerners' estates marked the latest phase in a long and convoluted history.

CONCLUSION

Although contemporaries saw plantation life as a product of conditions specific to the lowcountry, when viewed historically, northerners' activities reveal at least as much about broad currents in American culture as developments in coastal South Carolina. During the past quarter-century, historians have grown increasingly aware of northerners' role in nurturing the myth of a romantic Old South. Whether because of longings for a bygone era of agrarian virtue, the appeal of a society with comparatively low levels of class conflict and racial strife, or promising commercial prospects, northerners, especially members of the middle and upper classes, looked upon the South through rose-tinted glasses. Even as popular discourses cast the South as a backward region plagued by poverty, religious fundamentalism, and poisonous race relations, some northerners saw the region as deeply enticing.[92] Plantation life makes this point clear. For monied sporting enthusiasts to not only remake plantations for leisure but to idealize them as exotic realms of endeavor shows how pervasive beliefs about the South and its past had become. Fanciful depictions of a romantic southland not only saturated film and fiction; for a few members of the upper classes, they lay within reach. For northerners who became deeply invested in their lowcountry estates, the fantasy became palpably real.

---

[92] On northerners' views of the South, see especially David W. Blight, *Race and Reunion: The Civil War in American Memory* (Cambridge: Belknap Press of Harvard University Press, 2001), chap. 7; Silber, *Romance of Reunion*; Reiko Hillyer, *Designing Dixie: Tourism, Memory, and Urban Space in the New South* (Charlottesville: University of Virginia Press, 2014); McIntyre, *Souvenirs of the Old South*; Susan-Mary Grant, *North over South: Northern Nationalism and American Identity in the Antebellum Era* (Lawrence: University Press of Kansas, 2000). On the South as a benighted region, see George Brown Tindall, *The Emergence of the New South, 1913–1945* (Baton Rouge: Louisiana State University Press, 1967), chap. 6; Natalie J. Ring, *The Problem South: Region, Empire, and the New Liberal State, 1880–1930* (Athens: University of Georgia Press).

Plantation life had deep roots. Although the phrase did not become part of public discourse in the 1930s, it reflected decades of labor on the part of people across the United States. The triumphant reunion of North and South, systematic marginalization of African Americans, and widespread acceptance of white southerners' narrative of the Civil War and Reconstruction laid the groundwork for celebratory portrayals of the lowcountry and its past. That most Americans of the early twentieth century looked back on slavery ambivalently if not nostalgically, without concern for human suffering, toil, and violence, speaks volumes about the perspectives that developed. Had African Americans and their allies retained greater influence, differing views of the past might have kept the brutality of slavery and the violence of Jim Crow at the forefront of public consciousness. Yet in early twentieth-century America, black inferiority seemed to explain the state of race relations and African Americans' apparent regression toward savagery. Moreover, sophisticated systems of racial control in and outside the South limited challenges to white supremacy. In this context, narratives emphasizing slavery's civilizing influence and the tragedy of Reconstruction proliferated.[93]

African Americans proved crucial to development of plantation life. Not only did black labor sustain northerners' activities and African Americans' physical presence have symbolic power, but black attitudes toward the rural lowcountry produced a population willing to accommodate rather than resist northerners' aims. Outmigration left rural areas populated by members of older generations and younger blacks who, for varying reasons, opted not to follow their peers to nearby cities or to Philadelphia, Harlem, or Boston. The reasons that African Americans remained are complex, varied, and imperfectly understood. Bonds of family and community, emotional attachments, and limited knowledge of the world beyond all played a role. Regardless, the overall pattern meant that African Americans who stayed in the rural lowcountry proved willing to acquiesce to northerners' expectations regarding labor, tenancy, and even racialized performances. Those inclined to resist northerners' directives had already departed.[94]

[93] Blight, *Race and Reunion*; Silber, *Romance of Reunion*; Caroline E. Janney, *Remembering the Civil War: Reunion and the Limits of Reconciliation* (Chapel Hill: University of North Carolina Press, 2013), chaps. 5–9; Bruce E. Baker, *What Reconstruction Meant: Historical Memory in the American South* (Charlottesville: University Press of Virginia, 2007).

[94] Woofter, *Black Yeomanry*, chap. 4; Kiser, *Sea Island to City*, chaps. 3–5. For an example of African Americans in rural Berkeley County living with limited knowledge of urban areas circa 1930, see Mamie Garvin Fields, *Lemon Swamp and Other Places: A Carolina Memoir* (New York: Free Press, 1983), 200–202.

For owners and their guests, plantation life provided distinction. By tying traditional sporting pursuits to a seemingly unique realm of experience, plantation life boosted the value of pastimes centered on exploit and challenging endeavor. The experiential dimensions of northerners' estates situated sportsmen's and sportswomen's activities in a region recognized for its heritage, its place within the South, and its role in the nation's beginnings. All proved crucial in differentiating northerners' estates from similar venues elsewhere and layering meaning upon a well-established set of pastimes. Although upper-class Americans enjoyed field sports in locations throughout the United States and Europe, only the lowcountry offered plantation life. The novelty of owning a true "plantation" coalesced in the activities that filled northerners' days.

Ultimately, plantation life made select varieties of southernness a consumer commodity. In defining the meaning and purpose of their estates, sportsmen and sportswomen created a mode of experience that could scarcely have been more novel. To be sure, they did not will it into being by themselves, nor did they exercise absolute authority over it. African Americans, white southerners, journalists, and a few others contributed to the form and substance of plantation life. Still, northerners bore primary responsibility for social and material conditions on their estates, and the meanings they ascribed to their activities formed the wellspring of their experiences. Most important, northerners marked the first group of people to expend large sums of money and effort on owning plantations with leisure as a primary goal. Plantations had long been capital-intensive enterprises, and owners had historically borne the costs involved with expectations of profits. Planters made investments in land, labor, equipment, and machinery in pursuit of monetary gain. Prospective rewards tempered the challenges, conflicts, and difficulties inherent in operating an enterprise subject to storms, droughts, and the vicissitudes of a labor force that worked only under threat of violence. By contrast, sportsmen and sportswomen took up plantation ownership purely for the sake of pleasure. They took on the difficulties involved because of the enjoyment they derived from recreation. Although few Americans had the resources to create or purchase a lowcountry estate, for those who did, ownership provided access to a remarkable range of experiences.

Epilogue

In the late 1930s, the winter colonists' world began to unravel. The New Deal delivered the first blow. Federally funded construction projects transformed communities throughout the lowcountry. New roads, bridges, school buildings, libraries, and courthouses brought energy and activity to formerly quiet towns and heralded the rise of an activist state. In many areas, the surge of activity proved short-lived. Most federal projects took weeks or months, not years, to complete. Yet even where work progressed swiftly and then passed, the New Deal signaled the beginning of a new era. By inaugurating a new phase in American governance and a more closely connected world, it announced an end to the isolation and stagnation that had long prevailed. Although conditions on northerners' estates did not collapse overnight, they quickly began to change. As they did, many of the qualities that northerners prized began to slip away.

Owners of large estates felt the effects of the New Deal at once. An exodus of black laborers constrained plantation operations and raised concerns about the future of the lowcountry sporting scene. African Americans eagerly seized opportunities for employment with federal relief programs. Better wages, improved working conditions, and greater autonomy provided strong incentives. As workers left, owners and superintendents complained. In January 1937, Clare Boothe Luce telegrammed President Franklin D. Roosevelt about her inability to get "able bodied negroes" to do "light farm and house work" at her Mepkin Plantation. In a plainly stated critique of New Deal policy, she blamed the problem on "Congress's relief legislation."[1] W. B. Seabrook, the superintendent at Boone Hall Plantation, described similar conditions. Construction of a new bridge across the Wando River had "taken quite a few of the good young men," he reported, and "most of the other young negroes" had left for Works Progress Administration (WPA) projects. With relief work paying "common negro laborers $2.00 per day," Seabrook found keeping workers virtually impossible.[2]

Luce's and Seabrook's complaints revealed deeply felt frustrations. Owners and superintendents did not take the loss of labor lightly, for the outflow raised fears of long-term shortages and the inability to carry on as before. Owners who had recently made large investments expressed particular concern. In this sense, the comments of Luce and Seabrook are telling. Luce and her husband had recently poured more than $400,000 into Mepkin, and Seabrook, the superintendent of Thomas and Alexandra Stone's Boone Hall Plantation, presided over an operation that sought to be self-supporting through a combination of truck farming and pecan production. For these owners, labor shortages raised difficult choices. At the same time, other winter colonists also felt the pinch. Throughout the lowcountry, the loss of black workers raised apprehensions. As the years passed and the problem became acute, many owners recognized that conditions had changed for good.[3]

African Americans' enthusiasm for federal relief jobs not only troubled owners; it showed eagerness to abandon plantations. Even though work on northern-owned estates compared favorably with most varieties of agricultural labor, it left a great deal to be desired. Workers who accepted employment with federal projects struck out in search of better lives. Those who remained tended to be aged and unwilling to abandon familiar circumstances. Younger African Americans eagerly seized opportunities for income and relief from the monotony, isolation, and paternalism of the countryside. Just as their predecessors had during World War I, working-age blacks left plantation communities in large numbers.[4]

---

[1] Clare Boothe Luce to Franklin Delano Roosevelt, Jan. 9, 1937, quoted in Sylvia Jukes Morris, *Rage for Fame: The Ascent of Clare Boothe Luce* (New York: Random House, 1997), 299.

[2] W. B. Seabrook to Thomas A. Stone, Sept. 30, 1938, Boone Hall Scrapbook no. 3, South Carolina Historical Society, Charleston, SC (hereafter SCHS).

[3] Mepkin Plantation, Expenses and Commitments, Apr. 1, 1936–Jan. 31, 1938, folder 7, box 98, Henry R. Luce Papers, Library of Congress, Washington, DC; "Tide Harnessed at Boone Hall to Furnish Power for Entire Plantation," *News and Courier*, Nov. 8, 1936, C3.

[4] John Scott Strickland, "Traditional Culture and Moral Economy: Social and Economic Change in the South Carolina Low Country, 1865–1910," in *The Countryside in the Age*

The loss of plantation laborers not only compromised owners' ability to maintain their plantations, it also undermined conditions on their estates. As workers departed, the racial imbalance that northerners saw as characteristic of plantations ebbed. Decreasing black-white ratios muted one of the qualities northerners prized most. Since northerners viewed throngs of black bodies at work as characteristic of plantations and "the South," diminished levels of activity stood out. In addition, the symbolism inherent in owners' and superintendents' supervision of black workers lost potency. Fewer workers and the loss of able-bodied adult males left superintendents directing small groups of young boys and old men. In this context, white supervision of black labor became less an exercise in authority than part of a losing struggle carried out amid difficult conditions.[5]

Changes wrought by the New Deal also undermined plantation life. WPA and Public Works Administration (PWA) projects produced a surge of activity that lessened contrasts between old plantation districts and the "modern" world. New roads and bridges, school buildings, courthouses, post offices, and sewer and water systems brought modern buildings and infrastructure to communities throughout the lowcountry. Rural electrification, agricultural assistance, and soil conservation reached even the most remote areas. Collectively, New Deal programs signaled the arrival of improved public services; closer ties between federal, state, and local authorities; transportation improvements; and better communications within the lowcountry and with the world beyond. They also made the rural lowcountry seem less anachronistic and thus destroyed conditions that northerners prized.[6]

By the late 1930s, signs of progress abounded. Per capita income had improved and relief jobs had put money in the pockets of tens of

of Capitalist Transformation: Essays in the Social History of Rural America, ed. Steven Hahn and Jonathan Prude (Chapel Hill: University of North Carolina Press, 1985), 147–148; T. J. Woofter, *Black Yeomanry: Life on St. Helena Island* (New York: Henry Holt and Co., 1930), 90–95, 101, 266–267, 280–281; Clyde Vernon Kiser, *Sea Island to City: A Study of St. Helena Sea Islanders in Harlem and Other Urban Centers* (New York: Columbia University Press, 1932), 81–144; Lawrence S. Rowland and Stephen R. Wise, *The History of Beaufort County, South Carolina*, vol. 3: *Bridging the Sea Islands' Past and Present, 1893–2006* (Columbia: University of South Carolina Press, 2015), 189–195.

5 Gertrude S. Legendre, *The Sands Ceased to Run* (New York: William-Frederick Press, 1947), 4–5.

6 Jack Irby Hayes, *South Carolina and the New Deal* (Columbia: University of South Carolina Press, 2001), chaps. 3–5 and 11; Tara M. Mielnik, *New Deal, New Landscape: The Civilian Conservation Corps and South Carolina's State Parks* (Columbia: University of South Carolina Press, 2011), 1–16, 22–23, 39–40, 45, 87–89, 119–123, 141–143; Rowland and Wise, *Bridging the Sea Islands' Past and Present*, 329–348.

thousands of workers. Rural electrification had transformed the lives of thousands. Literacy programs had given white and black workers new confidence and opportunities. Government aid had provided especially needy citizens with food, clothing, and household goods. Although scholars have emphasized that World War II, not the New Deal, pulled South Carolina out of the Great Depression, Roosevelt's programs had lasting effects. By restoring confidence and creating a sense of progress where only despair had existed, the New Deal helped South Carolinians weather extraordinary hardships.[7]

For owners of large estates, the New Deal marked the first in a series of developments that eroded conditions throughout the region. As its effects set in, other factors also took a toll. The rise of "cafe society," the cosmopolitan lifestyle that became part of upper-class culture in the late 1920s, profoundly affected the lowcountry sporting scene. Cafe society marked a turn away from ostentatious display and self-indulgent behavior. It developed when a few socialites became "bored" with Newport and began searching for entertainment and amusement. Soon they began mingling with celebrities, artists, and writers in the cafes and restaurants of Times Square and Hollywood. The resulting mix of "old socialites and new celebrities" became known as cafe society because of its preferred venues. Participants gathered in "public hotels or cafes" instead of Fifth Avenue residences and "great houses." The change not only reflected shifting tastes but also showed a change in attitude. More public, more attuned to popular culture, and more adventuresome, cafe society expressed desires to "see and be seen," to mingle in fluid social conditions, and to do away with the rigid social protocols that had long been the hallmark of society.[8]

Cafe society affected the lowcountry sporting scene in three ways. First, its cosmopolitanism showed declining enthusiasm for exclusivity and yearnings for livelier, wittier modes of behavior. The presence of writers, journalists, and playwrights proved significant. Men and women

---

[7] Hayes, *South Carolina and the New Deal*, 185–186, 195–196, 203–206.

[8] "The Yankee Doodle Salon," *Fortune*, Dec. 1937, 123–127. See also Cleveland Armory, *Who Killed Society?* (New York: Harper and Brothers, 1960), 20, 107–108; Frederic Cople Jaher, *The Urban Establishment: Upper Strata in Boston, New York, Charleston, Chicago, and Los Angeles* (Urbana: University of Illinois Press, 1982), 280; Thierry Coudert, *Café Society: Socialites, Patrons, and Artists, 1920–1960* (Paris: Flammarion, 2010). Greater sensitivity to public sentiment also contributed to the rise of cafe society. As *Fortune* noted, the 1929 stock market crash led the upper classes to discontinue "vulgar" and "dangerous ostentation."

from these backgrounds added intellectual vitality to a social realm that had traditionally eschewed learning and erudition. Cafe society thus revealed growing fatigue with pretention and desires for challenging exchanges of words and ideas. Masculine posturing and trophies gained through sport, travel, and adventure became less important. Although still valued, their significance declined.[9]

Cafe society also marked a turn away from self-conscious emulation of European nobles. Whereas wealthy Americans had traditionally modeled themselves after Old World elites, men who made money in the entertainment industries – in movies, radio, and publishing – saw themselves as charting a new course. They occupied a distinctively American world. Fueled by the energy of mass culture, electricity, and the automobile, it rejected the stuffy elitism of gentleman's clubs and cloistered privacy of upper-class residences. Thus the favor for hotels and cafes. Although cafe society relied on patterns of inclusion and exclusion for status, it broke with the upper classes' traditional aversion to publicity. In fact, since many of its stalwarts worked in publishing and entertainment, cafe society welcomed attention. By cultivating interest from the public and the press alike, it showed comfort with the gaze of the masses.[10]

Third, cafe society marked the rise of what would eventually become known as the "jet set" and the fragmentation of upper-class life that came with it. Cafe society quickly assumed an international character and avoided developing clear-cut social boundaries. Abandoning the traditional social calendar, it seized upon airplanes and express trains to achieve unmatched spontaneity. The result was a lively, unpredictable blur of activity that took place on multiple continents and contrasted sharply with the well-rehearsed pageantry of society balls and dinners.[11]

Owners of lowcountry estates stood simultaneously on the margins and at the center of cafe society. On the one hand, their estates and enthusiasm for sport hunting showed commitment to established modes of behavior. Favor for "gentlemanly" field sports aligned estate owners with longstanding traditions. Moreover, northerners' "plantations" unabashedly emulated the aristocratic seats of the Old World. With their emphasis

[9] Jaher, *The Urban Establishment*, 269; "The Yankee Doodle Salon," 128. See also Armory, *Who Killed Society?*, 120–121.

[10] "The Yankee Doodle Salon," 128, 184, 186.

[11] Ibid.; Jaher, *The Urban Establishment*, 289; Coudert, *Café Society*. On the importance of lists such as Ward McAllister's "Four Hundred" and the *Social Register*, see Eric Homberger, *Mrs. Astor's New York: Money and Social Power in a Gilded Age* (New Haven: Yale University Press, 2002), chaps. 1 and 5.

on privacy, seclusion, and pretention, they owed obvious debts to the great houses of the English countryside and continental Europe. In short, northerners had made their preferences resoundingly clear.

On the other hand, nothing prevented estate owners from moving between the two worlds – or, for that matter, abandoning one for the other. In fact, the porous boundaries of cafe society readily accommodated such behavior. In December 1937, *Fortune* published lists of people known to be definitely "in" cafe society and others known as intermittent participants ("in and out"). The former category included two couples who owned lowcountry estates – George and Jessie Widener of Mackey Point Plantation and Marshall and Ruth Field of Chelsea Plantation – and close relatives of several other owners. The in-and-out list included another pair of owners and one relative. In short, a handful of plantation owners "belonged" to cafe society and others had familial connections. Moreover, other owners might readily have joined, had they been so inclined. Henry Luce and Clare Boothe Luce supply an obvious example. The renowned publisher and his glamorous wife would likely have thrived in cafe society, had they sought to participate.[12]

Cafe society did more than simply indicate changing tastes and behaviors; it identified northerners' estates as belonging to a particular historical moment. The high-water mark of northerners' activities came during an era when the lowcountry remained anachronistic but had become readily accessible and occupied a place on a transportation corridor traveled by upper-class Americans at select times of the year. Precariously balanced between an earlier time and a radically new world, the lowcountry offered rare opportunities. The authenticity and authority that northerners associated with lowcountry plantations offered a sense of security amid fast-changing circumstances and the chance for vital, meaningful experiences of a kind that seemed increasingly rare. The "primitiveness" of the rural lowcountry and the "ancient" remains that dotted the countryside thrilled, enlivened, and entertained. In combination, these attributes proved deeply compelling. By offering exceptional sensory and experiential possibilities, the lowcountry provided distinctive options. For social elites fatigued by familiar routines, the lowcountry offered a liminal

---

[12] Mr. and Mrs. Franklyn L. Hutton, owners of Prospect Hill Plantation, appeared on the "in and out" list. See "The Yankee Doodle Salon," 124–125. On the Luces' activities of the era, see especially Alan Brinkley, *The Publisher: Henry Luce and His American Century* (New York: Alfred A. Knopf, 2010), chaps. 5–9; Morris, *Rage for Fame*, chap. 15–34.

space, a refuge, a realm of pleasure and endeavor, and a place simultaneously apart yet not completely removed. In this sense, the lowcountry fed
yearnings for the real, the vital, and the uncommon without requiring
travel far afield or any of the discomforts and risks typically associated
with stays in truly remote territories.[13]

The circumstances surrounding the rise and fall of northerners' estates
highlight upper-class Americans' struggles to maintain status and authority in a democratizing age. Although wealthy Americans appeared all but
immune to the upheavals of the interwar era, in fact, they faced unprecedented challenges. The advent of income tax in 1915 began the slow
erosion of Gilded Age fortunes. Less than two decades later, the Great
Depression hit many upper-class Americans hard. Although not all saw
their fortunes shrivel, some did, and even those who did not nonetheless
found themselves constrained by the leftward shift in American politics.
Great wealth became a liability, pretention dangerous. Constant vying for
status and challenges from new wealth also took their toll. In the end,
wealthy Americans found it harder than ever to maintain the appearance
of a hereditary elite. Continually shifting social boundaries and faddish
varieties of distinction decreased the importance of overt displays of
material wealth.[14]

Northerners' estates developed as part of a short-lived bid to retain
the guise of a landed aristocracy amid economic and cultural upheavals.
In addition to seizing the potential of the lowcountry, the new estates
responded to seismic shifts in upper-class life. The confluence of the
estate-building boom and end of estate development on Long Island is
hardly coincidental. During the mid-1920s, construction of new estates
on Long Island slowed dramatically. It declined further during the Great
Depression and ended completely by World War II. At the same time,
the associated social scene also ebbed. Although historians have tended
to attribute these developments to economic conditions, changing tastes
also had a part. The surging popularity of the lowcountry did not stem
directly from developments on Long Island but nonetheless shared a
close relationship. As the premier playground for America's sporting set

---

[13] On yearnings for vital experience and authenticity, see T. J. Jackson Lears, *No Place of Grace: Antimodernism and the Transformation of American Culture, 1880–1920* (Chicago: University of Chicago Press, 1981), esp. 4–5, 56–58, 221–223, 301–307.

[14] Amory, *Who Killed Society?*, 9–24. Rowland and Wise emphasize the effects of the 1929 crash on northerners who owned estates in Beaufort County in *Bridging the Sea Islands' Past and Present*, 291–293.

reached the point of saturation, some of its devotees struck out for new horizons. For a time, a few of them found what they wanted on the South Carolina coast.[15]

World War II accelerated trends started by the New Deal. Military service and defense industry jobs prompted African Americans to leave plantations in sizable numbers. Neal Cox of Arcadia Plantation recalled losing at least nine men to the armed services.[16] At Medway Plantation, Gertrude and Sidney Legendre watched their "plantation Negroes" depart "one by one." Eventually, only three remained.[17] At Mepkin, the Luces found themselves even more shorthanded than before. "It seems that all Berkeley County negroes are working in war industries except the aged, the crippled and the young," observed one of Henry Luce's assistants. By the fall of 1942, Mepkin had been "almost completely stripped of field hands."[18]

The war also cut into owners' ability to enjoy their estates. Some joined the military and others contributed to the war effort as civilians. Benjamin R. Kittredge, Jr., and Sidney Legendre, for example, served in the US Navy. Jesse Metcalf, a veteran of World War I, became an officer in the US Army Reserve. Gertrude Legendre worked for the Office of Strategic Services. Charles L. Stevens served as an advisor to the War Production Board.[19] Most owners continued their private lives, largely because of age. The majority had already reached their fifties or sixties and thus had no choice but to watch events play out on newsreels and in the newspapers.[20]

---

[15] On developments on Long Island during the 1920s and 1930s, see Robert B. MacKay, Anthony K. Barber, and Carol A. Traynor, eds., *Long Island Country Houses and Their Architects* (New York: Society for the Preservation of Long Island Antiquities in association with W. W. Norton and Co., 1997), 26, 32–33.

[16] Neal Cox, *Neal Cox of Arcadia Plantation: Memoirs of a Renaissance Man* (Georgetown, SC: Alice Cox Harrison, 2003), 111.

[17] Legendre, *The Sands Ceased to Run*, 4–5.

[18] Alexander Hehmeyer to Henry R. Luce, Nov. 5, 1942, folder 11, box 85, Clare Boothe Luce Papers, Library of Congress, Washington, DC. See also V. C. Barringer to Harry F. Guggenheim, Sept. 6, 1943, in folder "Barringer, Victor, C., 1942–43," box 133, Harry F. Guggenheim Papers, Library of Congress, Washington, DC.

[19] Gertrude Legendre, *The Time of My Life* (Charleston: Wyrick and Co., 1987), chap. 10; David G. De Long, *Auldbrass: Frank Lloyd Wright's Southern Plantation* (New York: Rizzoli, 2003), 124; "Nicholas Roosevelt of Brokerage House," *New York Times*, June 30, 1965, 37. See also Richard S. Emmet, "Memories of Cheeha Combahee Plantation, 1929–1991," *Carologue*, autumn 1999, 19.

[20] A prosopographical study of estate owners has yet to be completed. The above analysis is based on a sample of twenty-five owners for whom dates of birth and death are available. Of this group, eleven had reached the age of fifty by 1940 and seven were sixty or older.

Owners who did not participate directly in the war nonetheless felt its effects. In Charleston, an expansion of the Navy Yard and construction of other defense-related facilities triggered a building boom. Employment at the Navy Yard doubled during the early stages of the war and reached 28,000 by 1943. The population of Charleston County grew by more than 41 percent and construction in outlying areas saw the city spill beyond its municipal boundaries. New military installations at Myrtle Beach, Conway, and Georgetown sparked similar developments at the northern end of the lowcountry. Elsewhere, wartime preparedness drills, coastal patrols, and training flights by military aircraft announced new priorities. Even in the most remote areas, the war made its presence felt.[21]

In May 1944, the New York *Daily News* reported declining interest in the South Carolina lowcountry. Beneath the headline "The Yankee Invasion of S. Carolina Ebbs," the newspaper explained that demand for large estates had all but evaporated and that several owners had recently sold their properties. In a few cases, South Carolinians had purchased plantations developed by wealthy northerners. "Reverting back to the natives," the *Daily News* observed, is "the very latest trend in plantations." The "Donald Allstons" now owned Mount Hope Plantation, formerly part of Arthur Whitney's estate at Willtown Bluff, and "a couple of South Carolinians" had bought Franklyn Hutton's Prospect Hill. These and other purchases signaled a reversal that would have been unthinkable only a few years before. In the mid-1930s, wealthy outsiders' enthusiasm for old plantations and coastal land had seemed insatiable. Now it had come to a screeching halt. Good land remained available, and, according to some sources, a few buyers remained interested. Still, the "invasion" had run its course. Northerners' desire for large estates had reached an end.[22]

The *Daily News* report announced what lowcountry residents already knew. Demand for lowcountry estates had dropped off sharply after

[21] Walter J. Fraser, Jr., *Charleston! Charleston!: The History of a Southern City* (Columbia: University of South Carolina Press, 1989), 387–393; Fritz P. Hamer, *Charleston Reborn: A Southern City, Its Navy Yard, and World War II* (Charleston: History Press, 2005), 39; Barbara J. Stokes, *Myrtle Beach: A History, 1900–1980* (Columbia: University of South Carolina Press, 2007), 50–8l; Rowland and Wise, *Bridging the Sea Islands' Past and Present*, chap. 14; Walter B. Edgar, *South Carolina: A History* (Columbia: University of South Carolina Press, 1998), 510–515; Lee G. Brockington, *Plantation between the Waters: A Brief History of Hobcaw Barony* (Charleston: History Press, 2006), 108–109.

[22] Robert Sullivan, "The Yankee Invasion Ebbs," *Daily News* (New York, NY), May 21, 1944, 44–45, 47.

1936 and reached a virtual standstill by the eve of World War II. Estate-making also slowed dramatically. Incremental development continued but grandiose building campaigns ended. Although estate owners continued wintering in the lowcountry, the fervor of earlier years had passed. In many ways, northerners had undermined the qualities they came to enjoy. By making the lowcountry a focus of popular attention, they had diminished the isolation and seclusion they coveted. By giving new life to old plantations, northerners had erased many of the anachronisms that marked the region as a vestige of an earlier time. By touting qualities they found appealing, northerners had made the lowcountry part of discourses about "the South." In sum, northerners aided the lowcountry's transformation from forgotten backwater to leading destination. By giving the region new status and celebrity, northerners had turned it into a different place. Although the transition represented a boon for the regional economy, it meant the downfall of plantation life. No longer a distinctive realm of experience, it became part of the past, just like so many other forms of activity associated with lowcountry plantations.[23]

* * *

Sportsmen's and sportswomen's plantations left an extraordinary legacy. Although they had their greatest influence in the 1930s, at the height of northerners' enthusiasm for the lowcountry, they continued to shape popular views of plantations for decades. Historical memory of slavery and contemporary views of plantations continue to show their influence. Moreover, grand "plantations" helped secure the lowcountry's claim to an extraordinary history. By suggesting continuities across time, legitimating the myth of an aristocratic past, and setting the region at the center of narratives of the nation and the South, new sporting estates affirmed celebratory accounts of regional history.

Northerners' estates shaped popular views of plantations in several ways. The transformation of architecture, landscape, and space involved in their making demanded reconsideration of what plantations had been historically and how new estates compared. The concentrated form of the new estates not only differed profoundly from plantations of earlier periods, it encouraged misapprehensions. Well-developed domestic cores

---

[23] Sidney Legendre explicitly identified America's entry into World War II as "the end of plantation life." See Sidney J. Legendre, "Diary of Life at Medway Plantation, Mt. Holly, South Carolina," private collection (copy in author's possession), undated entry [ca. Dec. 1941 or Jan. 1942].

set amid forests, pastures, and unimproved land gave northerners' plantations a two-part organization that approximated popular conceptions of landed estates, the plantations of imagined Old Souths included. Whereas antebellum plantations had diffuse material environments, the new estates followed a different model.[24] In practical terms, the change had two consequences. First, it made contemporary plantations the apparent counterparts of the fabled estates of the Old South. Diminution of spaces and structures associated with slavery, labor, and commercial enterprise left handsome houses and outbuildings standing amid large acreages, as stately mansions in bucolic circumstances. Material development of domestic spaces foregrounded elegant architecture and carefully maintained landscaping. Settings easily read as successors to the supposedly idyllic plantations of the antebellum era resulted. In an era when film, fiction, and most non-fiction accounts depicted plantations as columnar mansions occupied by aristocratic families, northerners' estates appeared cut from the same cloth. The loss of features associated with labor, slavery, and commercial enterprise muted the historical significance of productive activities. Even where rice fields, barns and sheds, and slave houses remained, they had been reworked, recast, and set amid seemingly "natural" landscapes that bore little resemblance to their historical predecessors. As remnants, they offered suggestive hints but nothing more.

Second, intermingling of old and new made it virtually impossible to discern, even generally, patterns of historical development through first-hand observation. Blurring of boundaries between old and new created an apparently unbroken record of material change. The variations among northerners' estates, the presence of still-decaying plantations in scattered locations, and subtle differences between "restored," rehabilitated, and refurbished buildings and landscapes all but destroyed authenticity as a recognizable quality. Selective retention of still-distressed elements and use of salvaged building materials compounded the problem. So, too, did claims about "authentic" and "restored" plantations, few of which

---

[24] On pre–Civil War material environments, see Philip D. Morgan, *Slave Counterpoint: Black Culture in the Eighteenth-Century Chesapeake and Lowcountry* (Chapel Hill: Omohundro Institute of Early American History and Culture by the University of North Carolina Press, 1998), 104–107, 110–123; S. Max Edelson, *Plantation Enterprise in Colonial South Carolina* (Cambridge: Harvard University Press, 2006); James L. Michie, *Richmond Hill Plantation, 1810–1868: The Discovery of Antebellum Life on a Waccamaw Rice Plantation* (Spartanburg, SC: Reprint Co., 1990), chap. 4; Charles Joyner, *Down by the Riverside: A South Carolina Slave Community* (Urbana: University of Illinois Press, 1984), 117–126, 139–140.

expressed discerning judgments. By the mid-1930s, plantations that retained most of their pre–Civil War form not only had become rare but, in all but a handful of cases, appeared virtually indistinguishable from others. Most plantations exhibited a mélange of elements from multiple periods. Parsing out temporal distinctions became all but impossible. Impulses to restore, renew, and preserve had created a complex tapestry that emphasized multiple phases of development yet told little about the underlying processes involved.

Discourses surrounding sportsmen's and sportswomen's estates reinforced romanticized readings of plantations and their history. The proliferation of writing about the new estates and historical plantations created a body of knowledge that went unchallenged for decades. The apparent authority and documentary power of books such as *Plantations of the Carolina Low Country* gave ennobling views of the region's plantations the status of near facts. At the same time, the absence of countervailing perspectives gave tacit legitimacy to the overriding focus on grand houses, genteel families, and material refinement. Visual imagery augmented written descriptions. Photographs included in newspaper and magazine features also emphasized planter's dwellings, groomed landscapes, and elite families. Only rarely did accounts published in the 1930s suggest that plantations had historically encompassed more than big houses and bountiful fields. Even when they did, none conveyed the scale and severity of the activities that had historically taken place, let alone the views and experiences of the people who carried them out. The contemporary role of African Americans and their labor also received limited attention. Although a few articles noted butlers and housekeepers, none showed the general laborers and guides that proved crucial to the operation of large sporting estates. As with historical memory of slavery, representations of northerners' plantations obscured the importance of African Americans.

As northerners' estates grew in number, pilgrimages to old plantation districts became annual rituals. In the spring of each year, as warming temperatures set the lowcountry ablaze in a sea of color, visitors flocked to the region. The famed gardens of Charleston and a host of other attractions drew large crowds. During the late 1920s, "tourist season" became an increasingly important source of revenue for communities throughout the lowcountry. Merchants and civic leaders took steps to increase visitation, partly by creating attractions designed to appeal to a wide range of interests. As Stephanie E. Yuhl has shown, in Charleston the major thrust of activity focused on historic sites and historical programs.

Plantation tours assumed a small but important role alongside tours of historic houses, historical pageants, and spirituals performances. From the beginning, plantation tours attracted a mix of tourists and lowcountry people. The first took place in February 1932, when members of the Charleston "Study Club" traveled to the Pinopolis area to visit several plantations and a handful of other sites. Of the seven plantations the group toured, one, Medway, had become a sporting estate. In 1934, the Society for the Preservation of Old Dwellings (SPOD), Charleston's leading historic preservation organization, organized three tours of plantations along the Cooper River. Held in conjunction with the second annual Azalea Festival, a week-long celebration of lowcountry history and culture, the tours raised funds for the Heyward House, a then-threatened dwelling associated Thomas Heyward, Jr., a signer of the Declaration of Independence. The itinerary included eight plantations, among them Mulberry and Gippy. In early April, the Agricultural Society of South Carolina took between seventy-five and a hundred people to see five plantations in lower Charleston and Beaufort counties, all of them owned by northerners. According to the *News and Courier*, the chance to see recent "renovations and improvements" accounted for the popularity of the outing. In 1936, SPOD organized a tour of three Edisto River plantations, including Franklyn Hutton's Prospect Hill and the Grove, then owned by Owen Winston and W. H. Barnum.[25]

Tours of plantations near Georgetown began after World War II. In 1947, the Women's Auxiliary of Prince George Winyah Parish organized three all-day excursions to plantations along the Santee, the Black, and the Pee Dee rivers and on the Waccamaw Neck. Of the twenty-three plantations visited, northerners owned fifteen. The itinerary included some of the most elegant estates in the lowcountry, among them Arcadia, the Wedge, Windsor, Mansfield, and Wedgefield. The 1947 outings inaugurated an annual tradition that continues to the present day. Meanwhile, a few owners opened their plantations for public visitation, and organizations

---

[25] "Study Club Visits Old Homes in Trip to Berkeley County," *News and Courier*, Feb. 19, 1932, 6; "Old Plantations to Be Seen on Lowcountry Tour Today," *News and Courier*, March 20, 1934, 3; "Agricultural Society to Visit Plantations," *News and Courier*, Apr. 9, 1934, 10; "Tour to Willtown and Plantations This Afternoon to Benefit Heyward-Washington House," *News and Courier*, March 28, 1936, 8; Stephanie E. Yuhl, *A Golden Haze of Memory: The Making of Historic Charleston* (Chapel Hill: University of North Carolina Press, 2005). On efforts to promote tourism, see also Edgar, *South Carolina: A History*, 493; Rowland and Wise, *Bridging the Sea Islands' Past and Present*, 232–233, 338–341.

in Charleston launched new tours. In 1946, Lawrence A. Walker, Jr., of Summerville purchased Mulberry Plantation from Clarence and Adelaide Chapman and immediately announced plans to open it to the public. The *News and Courier* noted that the move would make available "another type of fine colonial plantation property." Already Harrietta had been opened on a similar basis, and the gardens at Middleton Place and Magnolia Plantation had attracted tourists for decades. In 1958, the Women's Auxiliary of St. Phillip's Protestant Episcopal Church of Charleston led tours of Cooper River plantations where visitors could "brush aside the centuries and wander through plantation houses that have been Lowcountry landmarks for generations." All four of the plantations visited – Gippy, the Bluff, Medway, and Mulberry – served as winter residences for northerners.[26]

Limited sources make it difficult to know what tour participants learned on these excursions, but they likely heard information that mirrored prevailing discourses. Participants probably learned about once-beautiful houses where virtuous families had presided over prosperous estates until the downfall of the lowcountry's plantation empire. They undoubtedly heard about the benevolence of slavery and its benefits to masters and slave alike. They likely learned about the collapse of a great

---

[26] "Hopeswee Plantation," newspaper clipping cited as *The State* (Columbia, SC), March 23, 1973, in Hopeswee (Georgetown Co.) file, South Carolina Vertical File Collection, Charleston County Public Library, Charleston, SC (hereafter SCVF); "Rice Hope to Have First Public Showing on Moncks-Corner Pinopolis Club Tour," newspaper clipping cited as *News and Courier*, Apr. 1, 1951, in Rice Hope Plantation (Moncks Corners), SCVF; "Georgetown Tour Features Birthplace of Declaration of Independence Signer," newspaper clipping identified as *Charleston Evening Post* (Charleston, SC), Apr. 6, 1953, in Hopeswee (Georgetown Co.) file, SCVF; "Tour of Plantations to Begin at 1:30 p.m.," newspaper article cited as *News and Courier*, Apr. 4, 1954, in Otranto Hunting Club file, SCVF; "Mulberry Sale to Summerville Man Negotiated," newspaper clipping cited as *News and Courier*, Nov. 8, 1946, Mulberry Plantation file, SCVF; "Another Lowcountry Landmark Is Open to Public Inspection," unidentified newspaper clipping, Mulberry Plantation (Berkeley County) vertical file, SCHS; "Litchfield Plantation on Georgetown Tours," newspaper clipping cited as *News and Courier*, March 14, 1954, Litchfield Plantation file, SCVF; "St. Phillips Annual Tour Scheduled This Afternoon," newspaper clipping cited as *News and Courier*, March 22, 1958, Gippy Plantation file, SCVF; "S.C. Historical Society Fall Tour Is Set Sunday," newspaper clipping cited as *News and Courier*, Nov. 14, 1960, Litchfield Plantation file, SCVF. For information about recent plantation tours offered by Prince George Winyah Episcopal Church, see http://pgwi nyah.com/plantation-tours (accessed March 26, 2017). In 1960, the South Carolina Historical Society organized a fall tour of plantations on the Waccamaw Neck. Billed as "one of the tours sponsored annually to maintain interest in the history of the state," the outing has since become part of the organization's regular programming.

slaveholding civilization and the decline that set in afterward. Doubtless they heard about the wealthy people who had rediscovered the lowcountry and returned ruined estates to their former grandeur. In short, tourgoers probably heard narratives that told little about the history of the region while portraying plantations as magnificent country seats. In all probability, they heard nothing about the sharp differences between plantations of the past and new sporting estates. In sum, plantation tours likely disseminated the same romantic vision proffered by newspapers, magazines, and historical writing.[27]

Tours of plantations marked the last in a long sequence of developments that made northerners' estates widely known. In combination with newspaper and magazine reporting, plantation tours offered intimate glimpses of private worlds of pleasure and sumptuous display. Direct access allowed non-elites to experience the majesty and grandeur of the northerners' plantations under circumstances that foregrounded owners' status and wealth. As important, tours underscored the continuing symbolism of plantations. By the 1930s, the fiction of a bygone plantation world of romance and gentility had become a mainstay of American culture, its reach and power continually enlarged and extended by the steady advance of mass culture. Fiction, film, advertising, theater, tourist brochures, and travel literature celebrated an imagined land of gentlemanly planters, demure belles, and happy darkies.[28] Northerners' estates brought the fantasy into the present. By creating material environments that mirrored the fabled plantations of popular lore, the new plantations created real-world successors to the revered estates of yesteryear. By making plantations settings for traditional field sports and upper-class social life, they affirmed the myth of an aristocratic past and gave it new

---

[27] Narratives of lowcountry history remained remarkably consistent for decades. See, for example, Herbert Ravenel Sass, "South Carolina Rediscovered: A Native Son Finds Spectacular Changes in the 'Moonlight and Magnolia' State, Scene of Huge H-Bomb Project," *National Geographic* 103, no. 3 (March 1953): 286–288, 312; Herbert Ravenel Sass, *The Story of the South Carolina Lowcountry*, 3 vols. (West Columbia, SC: J. F. Hyer, 1956), I; Anthony Q. Devereux, *The Rice Princes: A Rice Epoch Revisited* (Columbia: State Printing Co., 1973).

[28] On portrayals of the South in mass culture during the era, see especially Karen L. Cox, *Dreaming of Dixie: How the South Was Created in American Popular Culture* (Chapel Hill: University of North Carolina Press, 2011), 5–8, 27–33, 37–80, 86–100, 126–129, 146–162; Grace Elizabeth Hale, *Making Whiteness: The Culture of Segregation in the South, 1890–1940* (New York: Pantheon Books, 1998), chaps. 3 and 4; M. M. Manring, *Slave in a Box: The Strange Career of Aunt Jemima* (Charlottesville: University Press of Virginia, 1998).

meaning. By reinforcing the social hierarchies of Jim Crow, northerners ensured that plantations remained sites of white authority. In refiguring old plantations for their purposes, sportsmen and sportswomen tied the majesty of a genteel South to the practices of a contemporary elite while legitimating the fiction of a glorious past. In the process, they curtailed opportunities for recovering the history of plantation slavery and facing its legacies head on.

* * *

Although northerners' plantations are now but a distant memory, they continue to influence views of slavery and its legacy in powerful ways. One recent example comes from the discovery of First Lady Michelle Obama's ancestral ties to slavery. In the weeks leading up to the 2008 presidential election, Americans learned that the wife of then-Democratic Party nominee Barack Obama had slave ancestors. On October 2, 2008, the *Washington Post* reported that Michelle Obama's great-great-grandfather, a man named Jim Robinson, had been a slave at Friendfield Plantation, on the coast of South Carolina. Born about 1850, Robinson came of age in the years when the Civil War drew to a close and Reconstruction began. He remained at Friendfield his entire life, working as a sharecropper. He married, had two sons, suffered the loss of his first wife while still relatively young, remarried, and lived out his days at the place where he had been born.[29]

Friendfield lies on the Sampit River, immediately west of Georgetown. During the antebellum era, it ranked among the most productive plantations in the Georgetown area. Its owner, Francis Withers, lived the life of a wealthy grandee. Withers's plantations – he owned several – produced as much as 720,000 pounds of rice in some years. In the 1850s his labor force numbered 472 men, women, and children, young Jim Robinson among them.[30]

The news of Michelle Obama's slave ancestry underscored the centrality of slavery in American history with uncommon power. On the campaign

[29] Shailagh Murray, "A Family Tree Rooted in American Soil," *Washington Post* (Washington, DC), Oct. 2, 2008, C1; Dahleen Glanton and Stacy St. Clair, "Michelle Obama's Family Tree Has Roots in a Carolina Slave Plantation," *Chicago Tribune* (Chicago, IL), Dec. 1, 2008, 1.

[30] George C. Rogers, Jr., *The History of Georgetown County, South Carolina* (Columbia: University of South Carolina Press, 1970), 285–287, 324.

trail, Barack Obama described his wife as "the most quintessentially American woman I know" and noted that her veins carried "the blood of slaves and slave owners." At the time, no connection to a white ancestor had been identified, although the Robinson family had long suspected one might exist.[31] As the fall wore on and Obama traded barbs with his Republican opponent, Senator John McCain of Arizona, people around the globe wondered what it would mean for a nation forged in slavery to elect a black man to its highest office. Yet even as new attention to race held the spotlight, news reports employed rhetorical conventions that downplayed the significance of slavery in American history and the severity of its legacies. The meaning of the term "plantation" offers a case in point. Although plantations are increasingly recognized as international symbols of racial slavery, the origins of their symbolism and the reasons for its historical strength and durability are poorly understood. Historians have tended to privilege representations over material realities – or, rather, what people saw as material realities – and to overlook the complexity of plantation histories. The results include an impoverished understanding of plantations, their symbolism, and the cultural significance of associated discourses.

Jim Robinson entered the world on a plantation of a particular type. During his lifetime it became markedly different. Changes in use, purpose, and material form changed the way people viewed Friendfield and its past. Those same changes compelled people to see its symbolism differently, for the Friendfield of the late nineteenth century differed dramatically from its antebellum predecessor. During the twentieth century, Friendfield changed even more. In the 1850s, the plantation had hummed with efficiency. Two hundred slaves worked its fields, cultivating rice along the Sampit, growing provision crops, and carrying out countless other tasks. Withers and his family lived in a handsome dwelling and owned a house in Charleston where they passed their summers. The Civil War and emancipation brought chaos to Friendfield. Plantation discipline collapsed, productivity declined, and decay set in. By the spring of 1865, few Georgetown District plantations remained in operation.

---

[31] Barack Obama quoted in Murray, "A Family Tree Rooted in American Soil." Further research discovered a white ancestor in Michelle Obama's maternal lineage. See Rachel L. Swarns and Jodi Kantor, "First Lady's Roots Reveal Slavery's Tangled Legacy," *New York Times*, Oct. 8, 2009, A1, A24; Rachel L. Swarns, *American Tapestry: The Story of the Black, White, and Multiracial Ancestors of Michelle Obama* (New York: Amistad, 2012).

Friendfield's status is unknown. If it remained in operation, it would have lost much of its productive capacity and most of its traditional order and efficiency. Chances are it lay idle.[32]

In the immediate aftermath of the war, Friendfield fared better than many of its counterparts. Cultivation of rice and other crops resumed quickly and reached moderate levels of output by the 1870s. During the 1880s, however, Friendfield declined precipitously. Decreased production led to diminished activity and material deterioration. As with other plantations, physical distress evinced the collapse of a once-productive social and economic system.[33]

The date of Robinson's death is unknown. He likely witnessed the early stages of Friendfield's decline and the material decay that occurred along with it. Removal of select buildings, flooding of rice fields, reforestation of upland territory, and general deterioration underscored the drama and severity of the changes taking place. If Robinson lived into the early twentieth century, he saw B. Walker Cannon, an absentee owner, lease Friendfield to a combination of black tenants and white operators. Full-time leisure use arrived in 1919, when a group of duck hunters from Florence, South Carolina, leased the property. Then, in the spring of 1930, Radcliffe and Frances Cheston of Philadelphia purchased Friendfield. Radcliffe Cheston had hunted the previous fall with friends in Williamsburg County and immediately became enamored with the lowcountry. In purchasing Friendfield, the Chestons sought exactly what so many other well-to-do northerners did: a winter residence and sporting estate steeped in the charm of time.[34]

During the next several years, the Chestons turned Friendfield into a showplace. A new dwelling designed by Philadelphia architect Arthur I. Meigs, a grounds and garden plan crafted by Umberto Innocenti, and extensive acreage made Friendfield a new kind of plantation. Agriculture became part of its past, leisure its foremost purpose. Simultaneously a refuge, a playground, and a site of spectacular beauty, Friendfield

---

[32] "Gerard Purchases Pee Dee Gun Club," *News and Courier*, Dec. 19, 1925, 2; Chalmers S. Murray, "Cheston Builds on Old Plantation," *News and Courier*, Aug. 2, 1931, B2; Friendfield Plantation (Georgetown County, South Carolina), National Register of Historic Places nomination, 1996, South Carolina Department of Archives and History, Columbia, SC.

[33] Friendfield Plantation, National Register of Historic Places nomination.

[34] Ibid.; Frances Cheston Train, *A Carolina Plantation Remembered: In Those Days* (Charleston: History Press, 2008), 31.

supplied the Chestons and their five children with an enviable setting for pleasure, recreation, and seasonal habitation.[35]

Friendfield exhibited in microcosm changes that occurred throughout the lowcountry and in a few places farther afield. Changes in use, purpose, and material form not only turned plantations into handsome estates; they changed the way people understood plantations and their pasts. They changed the symbolism and significance of plantations and their social and cultural roles. No longer simply former sites of slavery and commercial agriculture, plantations such as Friendfield embodied new purposes, new priorities, and new meanings. People who saw and experienced Friendfield during the Chestons' ownership recognized it as a product of their time. Although possessed of deep roots, it belonged to a variety of plantations with recent origins. Moreover, it embodied a valued history. Recognition of Friendfield as important in lowcountry history entailed more than age. It resulted from new perspectives on the past prompted by the Chestons' actions, the purposes for which they used Friendfield, and their social status. Memorialization of plantations and contemporary uses and associations profoundly influenced views of Friendfield and its past.

Friendfield's transformation grew out of ambitions and desires common to upper-class Americans of the era. Inspired by neither an imagined past nor the appeal of an idyllic southland, it demonstrated the cultural significance of vigorous recreation, immersion in unspoiled nature, and the revitalizing influence of both. The Chestons fit squarely within the norm. Radcliff Cheston's devotion to recreational hunting and the family's desire to escape what one of the Cheston daughters called "the drab, gray Northern winter" gave restorative, rejuvenating activities a central role in their lives.[36] Moreover, ties to an authentic, important history connoted status and authority. Like other Americans of the era, the Chestons prized associations with vestiges of the American past. Ownership of Friendfield made them stewards of history.

What Friendfield's transformation meant for the Robinson family and their descendants is less clear. Because the date of Jim Robinson's death is unknown, the date his descendants' association with Friendfield ended is also uncertain. The available evidence suggests that Robinson's kin left Friendfield during the first or second decade of the twentieth century.

[35] Murray, "Cheston Builds on Old Plantation"; Friendfield Plantation, National Register of Historic Places nomination; Train, *A Carolina Plantation Remembered*, 31–37.
[36] Train, *A Carolina Plantation Remembered*, 51–97.

Whatever the case, Robinson's sons made lives for themselves elsewhere. His eldest, Frasier Robinson, Sr., worked at a lumber mill, as a shoe repairman, and as a newspaper salesman in nearby Georgetown. With his wife, Rose Ella Cohen, Frasier Sr. had six children. The couple had their first son, Frasier Robinson, Jr., in 1912. In the early 1930s, Frasier Jr. traveled to Chicago in search of employment. As part of the Great Migration of African Americans from the South, he sought opportunities in the less racist, less segregated North. He obtained work with the US Postal Service, married, and raised a family in a cinderblock apartment in a public housing complex – a humble existence but an impressive step upward, given his father's beginnings in slavery and his own roots.[37]

The news articles that broke the story of Michelle Obama's ancestry depicted Friendfield in a manner typical of portrayals of former slave plantations since the early 1990s. The idiom casts plantations as solemn, mysterious places suffused by a palpable sense of foreboding. Depictions of this sort implicitly reference slavery, its tragedy, and its contested legacy; they hint at ongoing debates and seemingly unresolvable differences. In effect, they acknowledge the tensions that invariably surround discussions of race and slavery without taking up questions about their roots and the reasons for their persistence. Meanwhile, authors leave unaddressed the severity of slavery and its lasting influence. Tiptoeing around these and other subjects invokes a guise of neutrality while suggesting that truly compelling, convincing answers are unattainable. The message is that no amount of research and study can yield conclusive judgments; despite volumes of scholarship and extensive data, answers are shrouded somewhere in the mists of time.[38]

The appeal of "solemn-yet-neutral" portrayals is easily understood – they seek to sidestep controversy by offending no one. They cast slavery as a continuing problem without explaining why or naming its most

---

[37] Glanton and St. Clair, "Michelle Obama's Family Tree Has Roots in a Carolina Slave Plantation."

[38] See, for example, Ginger Thompson, "Reaping What Was Sown on the Old Plantation," *New York Times*, June 22, 2000, www.nytimes.com/2000/06/22/us/reaping-what-was-sown-old-plantation-landowner-tells-her-family-s-truth-park.html. Leigh Anne Duck notes an emphasis on "historical trauma and loss" in recent scholarship and literature in her "Plantation Cartographies and Chronologies," *American Literary History* 24, no. 4 (winter 2012): 848. For examples, see Jessica Adams, *Wounds of Returning: Race, Memory, and Property on the Postslavery Plantation* (Chapel Hill: University of North Carolina Press, 2007); Elizabeth Christine Russ, *The Plantation in the Postslavery Imagination* (New York: Oxford University Press, 2009).

damaging effects. They ignore vital questions, including many for which answers have long been available. Although the general stance is understandable, especially in an age of hyper-impartial news reporting, the consequences are enormous, for the end result is avoidance of slavery's role in the American past. To ignore what several generations of scholarship have made plainly clear denies the value of historical study and the power of history to chart paths toward understanding, healing, and reconciliation.

As scholars continue to examine the history of racial slavery and its role in shaping societies throughout the Western world, the need to understand the amnesia and mythmaking that took place afterward increases exponentially. That popular memory saw an exploitive and damaging institution as benign for so long speaks not only to human beings' capacity to rationalize injustice but to the erasure and forgetting that followed its downfall. The sanguine portrayals of slavery that acquired extraordinary strength and durability during the early twentieth century reflect more than romanticization and widespread refusal to face a sordid and shameful past. They also demonstrate the interconnectedness of narratives and place, the evocative power of material remains, and the implicit authority of the category "historic." The histories that became associated with plantations throughout the lowcountry offer a potent example. Although northerners' estates proved unusual in many respects, their influence is difficult to overstate. By the late 1930s, a region that many Americans had once vilified for political radicalism and extraordinary inequality consistently won praise for a glorious past. Sites where enslaved men, women, and children had lived and labored under deplorable conditions had become bucolic stage sets for the choreographed pageantry of an elite. The transformation of plantations throughout the region marked a profound reversal. By dramatically limiting opportunities for understanding the lowcountry past, northerners' estates became central to debates over slavery and its legacy. Like monuments to the Confederacy, veterans' memorials, and Civil War battlefields, northerners' "plantations" articulated a brand of remembering and forgetting that obscured more than it revealed.

Northerners' plantations underscore histories that have yet to receive due attention. Although the influence of deeply romanticized portrayals of the South and its past is well known, other developments have elicited little attention. The memorialization of plantations across the South offers a case in point. Identification of select plantations as remnants of a valued past not only laid the groundwork for northerners' efforts; it created a landscape of seemingly majestic monuments to lost grandeur. The decaying mansions that stood across large portions of the South during the early twentieth century not only demonstrated the impoverishment of the region after the Civil War; they legitimated the myth of a genteel past. Weathered houses, vine-covered barns, and untilled fields served as touchstones of remembrance, anchoring narratives of tragic loss and decline. Americans did not need *Gone with the Wind* and its many counterparts to learn about the southern past; they saw it as embodied presence. Although views of authentic remains proved no more impartial than the most fanciful portrayals of the era, they possessed unmatched power and durability. Physical presence and ties to seemingly tailor-made narratives made the difference. As the years passed and Americans grew circumspect of plainly romanticized portrayals, they increasingly saw surviving edifices as fundamental to knowledge of the South and its past.[39]

As important, plantations continued to take on new meanings. People saw plantations as living spaces that continued to change and develop, even after large-scale agriculture ended. Historians' emphasis on staple crops and coerced labor has left important examples little unexplored. No matter how unusual northerners' estates may seem in retrospect, contemporaries viewed them as plantations unequivocally. They saw northerners' activities as fundamentally aristocratic and thus reinforcing longstanding ties between plantations and social and cultural elites. Grappling with northerners' "plantations" and the discourses they inspired thus brings neglected histories to light, with powerful implications for perspectives on the cultural history of early twentieth-century America.

---

[39] Memorialization, in this usage, refers to the identification of plantations as historical monuments in the manner described in Alois Riegl, "The Modern Cult of Monuments: Its Essence and Its Development," *Oppositions* 25 (fall 1992): 21–51. See also David Lowenthal, *The Past Is a Foreign Country* (Cambridge: Cambridge University Press, 1985), 238–243; Françoise Choay, *The Invention of the Historic Monument*, trans. Lauren M. O'Connell (Cambridge: Cambridge University Press, 2001), 1–16, 82–116. For 1930s accounts that differentiate between plantation romances and history, see Charles McD. Puckette, "Charleston and the Carolina Lowcountry," *New York Times*, Dec. 13, 1931, book review section, 3; Herbert Ravenel Sass, "The Low-Country," in *The Carolina Low-Country*, ed. Augustine T. Smythe et al. (New York: Macmillan Co., 1931), 4. On interest in authentic remnants of the pre–Civil War South during the 1930s, see Catherine A. Stewart, *Long Past Slavery: Representing Race in the Federal Writers' Project* (Chapel Hill: University of North Carolina Press, 2016), 1–61; Yuhl, *A Golden Haze of Memory*; W. Fitzhugh Brundage, *The Southern Past: A Clash of Race and Memory* (Cambridge: Belknap Press of Harvard University Press, 2005), chap. 5.

Ultimately, the significance of northerners' lowcountry estates lies less in their unusualness than in what they reveal about American culture during an era of tumult and instability. Once plantations became objects of the imagination, cultural norms became the sole constraints on use and material expression. During the early twentieth century, broad commitments to white supremacy and pervasive denial of a shameful past created open-ended possibilities. As popular interest in the South surged, sportsmen and sportswomen seized opportunities for self-definition that quickly became intertwined with existing cultural practices and new rituals. During World War II, ideological shifts began to limit celebrations of a romantic Old South. The contradictions inherent in fighting totalitarian regimes abroad while supporting Jim Crow at home, military service for more than one million black men, and concern for fundamental human rights turned Americans' attention to global concerns and stoked growing unease about racial oppression. In the mid-1950s, when Chlotilde R. Martin recalled "those fabulous days of the early 1930s," she recalled a fleeting moment when the grandeur of a seemingly remote past had created rare possibilities. Although a few people of wealth and privilege continued spending their winters at "plantations," their estates no longer served as realms of intrigue and exoticism. Nor did "wealthy northerners" come with aims of turning old plantations into "hunting preserves and 'playthings'" anymore.[40] Yet the legacy of "those fabulous days" lived on – in majestic landscapes where vestiges of a seemingly noble past commingled with stately buildings and verdant gardens, all steeped in the charm of time, all evoking a history the lowcountry never had.

---

[40] Chlotilde R. Martin, "Beaufort Plantation Turnover Growing Larger," undated newspaper clipping [ca. 1952] in Mabel FitzSimons Scrapbook, SCHS. On the culture of the 1920s and 1930s, see especially Warren I. Susman, *Culture as History: The Transformation of American Society in the Twentieth Century* (Washington, DC: Smithsonian Institution Press, 2003), chaps. 7 and 9; Susan Hegeman, *Patterns for America: Modernism and the Concept of Culture* (Princeton, NJ: Princeton University Press, 1999); Terry A. Cooney, *Balancing Acts: American Thought and Culture in the 1930s* (New York: Twayne Publishers, 1995); Michael Kammen, *Mystic Chords of Memory: The Transformation of Tradition in American Culture* (New York: Knopf, 1991), 299–309, 407–415.

Bibliography

PRIMARY SOURCES

Newspapers

Beaufort Gazette (Beaufort, SC)
Chicago Tribune (Chicago, IL)
Georgetown Times (Georgetown, SC)
News and Courier (Charleston, SC)
New York Daily News (New York, NY)
New York Times (New York, NY)
Wall Street Journal (New York, NY)
Washington Post (Washington, DC)

Manuscript Collections

American Institute of Architects Archives, Washington, DC
    Benjamin Judah Lubschez Membership File
Archibald S. Alexander Library, Rutgers University, New Brunswick, NJ
    Joseph Sherman Frelinghuysen Papers
Charleston County Public Library, Charleston, SC
    South Carolina Vertical File Collection
Hagley Library, Wilmington, DE
    Kinloch Gun Club Records
    Miscellaneous Papers of Eugene du Pont
Historic Charleston Foundation, Charleston, SC
    Mulberry Plantation History File
Library of Congress, Washington, DC
    Clare Boothe Luce Papers
    Harry Frank Guggenheim Papers
    Henry Robinson Luce Papers

Private Collection
    Sidney J. Legendre, Diary of Life at Medway Plantation, Mt. Holly, South
        Carolina, May 1937–April 1941
South Carolina Historical Society, Charleston, SC
    Boone Hall Scrapbooks
    Henry O. Marcy Diary, 1864–1899
    Kittredge Family Papers
    Mabel FitzSimons Scrapbook
    News and Courier Records
    Peter Gaillard Stoney, List and Memorandum Book, 1824–1833
    Plat of Medway Plantation, 1792
    Records of the Society for the Preservation of Spirituals
    Stoney Family Papers
    Vertical File Collection
South Caroliniana Library, University of South Carolina, Columbia, SC
    Berkeley County Photo Collection
Southern Historical Collection, University of North Carolina, Chapel Hill, NC
    Stoney and Porcher Family Papers (microfilm)

Published Materials

Allen, George Marshall. "Charleston: A Typical City of the South." *Magazine of Travel* 1, no. 2 (Feb. 1895): 99–129.

Andrews, Sidney. *The South since the War: As Shown by Fourteen Weeks of Travel and Observation in Georgia and the Carolinas*. Boston: Ticknor and Fields, 1866.

Architects' Emergency Committee. *Great Georgian Houses of America*. 2 vols. 1937; reprint, New York: Dover Publications, 1970.

Articles of Agreement and Rules of the Pineland Club, Robertsville, Hampton County, South Carolina. N.p.: n.p., 1900.

Baker, John Cordis, ed. *American Country Homes and Their Gardens*. Philadelphia: House and Garden, 1906.

Ball, William Watts. *The State That Forgot: South Carolina's Surrender to Democracy*. Indianapolis: Bobbs-Merrill Company, 1932.

Bennett, John. "Gullah: A Negro Patois." *South Atlantic Quarterly* 7, no. 4 (Oct. 1908): 332–347.

———. "Gullah: A Negro Patois, Part II." *South Atlantic Quarterly* 8, no. 1 (Jan. 1909): 39–52.

"Carolina Classic." *House and Garden* 73, no. 2 (Feb. 1938): 34–35.

Carter, Henry H. *Early History of the Santee Club*. Boston [?]: n.p., 1934.

Christensen, A. H. M. *Afro-American Folk Lore: Told Round Camp Fires on the Sea Islands of South Carolina*. Boston: J. G. Cupples Company, 1892.

Cram, Mildred. *Old Seaport Towns of the South*. New York: Dodd, Mead and Company, 1917.

Cross, J. Russell. *Historic Ramblin's through Berkeley*. Columbia: R. L. Bryan Company, 1985.

Crowley, Herbert. "Rich Men and Their Houses." *Architectural Record* 12, no. 1 (May 1902): 27–32.

Curtis, Paul A. "Will We Have Good Duck Shooting Again?" *Country Life in America* 69, no. 1 (Nov. 1935): 34–35, 78–79.

Cyclopedia of Eminent and Representative Men of the Carolinas of the Nineteenth Century. 2 vols. Madison, WI: Brant and Fuller, 1892.

Daniels, Jonathan. *A Southerner Discovers the South*. New York: Macmillan, 1938.

Daniels, Winthrop M. "The Slave Plantation in Retrospect." *Atlantic Monthly* 107, no. 3 (March 11, 1911): 363–369.

Deas, Anne Simmons. *Points of Colonial Interest around Summerville. Dorchester, Newington, Ingleside, St. James, Goose Creek*. Summerville, SC: S. P. Driggers, 1905.

"The Deep South." *House and Garden* 76, no. 1 (Nov. 1939): 28–49.

Derieux, James C. "The Renaissance of the Plantation." *Country Life* 41, no. 3 (Jan. 1932): 34–39.

Desmond, Harry W., and Herbert Croly. *Stately Homes in America*. New York: D. Appleton and Company, 1903.

Devereux, Anthony Q. *The Rice Princes: A Rice Epoch Revisited*. Columbia: State Printing Company, 1973.

Doar, David. *Rice and Rice Planting in the South Carolina Low Country*. Charleston: Charleston Museum, 1936.

Dwight, H. R. *Some Historic Spots in Berkeley*. Moncks Corner, SC: Monck's Corner Drug Company, 1921.

Eberlein, Howard Donaldson. *The Architecture of Colonial America*. 1915; reprint, New York: Johnson Reprint Corporation, 1968.

"Preserving Our Architectural Birthright." *Country Life* 38, no. 6 (Oct. 1920): 70–72.

Elliman, Huyler, and Mullally, Inc. *Romantic Charleston*. Charleston: n.p., [1935?].

Elliott, William. *Carolina Sports by Land and Water*. New York: Derby and Jackson, 1859.

Ferree, Barr. *American Estates and Gardens*. New York: Munn and Company, 1904.

Fletcher, Coyne. "In the Lowlands of South Carolina." *Frank Leslie's Popular Monthly* (March 1891): 280–288.

"From Thomasville to Tallahassee." *Country Life* 17, no. 4 (Feb. 1935): 11–14.

Furman, Sara. "Harrietta: An Old Plantation House on the Santee River." *House Beautiful* 70, no. 6 (1931): 475–480.

G., A. G., "A Southern Coast Home." *Southern Cultivator* 52, no. 8 (Aug. 1894): 402.

Gaines, Francis Pendleton. *The Southern Plantation: A Study in the Development and Accuracy of a Tradition*. New York: Columbia University Press, 1924.

Gonzales, Ambrose Elliott. *The Black Border: Gullah Stories of the Carolina Coast*. Columbia: The State Company, 1922.

Hallock, Charles. *Hallock's American Club List and Sportsman's Glossary.* New York: Forest and Steam Publishing Company, 1878.

Hammond, Harry. *South Carolina: Resources and Population, Institutions and Industries.* Charleston: Walker, Evans, and Cogswell, 1883.

Hart, D. J. "Wild Turkey Hunting in South Carolina: The Ways and Habits of Meleagris Gallapavo." *Field and Stream* 20, no. 8 (Dec. 1915): 775–781.

Herbert, William. *Houses for Town or Country.* New York: Duffield and Company, 1907.

Heyward, DuBose. "Charleston: Where Mellow Past and Present Meet." *National Geographic Magazine* 75, no. 3 (March 1939): 273–312.

   Porgy. New York: George H. Doran Company, 1925.

Hewyard, Duncan Clinch. *Seed from Madagascar.* 1937; reprint, Columbia: University of South Carolina Press, 1993.

Hooper, Charles Edward. *Reclaiming the Old House: Its Modern Problems and Their Solutions as Governed by the Methods of Its Builders.* New York: McBride, Nast and Company, 1913.

Horton, Mrs. Thaddeus. "Romances of Some Southern Homes." *Ladies' Home Journal* 27, no. 10 (Sept. 1900): 9–10.

Howells, William D. "In Charleston." *Harper's Magazine* 131, no. 785 (Oct. 1915): 747–757.

Hungerford, Edward. "Charleston of the Real South." *Travel* 11, no. 6 (Oct. 1913): 32–35, 57–58.

Irvin, Willis. *Selections from the Work of Willis Irvin – Architect.* N.p.: n.p., 1937.

Irving, John Beaufain. *A Day on Cooper River.* Charleston: A. E. Miller, 1842.

Jones, Charles Colcock. *Negro Myths from the Georgia Coast Told in the Vernacular.* Boston: Houghton, Mifflin, and Company, 1888.

Jones, J. Roy. *Year Book of the Department of Agriculture, Commrce and Industries of the State of South Caroilna.* Columbia: General Assembly of South Carolina, 1936.

Kimball, Fiske. "The American Country House." *Architectural Record* 46, no. 4 (Oct. 1919): 291–327.

King, Edward. *The Great South.* 2 vols. Hartford, CT: American Publishing Company, 1875.

Kirk, Francis Marion. *A History of the St. John's Hunting Club.* N.p.: St. John's Hunting Club, 1950.

Legendre, Gertrude Sanford. *The Sands Ceased to Run.* New York: William-Frederick Press, 1947.

   The Time of My Life. Charleston: Wyrick and Company, 1987.

Legendre, Sidney Jennings. *Land of the White Parasol and the Million Elephants.* New York: Dodd, Mead, and Company, 1936.

   Okovango, Desert River. New York: J. Messner, 1939.

Lieding, Harriette Kershaw. *Historic Homes of South Carolina.* Philadelpha: J. B. Lippincott and Company, 1921.

"Life Goes to a Party with the Sidney Legendres on a Deer Hunt in South Carolina." *Life*, Jan. 24, 1938, 54–57.

"The Low Country." *Life*, Dec. 25, 1939, 38–45.

Luce, Clare Boothe. "The Victorious South." *Vogue*, June 1, 1937, 79–81, 120–122.

Mahoney, Nell Savage. "The Melody Lingers On." *Country Life* 67, no. 6 (Apr. 1935): 10–15.

Major, Mrs. Howard. "Southern Plantation Homes." *House and Garden* 53 (Nov. 1926): 112–113, 126, 130.

Mechlin, Leila. "A Glimpse of Old Charleston and the Nearby Rice Plantations." *American Magazine of Art* 14, no. 9 (Sept. 1923): 475–485.

"Medway Plantation." *Town and Country* 103, no. 4318 (March 1949): 76–79.

"Mepkin Plantation, Moncks Corner, S.C." *Architectural Forum* 66, no. 6 (June 1937): 515–522.

Mills, Robert. *Atlas of the State of South Carolina*. Baltimore: F. Lucas, Jr., 1825.

"Modern in South Carolina, Winter Home of Mr. and Mrs. Henry R. Luce, Moncks Corners, South Carolina." *House and Garden* 72 (Aug. 1937): 36–39.

Oakland Club. *Oakland Club, St. Stephens P.O., Berkeley County, South Carolina*. N.p.: n.p., 1908.

"The Oaks, a Restored Mansion of the South." *Country Life* 29, no. 2 (Dec. 1915): 53–55.

Parrish, Lydia. "A Heritage We Must Not Lose." *Country Life* 69, no. 2 (Dec. 1935): 50–55, 62.

Pepper, George W. *Personal Recollections of Sherman's Campaigns in Georgia and the Carolinas*. Zanesville, OH: Hugh Dunne, 1866.

Perrett, Antoinette. "History of the Country Estate – I. Egypt." *Country Life* 68, no. 5 (Sept. 1935): 27–29, 74.

"History of the Country Estate – II. Babylonia." *Country Life* 68, no. 6 (Oct. 1935): 34–36, 72, 75.

"History of the Country Estate – III. The Roman Farm." *Country Life* 69, no. 1 (Nov. 1935): 37–39, 72–76.

"History of the Country Estate – IV. The Roman Pleasure Villa." *Country Life* 69, no. 2 (Dec. 1935): 45–47, 76–77.

"History of the Country Estate – V. Moorish Empire in Spain." *Country Life* 69, no. 5 (March 1936): 49–50, 70–71.

"History of the Country Estate – XIII. Southern Estates." *Country Life* 71, no. 2 (Dec. 1936): 45–46, 115–118.

Peterkin, Julia Mood. *Black April: A Novel*. Indianapolis: Bobbs-Merrill Company, 1927.

Roll, Jordan, Roll. New York: Robert B. Ballou, 1933.

Scarlet Sister Mary. Indianapolis: Bobbs-Merrill Company, 1928.

Ravenel, Henry Edmund. *Ravenel Records*. Atlanta: Franklin Printing and Publishing Company, 1898.

Rice, James Henry. *Glories of the Carolina Coast*. Columbia: R. L. Bryan Company, 1925.

Richards, T. Addison. "The Rice Lands of the South." *Harper's New Monthly Magazine* (Nov. 1859): 721–738.

Salley, A. S. *The Happy Hunting Ground: Personal Experiences in the Low-Country of South Carolina*. Columbia: The State Company, 1926.

Sass, Herbert R. *Look Back to Glory*. Indianapolis: Bobbs-Merrill Company, 1933.

"South Carolina Rediscovered: A Native Sun Finds Spectacular Changes in the 'Moonlight and Magnolia' State, Scene of Huge H-Bomb Project." *National Geographic* 103, no. 1 (March 1953): 281–321.

The Story of the South Carolina Lowcountry. 3 vols. West Columbia, SC: J. F. Hyer, 1956.

"The Ten Rice Rivers." *Saturday Evening Post* (Dec. 13, 1941): 20–21, 105–108.

Saxxon, Lyle. "Vanished Paradise." *Country Life* 67, no. 1 (Nov. 1934): 35–41, 84, 106.

Shaffer, E. T. H. "The Ashley River and Its Gardens." *National Geographic* 49, no. 5 (May 1926): 524–532, 549–550.

"Shooting Lodge in Carolina: 'Richmond,' the Estate of George A. Ellis, Esq., Near Monks Corner, South Carolina." *Country Life* 65, no. 2 (Dec. 1933): 60–61.

Simms, William Gilmore. *The Geography of South Carolina.* Charleston: Babcock and Company, 1843.

Simons, Albert, and Samuel Lapham. *Charleston, South Carolina.* New York: Press of the American Institute of Architects, 1927.

Simons, Katherine Drayton. *Roads of Romance and Historic Spots Near Summerville, South Carolina.* Charleston: Southern Printing and Publishing Company, 1924.

Smith, Alice R. H., and D. E. Huger Smith. *The Dwelling Houses of Charleston, South Carolina.* Philadelphia: J. B. Lippincott Company, 1917.

Smith, Alice R. H., Herbert Ravenel Sass, and D. E. Huger Smith. *A Carolina Plantation of the Fifties.* New York: W. Morrow and Company, 1936.

Smith, Harry Worchester. *Life and Sport in Aiken and Those Who Made It.* New York: Derrydale Press, 1935.

Smith, Reed. *Gullah.* Columbia: University of South Carolina, 1926.

Smythe, Augustine T., et al. *The Carolina Low-Country.* New York: Macmillan, 1931.

"South Carolina Plantations." *Town and Country* 90, no. 4144 (Jan. 15, 1935): 30–33.

Starnes, Hugh. "The Rice-Fields of Carolina." *Southern Bivouac* 2, no. 6 (Nov. 1886): 329–342.

Stoney, Samuel Gaillard. *Plantations of the Carolina Low Country.* Charleston: Carolina Art Association, 1939.

    Plantations of the Carolina Low Country. Charleston: Carolina Art Association, 1945.

    Plantations of the Carolina Low Country. Charleston: Carolina Art Association, 1955.

    Plantations of the Carolina Low Country. Charleston: Carolina Art Association, 1964.

Stoney, Samuel Gaillard, and Gertrude Mathews Shelby. *Black Genesis: A Chronicle.* New York: Macmillan, 1930.

Todd, John R., and Francis M. Huston. *Prince Williams Parish and Plantations.* Richmond: Garrett and Massie, 1935.

Toombs, Frederick R. "Midwinter Hunting in the South." *Town and Country,* Jan. 13, 1906, 26.

Ware, William Roach, ed. *The Georgian Period: A Collection of Papers Dealing with "Colonial" or Eighteenth-Century Architecture in the United States.* 3 vols. New York: American Architect, 1908.

Watson, E. J. *Fourth Annual Report of Agriculture, Commerce and Immigration of the State of South Carolina.* Columbia: Gonzales and Bryan, 1908.

Handbook of South Carolina, 1907. Columbia: The State Company, 1907.

"Wedgefield Plantation." *Country Life* 75, no. 2 (Dec. 1938): 53–57, 101–102.

Whitson, William. "A South Carolina Hunt." *Forest and Stream* 53, no. 14 (Sept. 30, 1899): 264–265.

Willett, N. L. *Game Preserves and Game of Beaufort, Colleton and Jasper Counties, South Carolina: Hunters' Paradise, Manly Sports.* Beaufort, SC: Charleston and Western Carolina Railway Company, 1927.

Wilson, Robert. *An Address Delivered before the St. John's Hunting Club, at Indianfield Plantation, St. John's, Berkeley, July 4, 1907.* Charleston: Walker, Evans, and Cogswell Company, 1950.

Woolson, Constance. "Up the Ashley and Cooper." *Harpers New Monthly Magazine* 52 (Dec. 1975): 1–24.

"The Yankee Doodle Salon." *Fortune* 16 (Dec. 1937): 123–129, 180, 183–184.

Government Records

Berkeley County Register of Deeds, Moncks Corner, SC
    Deeds and plats pertaining to Medway and Mulberry plantations

Charleston County Register of Mense Conveyance, Charleston, SC
    Deeds and plats pertaining to Mulberry Plantation

National Archives and Records Administration, Washington, DC
    Records of the Bureau of Refugees, Freedmen, and Abandoned Lands, 1865–1872
    Seventh Census of the United States, Slave Schedule, 1850. St. John's Berkeley Parish, Charleston District, microfilm M432, roll 862.

South Carolina Department of Archives and History, Columbia, SC
    MS census returns, Charleston District, Seventh and Eight Censuses of the United States
    Preservation Consultants, Inc. Berkeley County Historical and Architectural Inventory, 1989
    National Register of Historic Places nominations
        Atalaya and Brookgreen Gardens, Georgetown County
        Friendfield Plantation, Georgetown County
        Mansfield Plantation, Georgetown County

United States Census Bureau. *Agriculture of the United States in 1860.* Washington, DC: Government Printing Office, 1864.

Fifteenth Census of the United States: 1930, Agriculture. Vol. 2. Washington, DC: Government Printing Office, 1932.

Fourteenth Census of the United States, Taken in the Year 1920. Vol. 1: *Population, 1920, Number and Distribution of Inhabitants*. Washington, DC: Government Printing Office, 1921.

Fourteenth Census of the United States, Taken in the Year 1920. Vol. 9: *Manufacturers, 1919*. Washington, DC: Government Printing Office, 1923.

Population of the United States in 1860; Compiled from the Original Returns of the Eight Census. Washington, DC: Government Printing Office, 1864.

Thirteenth Census of the United States, Taken in the Year 1910. Vol. 7: *Agriculture, 1909 and 1910*. Washington, DC: Government Printing Office, 1913.

Twelfth Census of the United States, Taken in the Year 1900: Agriculture. Pt. II. Crops and Irrigation. Washington, DC: Government Printing Office, 1902.

SECONDARY SOURCES

Articles and Essays

Alpern, Stanley B. "Did Enslaved Africans Spark South Carolina's Eighteenth-Century Rice Boom?" In *African Ethnobotany in the Americas*, ed. Robert A. Voeks and John Rashford, 35–66. New York: Springer, 2013.

Baum, Jack. "A History of Market Hunting in the Currituck Sound Area, Part 1." *Wildlife in North Carolina* 32, no. 11 (Nov. 1968): 13–15.

"A History of Market Hunting in the Currituck Sound Area, Part 2." *Wildlife in North Carolina* 32, no. 12 (Dec. 1968): 4–8, 31.

Brendan, Gill. "Frank Lloyd Wright's Auldbrass." *Architectural Digest* 50, no. 12 (Dec. 1993): 126–137, 180–181.

Brueckheimer, William R. "The Quail Plantations of the Thomasville-Tallahassee-Albany Regions." In *Proceedings: Tall Timbers Ecology and Management Conference*, no. 16, 141–165. Tallahassee: Tall Timbers Research Station, 1982.

Brundage, W. Fitzhugh. *The Southern Past: A Clash of Race and Memory*. Cambridge: Belknap Press of Harvard University Press, 2005.

ed. *Where These Memories Grow: History, Memory, and Southern Identity*. Chapel Hill: University of North Carolina Press, 2000.

da Cunha, Olivia Gomes. "Somewhere Close to Nashville: Plantation Cartographies." *Review* 34, nos. 1–2 (2011): 79–113.

Carney, Judith. "The African Antecedents of Uncle Ben in U.S. Rice History." *Journal of Historical Geography* 29, no. 1 (2003): 1–21.

"Landscapes of Technology Transfer: Rice Cultivation and African Continuities." *Technology and Culture* 37, no. 1 (Jan. 1996): 5–35.

Clifton, James M. "Twilight Comes to the Rice Kingdom: Postbellum Rice Culture on the South Atlantic Coast, 1820–1880." *Georgia Historical Quarterly* 62 (summer 1978): 146–52.

Coclanis, Peter A. "Distant Thunder: The Creation of a World Market in Rice and the Transformations It Wrought." *American Historical Review* 98, no. 4 (Oct. 1993): 1050–1078.

Courson, Maxwell Taylor. "Howard Earle Coffin, King of the Georgia Coast." *Georgia Historical Quarterly* 83, no. 2 (summer 1999): 322–341.

Davis, John E. "The Plantation Broker." *South Carolina Wildlife* 50, no. 6 (Nov.-Dec. 2003): 7–13.

Duck, Leigh Anne. "Plantation Cartographies and Chronologies." *American Literary History* 24, no. 2 (winter 2012): 842–852.

Easterby, J. H. "The St. Thomas Hunting Club, 1785–1801: Its Rules, Excerpts from Its Minutes, and a List of Members." *South Carolina Historical and Genealogical Magazine* 46, no. 3 (July 1945): 123–131.

"The St. Thomas Hunting Club, 1785–1801 (Continued)." *South Carolina Historical and Genealogical Magazine* 46, no. 4 (Oct. 1945): 209–213.

Edelson, S. Max. "Beyond 'Black Rice': Reconstructing Material and Cultural Contexts for Early Plantation Agriculture." *American Historical Review* 115, no. 1 (Feb. 2010): 125–135.

Eltis, David, Philip Morgan, and David Richardson. "Agency and Diaspora in Atlantic History: Reassessing the African Contributions to Rice Cultivation in the Americas." *American Historical Review* 112, no. 5 (2007): 1329–1358.

"Black, Brown, or White? Color-Coding American Commercial Rice Production." *American Historical Review* 115, no. 1 (Feb. 2010): 164–171.

Emmet, Richard S. "Memoires of Cheeha Combahee Plantation, 1929–1991." *Carologue* (autumn 1999): 18–22.

Fields-Black, Edda L. "Atlantic Rice and Rice Farmers: Rising from Debate, Engaging New Sources, Methods, and Modes of Inquiry, and Asking New Questions." *Atlantic Studies* 12, no. 3 (Sept. 2015): 276–295.

Gebhard, David. "The American Colonial Revival in the 1930s." *Winterthur Portfolio* 22, nos. 2–3 (summer/autumn 1987): 109–148.

Hart, John Fraser. "The Role of the Plantation in Southern Agriculture." *Proceedings: Tall Timbers Ecology and Management Conference, February 22–24, 1979, Thomasville, Georgia*. Tallahassee: Tall Timbers Research Station, 1982.

Hawley, Norman R. "The Old Rice Plantations in and around the Santee Experimental Forest." *Agricultural History* 23, no. 2 (Apr. 1949): 86–91.

Hemingaway, Theodore. "Prelude to Change: Black Carolinians in the War Years, 1914–1920." *Journal of Negro History* 65, no. 3 (1980): 212–227.

Hilliard, Sam B. "Antebellum Tidewater Rice Culture in South Carolina and Georgia." In *European Settlement and Development in North America: Essays on Geographical Change in Honour and Memory of Andrew Hill Clark*, ed. James R. Gibson, 91–115. Toronto: University of Toronto Press, 1978.

Hoelscher, Steven. "Making Place, Making Race: Performances of Whiteness in the Jim Crow South." *Annals of the Association of American Geographers* 93, no. 3 (Sept. 2003): 657–686.

"The White-Pillard Past: Landscapes of Memory and Race in the American South." In *Landscape and Race in the United States*, ed. Richard H. Schein, 39–73. New York: Routledge, 2006.

Hoffman, Edwin D. "The Gensis of the Modern Movement for Equal Rights in South Carolina, 1930–1939." *Journal of Negro History* 44, no. 4 (Oct. 1959): 346–369.

Hofstadter, Richard. "U. B. Phillips and the Plantation Legend." *Journal of Negro History* 29, no. 2 (Apr. 1944): 109–124.

Jaher, Frederic Cople. "The Gilded Elite: American Multimillionaires, 1865 to the Present." In *Wealth and the Wealthy in the Modern World*, ed. W. D. Rubinstein, 189–276. New York: St. Martin's Press, 1980.

"Style and Status: High Society in Late Nineteenth-Century New York." In *The Rich, the Well Born, and the Powerful: Elites and Upper Classes in History*, ed. Frederic Cople Jaher, 258–284. Urbana: University of Illinois Press, 1973

Kelly, Brian. "Black Laborers, the Republican Party, and the Crisis of Reconstruction of Lowcountry South Carolina." *International Review of Social History* 51 (2006): 375–414.

Kittredge, Carola. "Charleston's Grandest Dame." *Town and Country* 149, no. 5179 (Apr. 1995): 118–121.

Kovacik, Charles F. "Plantations and the Low Country Landscape." In *Snapshots of the Carolinas: Landscapes and Cultures*, ed. D. Gordon Bennett, 3–6. Washington, DC: Association of American Geographers, 1996.

"South Carolina Rice Coast Landscape Changes." In *Proceedings: Tall Timbers Ecology and Management Conference*. No. 16. Tallahassee: Tall Timbers Research Station, 1982, 47–65.

Kovacik, Charles F., and Robert E. Mason. "Changes in the South Carolina Sea Island Cotton Industry." *Southeastern Geographer* 25, no. 2 (Nov. 1985): 77–104.

Lewis, Carolyn Baker. "The World around Hampton: Post-Bellum Life on a South Carolina Plantation." *Agricultural History* 58, no. 3 (July 1984): 456–476.

Longstreth, Richard W. "Academic Eclecticism in American Architecture." *Winterthur Portfolio* 17, no. 1 (spring 1982): 55–82.

Lowenthal, David. "Past Time, Present Place: Landscape and Memory." *Geographical Review* 65 (Jan. 1975): 1–36.

May, Bridget A. "Progressivism and the Colonial Revival: The Modern Colonial House, 1900–1920." *Winterthur Portfolio* 26, nos. 2–3 (summer/autunm 1991): 107–122.

Moore, Jamie W. "The Lowcountry in Economic Transition: Charleston since 1865." *South Carolina Historical Magazine* 80, no. 2 (Apr. 1979): 156–171.

Moore, John Hammond. "Charleston in World War I: Seeds of Change." *South Carolina Historical Magazine* 86, no. 1 (Jan. 1985): 39–49.

Morgan, Philip D. "Work and Culture: The Task System and the World of Lowcountry Blacks, 1700 to 1880." *William and Mary Quarterly* 39, no. 4 (Oct. 1982): 563–599.

Oakes, Timothy. "Place and the Paradox of Modernity." *Annals of the Association of American Geographers* 87, no. 3 (Sept. 1997): 509–531.

Pred, Allan. "Place as a Historically Contingent Process: Structuration and the Time-Geography of Becoming Places." *Annals of the Association of American Geographers* 74, no. 2 (June 1984): 279–297.

Prunty, Merle, Jr. "The Renaissance of the Southern Plantation." *Geographical Review* 45, no. 4 (Oct. 1955): 459–491.

Rhoads, William B. "The Colonial Revival and American Nationalism." *Journal of the Society of Architectural Historians* 35, no. 4 (Dec. 1976): 239–254.

Roberts, Blain, and Ethan J. Kytle. "Looking the Thing in the Face: Slavery, Race, and the Commemorative Landscape in Charleston, South Carolina, 1865–2010." *Journal of Southern History* 73, no. 3 (Aug. 2012): 639–684.

Rothery, Mark. "The Shooting Party: The Associational Cultures of Rural and Urban Elites in the Late Nineteenth and Early Twentieth Centuries." In *Our Hunting Fathers: Field Sports in England after 1850*, ed. R. W. Hoyle, 96–118. Lancaster, UK: Carnegie Publishing, 2007.

Rowland, Lawrence S. "'Alone on the River': The Rise and Fall of the Savannah River Rice Plantations of St. Peter's Parish, South Carolina." *South Carolina Historical Magazine* 88, no. 3 (July 1987): 121–150.

Saville, Julie. "Grassroots Reconstruction: Agricultural Labour and Collective Action in South Carolina, 1860–1868." *Slavery and Abolition* 12, no. 3 (Dec. 1991): 173–182.

Schein, Richard H. "Normative Dimensions of Landscape." In *Everyday America: Cultural Landscape Studies after J. B. Jackson*, ed. Chris Wilson and Paul Groth, 199–218. Berkeley: University of California Press, 2003.

⸻. "The Place of Landscape: A Conceptual Framework for Interpreting an American Scene." *Annals of the Association of American Geographers* 87, no. 4 (Dec. 1997): 660–680.

Schmidt, Albert J. "Hyrne Family Letters." *South Carolina Historical Magazine* 63, no. 3 (July 1962): 150–157.

Shick, Tom W., and Don H. Doyle. "The South Carolina Phosphate Boom and the Stillbirth of the New South, 1867–1920." *South Carolina Historical Magazine* 86, no. 1 (Jan. 1985): 1–31.

Smith, Hayden R. "Knowledge of the Hunt: African American Hunting Guides in the South Carolina Lowcountry at the Turn of the Twentieth Century." In *Leisure, Plantations, and the Making of a New South: The Sporting Plantations of the South Carolina Lowcountry and Red Hills Region, 1900–1940*, ed. Julia Brock and Daniel Vivian, 131–148. Lanham, MD: Lexington Books, 2015.

⸻. "Reserving Water: Environmental and Technological Relationships with Colonial South Carolina Inland Rice Plantations." In *Rice: Global Networks and New Histories*, ed. Francesca Bray, Peter A. Coclanis, Edda L. Fields-Black, and Dagmar Schaefer, 189–211. New York: Cambridge University Press, 2015.

Stewart, Mart A. "Rice, Water, and Power: Domination and Resistance in the Lowcountry, 1790–1880." *Environmental History Review* 15 (fall 1991): 47–64.

Stoney, Samuel G., ed. "Recollections of John Stafford Stoney, Confederate Surgeon." *South Carolina Historical Magazine* 60, no. 4 (Oct. 1959): 208–220.

Strickland, John Scott. "'No More Mud Work': The Struggle for the Control of Labor and Production in Low Country South Carolina, 1863–1880." In *The Southern Enigma: Essays on Race, Class, and Folk Culture*, ed. Walter J. Fraser, Jr., and Winfred B. Moore, 43–62. Westport, CT: Greenwood Press, 1983.

"Traditional Culture and Moral Economy: Social and Economic Change in the South Carolina Low Country, 1865–1910." In *The Countryside in the Age of Capitalist Transformation*, ed. Steven Hahn and Jonathan Prude, 141–178. Chapel Hill: University of North Carolina Press, 1985.

Tindall, George B. "Mythology: A New Frontier in Southern History." In *The Idea of the South: Pursuit of a Central Theme*, ed. Frank E. Vandiver, 1–15. Chicago: William Marsh Rice University by University of Chicago Press, 1964.

Taylor, Zach. "Currituck's Grand Old Hunting Clubs." *Sports Afield* 180, no. 6 (Dec. 1978): 38–40.

Vlach, John Michael. "The Plantation Tradition in an Urban Setting: The Case of the Aiken-Rhett House in Charleston, South Carolina." *Southern Cultures* 5, no. 4 (winter 1999): 52–69.

Wayne, Lucy B. "'Burning Brick and Making a Large Fortune at It Too': Landscape Archaeology and Lowcountry Brickmaking." In *Carolina's Historical Landscapes: Archaeological Perspectives*, ed. Linda F. Stine et al., 97–111. Knoxville: University of Tennessee Press, 1997.

Waterman, Thomas T. "French Influence in Early American Architecture." *Gazette des Beaux Arts* 28 (1945): 87–112.

Wilson, Richard Guy. "Architecture and the Reinterpretation of the American Renaissance." *Winterthur Portfolio* 18, no. 1 (spring 1983): 69–87.

Wood, Peter H. "Slave Labor Camps in Early America: Overcoming Denial and Discovering the Gulag." In *Inequality in Early America*, ed. Carla Gardina Pestana and Sharon V. Salinger, 222–238. Hanover, NH: University Press of New England, 1999.

Books

Adams, Jessica. *Wounds of Returning: Race, Memory, and Property on the Postslavery Plantation*. Chapel Hill: University of North Carolina Press, 2007.

Aiken, Charles S. *The Cotton Plantation South since the Civil War*. Baltimore: Johns Hopkins University Press, 1998.

Akin, Edward N. *Flagler, Rockefeller Partner and Florida Baron*. Kent, OH: Kent State University Press, 1988.

Allen, Frederick Lewis. *Only Yesterday*. New York: Harper and Brothers, 1931.

Allsen, Thomas T. *The Royal Hunt in Eurasian History*. Philadelphia: University of Pennsylvania Press, 2006.

Armory, Cleveland. *Who Killed Society?* New York: Harper and Brothers, 1960.

Aslet, Clive. *The American Country House*. New Haven: Yale University Press, 1990.

Axlerod, Alan, ed. *The Colonial Revival in America*. New York: Norton, 1985.

Baker, Bruce E. *What Reconstruction Meant: Historical Memory in the American South*. Charlottesville: University Press of Virginia, 2007.

Baldwin, William P. *Lowcountry Plantations Today*. Greensboro, NC: Legacy Publications, 2002.

Baruch, Bernard M. *Baruch: My Own Story*. 2 vols. New York: Henry Holt and Company, 1957.

Beach, Virginia. *Medway*. Charleston: Wyrick and Company, 1999.

   Rice and Ducks: The Surprising Convergence That Saved the Carolina Low-country. Charleston: Evening Post Books, 2014.

Becker, Stephen D. *Marshall Field III: A Biography*. New York: Simon and Schuster, 1964.

Beckert, Sven. *The Monied Metropolis: New York City and the Consolidation of the American Bourgeois, 1850–1896*. New York: Cambridge University Press, 2001.

Bederman, Gail. *Manliness and Civilization: A Cultural History of Gender and Race in the United States, 1880–1917*. Chicago: University of Chicago Press, 1995.

Bender, Barbara, ed. *Landscape: Politics and Perspectives*. Providence, RI: Berg Publishers, 1993.

Berch, Bettina. *The Woman behind the Lens: The Life and Work of Frances Benjamin Johnston, 1864–1952*. Charlottesville: University Press of Virginia, 2000.

Berlin, Ira. *Many Thousands Gone: The First Two Centuries of Slavery in North America*. Cambridge: Belknap Press of Harvard University Press, 1998.

Blackburn, Robin. *The Making of New World Slavery: From the Baroque to the Modern, 1492–1800*. New York: Verso, 1997.

Blair, William A. *Cities of the Dead: Contesting the Memory of the Civil War in the South, 1865–1914*. Chapel Hill: University of North Carolina Press, 2004.

Blight, David W. *Beyond the Battlefield: Race, Memory, and the American Civil War*. Amherst: University of Massachusetts Press, 2002.

   Race and Reunion: The Civil War in American Memory. Cambridge: Belknap Press of Harvard University Press, 2001.

Brander, Michael. *Hunting and Shooting: From Earliest Times to the Present Day*. New York: G. P. Putnam's Sons, 1971.

Briggs, Loutrel W. *Charleston Gardens*. Columbia: University of South Carolina Press, 1951.

Brinkley, Alan. *The Publisher: Henry Luce and His American Century*. New York: Alfred A. Knopf, 2010.

Brock, Julia, and Daniel Vivian, eds. *Leisure, Plantations, and the Making of a New South: The Sporting Plantations of the South Carolina Lowcountry and Red Hills Region, 1900–1940*. Lanham, MD: Lexington Books, 2015.

Brockington, Lee. *Plantation between the Waters: A Brief History of Hobcaw Barony*. Charleston: History Press, 2006.

Brown, Thomas J. *Civil War Canon: Sites of Confederate Memory in South Carolina*. Chapel Hill: University of North Carolina Press, 2015.

Brundage, W. Fitzhugh. *The Southern Past: A Clash of Race and Memory.* Cambridge: Belknap Press of Harvard University Press, 2005.

ed. *Where These Memories Grow: History, Memory, and Southern Identity.* Chapel Hill: University of North Carolina Press, 2000.

Buck, Paul H. *The Road to Reunion, 1865–1900.* Boston: Brown, Little and Company, 1937.

Bullard, Mary R. *Cumberland Island: A History.* Athens: University of Georgia Press, 2003.

Campbell, Edward D. C. *The Celluloid South: Hollywood and the Southern Myth.* Knoxville: University of Tennessee Press, 1981.

Carney, Judith A. *Black Rice: The African Origins of Rice Cultivation in the Americas.* Cambridge: Harvard University Press, 2001.

Chaplin, Joyce. *An Anxious Pursuit: Agricultural Innovation and Modernity in the Lower South, 1730–1815.* Chapel Hill: Institute of Early American History and Culture by the University of North Carolina Press, 1993.

Childs, Arney R. *Rice Planter and Sportsman: The Recollections of J. Motte Alston, 1821–1909.* 1953; reprint, Columbia: University of South Carolina Press, 1999.

Choay, Françoise. *The Invention of the Historic Monument.* Trans. Lauren M. O'Connell. Cambridge: Cambridge University Press, 2001.

Cobb, James C. *Away Down South: A History of Southern Identity.* New York: Oxford University Press, 2005.

Coclanis, Peter A. *The Shadow of a Dream: Economic Life and Death in the South Carolina Low Country, 1670–1920.* New York: Oxford University Press, 1989.

Coit, Margaret L. *Mr. Baruch.* Boston: Houghton Mifflin, 1957.

Cooney, Terry A. *Balancing Acts: American Thought and Culture in the 1930s.* New York: Twayne Publishers, 1995.

Côté, Richard N. *Preserving the Legacy: Medway Plantation on Back River.* Mt. Pleasant, SC: the author, [1993?].

Cothran, James R. *Charleston Gardens and the Landscape Legacy of Loutrel Briggs.* Columbia: University of South Carolina Press, 2010.

Coudert, Thierry. *Café Society: Socialites, Patrons, and Artists, 1920–1960.* Paris: Flammarion, 2010.

Cox, Karen L. *Dixie's Daughters: The United Daughters of the Confederacy and the Preservation of Confederate Culture.* Gainesville: University Press of Florida, 2003.

Dreaming of Dixie: How the South Was Created in American Popular Culture. Chapel Hill: University of North Carolina Press, 2011.

Cox, Neal. *Neal Cox of Arcadia Plantation: Memories of a Renaissance Man.* Georgetown, SC: Alice Cox Harrelson, 2003.

Creel, Margaret Washington. *A Peculiar People: Slave Religion and Community-Culture among the Gullahs.* New York: New York University Press, 1988.

Curtin, Philip D. *The Rise and Fall of the Plantation Complex: Essays in Atlantic History.* 2nd ed. New York: Cambridge University Press, 1990.

Cuthbert, Robert B., and Stephen G. Hoffius, eds. *Northern Money, Southern Land: The Lowcountry Plantation Sketches of Chlotilde R. Martin.* Columbia: University of South Carolina Press, 2009.

Daniel, Pete. *Breaking the Land: The Transformation of Cotton, Tobacco, and Rice Cultures since 1880.* Urbana: University of Illinois Press, 1985.

Danielson, Michael N., and Patricia R. F. Danielson. *Profits and Politics in Paradise: The Development of Hilton Head Island.* Columbia: University of South Carolina Press, 1995.

Davidson, Chalmers Gaston. *The Last Foray: The South Carolina Planters of 1860, a Sociological Study.* Columbia: South Carolina Tricentennial Commission by the University of South Carolina Press, 1971.

De Long, David G. *Auldbrass: Frank Lloyd Wright's Southern Plantation.* New York: Rizzoli, 2003.

Devlin, George A. *South Carolina and Black Migration, 1865–1940.* New York: Garland Publishing, 1989.

Doyle, Don H. *New Men, New Cities, New South: Atlanta, Nashville, Charleston, Mobile, 1860–1910.* Chapel Hill: University of North Carolina Press, 1990.

Dunlap, Thomas R. *Saving America's Wildlife.* Princeton: Princeton University Press, 1988.

Dusinberre, William. *Them Dark Days: Slavery in the American Rice Swamps.* New York: Oxford University Press, 1996.

Edelson, S. Max. *Plantation Enterprise in South Carolina.* Cambridge: Harvard University Press, 2006.

Edgar, Walter B. *History of Santee Cooper, 1934–1984.* Columbia: R. L. Bryan Company, 1984.

    South Carolina: A History. Columbia: University of South Carolina Press, 1998.

Edgar, Walter B., N. Louise Bailey, et al., eds. *Biographical Directory of the South Carolina House of Representatives.* 5 vols. Columbia: University of South Carolina Press, 1974–1992.

Eichstedt, Jennifer L., and Stephen Small. *Race and Ideology in Southern Plantation Museums.* Washington, DC: Smithsonian Books, 2002.

Ellis, Clifton, and Rebecca Ginsburg, eds. *Cabin, Quarter, Plantation: Architecture and Landscapes of North American Slavery.* New Haven: Yale University Press, 2010.

Entrikin, J. Nicholas. *The Betweenness of Place: Towards a Geography of Modernity.* Baltimore: Johns Hopkins University Press, 1991.

Fahs, Alice, and Joan Waugh, eds. *The Memory of the Civil War in American Culture.* Chapel Hill: University of North Carolina Press, 2004.

Ferguson, Leland G. *Uncommon Ground: Archaeology and Early African America, 1650–1800.* Washington, DC: Smithsonian Institution Press, 1992.

Fields, Mamie Garvin. *Lemon Swamp and Other Places: A Carolina Memoir.* New York: Free Press, 1983.

Foner, Eric. *Nothing but Freedom: Emancipation and Its Legacy.* Baton Rouge: Louisiana State University Press, 1983.

    Reconstruction: America's Unfinished Revolution, 1863–1877. New York: Harper and Row, 1988.

Forman, Henry Chandlee. *The Architecture of the Old South: The Medieval Style, 1585–1850.* Cambridge: Harvard University Press, 1948.

Foster, Gaines M. *Ghosts of the Confederacy: Defeat, the Lost Cause, and the Emergence of the New South, 1865–1913.* New York: Oxford University Press, 1987.

Fraser, Walter J., Jr. *Charleston! Charleston!: The History of a Southern City.* Columbia: University of South Carolina Press, 1989.

  Lowcountry Hurricanes: Three Centuries of Storms at Sea and Ashore. Athens: University of Georgia Press, 2006.

Frazer, Susan Hume. *The Architecture of William Lawrence Bottomley.* New York: Acanthus Press, 2007.

Frederickson, George M. *The Black Image in the White Mind: The Debate on Afro-American Character and Destiny, 1817–1914.* New York: Harper and Row, 1971.

Gaston, Paul M. *The New South Creed: A Study in Southern Mythmaking.* Baton Rouge: Louisiana State University Press, 1970.

Gelernter, Mark. *A History of American Architecture: Buildings in Their Cultural and Technological Context.* Hanover, NH: University Press of New England, 1999.

Giltner, Scott E. *Hunting and Fishing in the New South: Black Labor and White Leisure after the Civil War.* Baltimore: Johns Hopkins University Press, 2008.

Goings, Kenneth W. *Mammy and Uncle Moses: Black Collectibles and American Stereotyping.* Bloomington: Indiana University Press, 1994.

Grant, James. *Bernard M. Baruch: The Adventures of a Wall Street Legend.* New York: Simon and Schuster, 1983.

Grant, Susan-Mary. *North over South: Northern Nationalism and American Identity in the Antebellum Era.* Lawrence: University Press of Kansas, 2000.

Gray, Lewis Cecil. *History of Agriculture in the Southeastern United States to 1860.* 2 vols. Washington, DC: Carnegie Institution of Washington, 1933.

Greenspan, Anders. *Creating Colonial Williamsburg: The Restoration of Virginia's Eighteenth-Century Capital.* 2nd ed. Chapel Hill: University of North Carolina Press, 2009.

Griffen, Sarah L., and Kevin D. Murphy, eds. *"A Noble and Dignified Stream": The Piscataqua Region in the Colonial Revival, 1860–1930.* York, ME: Old York Historical Society, 1992.

Griffin, Emma. *Blood Sport: Hunting in Britain since 1066.* New York: Yale University Press, 2007.

Guterl, Matthew Pratt. *The Color of Race in America, 1900–1940.* Cambridge: Harvard University Press, 2001.

Guthrie, Patricia. *Catching Sense: African American Communities on a South Carolina Sea Island.* Westport, CT: Bergin and Garvey, 1996.

Hale, Grace Elizabeth. *Making Whiteness: The Culture of Segregation in the South, 1890–1940.* New York: Pantheon Books, 1998.

Halfacre, Angela C. *A Delicate Balance: Constructing a Conservation Culture in the South Carolina Lowcountry.* Columbia: University of South Carolina Press, 2012.

Hamer, Fritz P. *Charleston Reborn: A Southern City, Its Navy Yard, and World War II.* Charleston: History Press, 2005.

Handler, Richard, and Eric Gable. *The New History in an Old Museum: Creating the Past at Colonial Williamsburg.* Durham: Duke University Press, 1997.

Hart, Emma. *Building Charleston: Town and Society in the Eighteenth-Century British Atlantic World.* Charlottesville: University of Virginia Press, 2010.

Hayes, Jack Irby, Jr. *South Carolina and the New Deal*. Columbia: University of South Carolina Press, 2001.

Hegeman, Susan. *Patterns for America: Modernism and the Concept of Culture*. Princeton: Princeton University Press, 1999.

Herman, Daniel Justin. *Hunting and the American Imagination*. Washington, DC: Smithsonian Institution Press, 2001.

Hewitt, Mark A. *The Architect and the American Country House, 1890–1940*. New Haven: Yale University Press, 1990.

Hilderbrand, Gary R., ed. *Making a Landscape of Continuity: The Practice of Innocenti and Webel*. Cambridge: Harvard University Graduate School of Design, 1997.

Hillyer, Reiko. *Designing Dixie: Tourism, Memory, and Urban Space in the New South*. Charlottesville: University of Virginia Press, 2014.

Holden, Charles J. *In the Great Maelstrom: Conservatives in Post–Civil War South Carolina*. Columbia: University of South Carolina Press, 2002.

Holmgren, Virginia C. *Hilton Head: A Sea Island Chronicle*. Hilton Head Island: Hilton Head Publishing Company, 1959.

Holt, Thomas C. *Black over White: Negro Political Leadership in South Carolina during Reconstruction, 1861–1877*. Urbana: University of Illinois Press, 1977.

Homberger, Eric. *Mrs. Astor's New York: Money and Social Power in a Gilded Age*. New Haven: Yale University Press, 2002.

Horton, James O., and Lois E. Horton, eds. *Slavery and Public History: The Tough Stuff of American Memory*. New York: New Press, 2006.

Hosmer, Charles B., Jr. *Presence of the Past: A History of the Preservation Movement in the United States before Williamsburg*. New York: G. P. Putnam's Sons, 1965.

Hudgins, Carter L., Carl R. Lounsbury, Louis P. Nelson, and Jonathan H. Poston, eds. *The Vernacular Architecture of Charleston and the Lowcountry, 1670–1990: A Field Guide*. Charleston: Historic Charleston Foundation, 1994.

Hurley, Suzanne Cameron Linder. *Anglican Churches in Colonial South Carolina* Charleston: Wyrick and Company, 2000.

Hutchisson, James M., and Harlan Greene, eds. *Renaissance in Charleston: Art and Life in the Carolina Low Country, 1900–1940*. Athens: University of Georgia Press, 2003.

Jacobson, Matthew Frye. *Whiteness of a Different Color: European Immigrants and the Alchemy of Race*. Cambridge: Harvard University Press, 1998.

Jacoby, Karl. *Crimes against Nature: Squatters, Poachers, Thieves, and the Hidden History of Conservation*. Berkeley: University of California Press, 2001.

Jaher, Frederic Cople. *The Urban Establishment: Upper Strata in Boston, New York, Charleston, Chicago, and Los Angeles*. Urbana: University of Illinois Press, 1982.

Janney, Carol E. *Burying the Dead but Not the Past: Ladies' Memorial Associations and the Lost Cause*. Chapel Hill: University of North Carolina Press, 2008.

Remembering the Civil War: Reunion and the Limits of Reconciliation. Chapel Hill: University of North Carolina Press, 2013.

Johnson, Guy B. *Folk Culture on St. Helena Island, South Carolina*. Chapel Hill: University of North Carolina Press, 1930.

Jones, Andrew. *Memory and Material Culture*. New York: Cambridge University Press, 2007.

Joyner, Charles. *Down by the Riverside: A South Carolina Slave Community*. Urbana: University of Illinois Press, 1984.

Kammen, Michael G. *Mystic Chords of Memory: The Transformation of Tradition in American Culture*. New York: Knopf, 1991.

Kathrens, Michael C. *Newport Villas: The Revival Styles, 1885–1935*. New York: W. W. Norton, 2009.

Kimball, Fiske. *Domestic Architecture of the American Colonies and of the Early Republic*. New York: Charles Scribners' Sons, 1927.

Kirby, Jack Temple. *Media-Made Dixie: The South in the American Imagination*. Rev. ed. Athens: University of Georgia Press, 1986.

Mockingbird Song: Ecological Landscapes of the South. Chapel Hill: University of North Carolina Press, 2006.

Rural Worlds Lost: The American South, 1920–1960. Baton Rouge: Louisiana State University Press, 1987.

Kiser, Clyde Vernon. *Sea Island to City: A Study of St. Helena Islanders in Harlem and Other Urban Centers*. New York: Columbia University Press, 1932.

Klein, Rachel N. *Unification of a Slave State: The Rise of a Planter Class in the South Carolina Backcountry, 1760–1808*. Chapel Hill: Institute of Early American History and Culture by University of North Carolina Press, 1990.

Kovacik, Charles F., and John J. Winberry. *South Carolina: A Geography*. Boulder: Westview Press, 1987.

Lachicotte, Alberta Morel. *Georgetown Rice Plantations*. Columbia: The State Printing Company, 1955.

Lane, Mills. *Architecture of the Old South: South Carolina*. Savannah: Beehive Press, 1984.

Lareau, Jane, and Richard Dwight Porcher. *Lowcountry: The Natural Landscape*. Greensboro, NC: Legacy Publications, 1988.

Lears, T. J. Jackson. *No Place of Grace: Antimodernism and the Transformation of American Culture, 1880–1920*. Chicago: University of Chicago Press, 1981.

Leepson, Marc. *Saving Monticello: The Levy Family's Epic Quest to Rescue the House That Jefferson Built*. New York: Free Press, 2001.

Lefebvre, Henri. *The Production of Space*. Trans. Donald Nicholson-Smith. Cambridge: Blackwell, 1991.

Legendre, Gertrude S. *Medway Plantation, 1686–1980*. Charleston: n.p., 1980.

Linder, Suzanne Cameron. *Historical Atlas of the Rice Plantations of the ACE River Basin – 1860*. Columbia: South Carolina Department of Archives and History Foundation, Ducks Unlimited, and the Nature Conservancy, 1995.

Linder, Suzanne Cameron, and Marta Leslie Thacker. *Historical Atlas of the Rice Plantations of Georgetown County and the Santee River*. Columbia: South Carolina Department of Archives and History for the Historic Ricefields Association, Inc., 2001.

Lindgren, James M. *Preserving the Old Dominion: Historic Preservation and Virginia Traditionalism.* Charlottesville: University Press of Virginia, 1993.

Long, Franklin Leslie, and Luce B. Long. *The Henry Ford Era at Richmond Hill, Georgia.* Darien, GA: Darien Graphics, 1998.

Lowenthal, David. *The Past Is a Foreign Country.* Cambridge: Cambridge University Press, 1986.

Lowenthal, David, and Marcus Binney. *Our Past before Us: Why Do We Save It?* London, England: T. Smith, 1981.

MacKay, Robert B., Anthony K. Baker, and Carol A. Traynor, eds. *Long Island Country Houses and Their Architects, 1860–1940.* New York: Society for the Preservation of Long Island Antiquities in association with W. W. Norton and Company, 1997.

Mandell, Richard. *Pinehurst: Home of American Golf.* Pinehurst, NC: T. Eliot Press, 2007.

Manring, M. M. *Slave in a Box: The Strange Career of Aunt Jemima.* Charlottesville: University of Virginia Press, 1998.

Margeson, Hank, and Joseph Kitchens. *Quail Plantations of South Georgia and North Florida.* Athens: University of Georgia Press, 1991.

Marks, Stuart A. *Southern Hunting in Black and White: Nature, History, and Ritual in a Carolina Community.* Princeton: Princeton University Press, 1991.

Marling, Karal Ann. *George Washington Slept Here: Colonial Revivals and American Culture, 1876–1986.* Cambridge: Harvard University Press, 1988.

Maynard, W. Barksdale. *Architecture in the United States, 1800–1850.* New Haven: Yale University Press, 2002.

McCash, William Barton, and June Hall McCash. *The Jekyll Island Club: Southern Haven for America's Millionaires.* Athens: University of Georgia Press, 1989.

McConnell, Stuart. *Glorious Contentment: The Grand Army of the Republic, 1865–1900.* Chapel Hill: University of North Carolina Press, 1992.

McCrady, Edward. *The History of South Carolina under the Proprietary Government, 1670–1719.* New York: Macmillan, 1897.

    The History of South Carolina in the Revolution, 1780–1783. New York: Macmillan, 1902.

    The History of South Carolina under the Royal Government, 1719–1776. New York: Macmillan, 1901.

McCurry, Stephanie. *Masters of Small Worlds: Yeoman Households, Gender Relations, and the Political Culture of the Antebellum South Carolina Low Country.* New York: Oxford University Press, 1995.

McElya, Micki. *Clinging to Mammy: The Faithful Slave in Twentieth-Century America.* Cambridge: Harvard University Press, 2007.

McInnis, Maurie D. *The Politics of Taste in Antebellum Charleston.* Chapel Hill: University of North Carolina Press, 2005.

McIntyre, Rebecca Cawood. *Souvenirs of the Old South: Northern Tourism and Southern Mythology.* Gainesville: University Press of Florida, 2011.

McKinley, Shepherd W. *Stinking Stones and Rocks of Gold: Phosphate, Fertilizer, and Industrialization in Postbellum South Carolina.* Gainesville: University Press of Florida, 2014.

Michie, James L. *Richmond Hill Plantation, 1810–1860: The Discovery of Antebellum Life on a Waccamaw Rice Plantation*. Spartanburg, SC: Reprint Company, 1990.

Mielnik, Tara M. *New Deal, New Landscape: The Civilian Conservation Corps and South Carolina's State Parks*. Columbia: University of South Carolina Press, 2011.

Mills, Cynthia J., and Pamela H. Simpson, eds. *Monuments to the Lost Cause: Women, Art, and the Landscapes of Southern Memory*. Knoxville: University of Tennessee Press, 2003.

Montgomery, Michael, ed. *The Crucible of Carolina: Essays in the Development of Gullah Language and Culture*. Athens: University of Georgia Press, 1994.

Moore, Alexander. *Poco Sabo Plantation: A Place in Time*. N.p.: H. Anthony Ittleson, 2005.

Moore, John Hammond, ed. and comp. *South Carolina in the 1880s: A Gazetteer*. Orangeburg, SC: Sandlapper Publishing, 1989.

Morgan, Philip D. *Slave Counterpoint: Black Culture in the Eighteenth-Century Chesapeake and Lowcountry*. Chapel Hill: Omohundro Institute of Early American History and Culture by the University of North Carolina Press, 1998.

Morris, Sylvia Jukes. *Rage for Fame: The Ascent of Clare Boothe Luce*. New York: Random House, 1997.

Morrison, William. *The Main Line: Country Houses of Philadelphia's Storied Suburb, 1870–1930*. New York: Acanthus Press, 2002.

Neff, John R. *Honoring the Civil War Dead: Commemoration and the Problem of Reconciliation*. Lawrence: University Press of Kansas, 2005.

Paisley, Clifton. *From Cotton to Quail: An Agricultural Chronicle of Leon County, Florida, 1860–1967*. Gainesville: University of Florida Press, 1968.

Perry, Grace Fox. *Moving Finger of Jasper*. N.p.: n.p., 1962.

Petty, Julian J. *The Growth and Distribution of Population in South Carolina*. Columbia: State Council for Defense, 1943.

Phillips, Ullrich B. *American Negro Slavery*. New York: D. Appleton and Company, 1918.

Pollitzer, William S. *The Gullah People and Their African Heritage*. Athens: University of Georgia Press, 1999.

Porcher, Richard Dwight, and Sara Fick. *The Story of Sea Island Cotton*. Charleston: Wyrick and Company, 2005.

Poston, Jonathan H. *The Buildings of Charleston: A Guide to the City's Architecture*. Columbia: University of South Carolina Press, 1997.

Pressly, Thomas J. *Americans Interpret Their Civil War*. Princeton: Princeton University Press, 1954.

Proctor, Nicholas W. *Bathed in Blood: Hunting and Mastery in the Old South*. Charlottesville: University Press of Virginia, 2002.

Ramsay, David. *Ramsay's History of South Carolina*. 2 vols. Newberry, SC: W. J. Duffie, 1858.

Reiger, John F. *American Sportsmen and the Origins of Conservation*. 3rd ed. Corvallis: Oregon State University Press, 2001.

Rhoads, William B. *The Colonial Revival*. New York: Garland Publishing, 1977.

Robb, Alex M. *The Sanfords of Amsterdam: The Biography of a Family in America*. New York: William-Frederick Press, 1969.

Rogers, George C., Jr. *The History of Georgetown County, South Carolina*. Columbia: University of South Carolina Press, 1970.

Rose, Willie Lee. *Rehearsal for Reconstruction: The Port Royal Experiment*. Indianapolis: Bobbs-Merrill, 1964.

Roth, Leland M. *Shingle Styles: Innovation and Tradition in American Architecture, 1874 to 1982*. New York: Henry N. Abrams, Inc., 1999.

Rowland, Lawrence S., and Stephen R. Wise. *The History of Beaufort County, South Carolina*, vol. 3: *Bridging the Sea Islands' Past and Present, 1893–2006*. Columbia: University of South Carolina Press, 2015.

Rubin, Anne Sarah. *Through the Heart of Dixie: Sherman's March and American Memory*. Chapel Hill: University of North Carolina Press, 2014.

Russ, Elizabeth Christine. *The Plantation in the Postslavery Imagination*. New York: Oxford University Press, 2009.

Saville, Julie. *The Work of Reconstruction: From Slave to Wage Laborer in South Carolina, 1860–1870*. New York: Cambridge University Press, 1994.

Schmitt, Peter J. *Back to Nature: The Arcadian Myth in Urban America*. New York: Oxford University Press, 1969.

Schwalm, Leslie A. *A Hard Fight for We: Women's Transition from Slavery to Freedom in South Carolina*. Urbana: University of Illinois Press, 1997.

Schwartz, Stuart B. *Sugar Plantations in the Formation of Brazilian Society: Bahia, 1550–1835*. New York: Cambridge University Press, 1985.

Severens, Kenneth. *Charleston Antebellum Architecture and Civic Destiny*. Knoxville: University Tennessee Press, 1988.

Severens, Martha R. *Alice Ravenel Huger Smith: An Artist, a Place, and a Time*. Charleston: Carolina Art Association, 1993.

 The Charleston Renaissance. Spartanburg, SC: Saraland Press, 1998.

Shaffer, Donald R. *After the Glory: The Struggles of Black Civil War Veterans*. Lawrence: University Press of Kansas, 2004.

Silber, Nina. *The Romance of Reunion: Northerners and the South, 1865–1900*. Chapel Hill: University of North Carolina Press, 2004.

Smith, John David. *An Old Creed for the New South: Proslavery Ideology and Historiography, 1865–1918*. Westport, CT: Greenwood Press, 1985.

Smith, John David, and J. Vincent Lowery. *The Dunning School: Historians, Race, and the Meaning of Reconstruction*. Lexington: University Press of Kentucky, 2013.

Smith, Julia Floyd. *Slavery and Rice Culture in Low Country Georgia, 1750–1860*. Knoxville: University of Tennessee Press, 1985.

Smith, Timothy B. *The Chickamauga Memorial: The Establishment of America's First Civil War National Military Park*. Knoxville: University of Tennessee Press, 2009.

 The Golden Age of Battlefield Preservation: The Decade of the 1890s and the Establishment of America's First Five Military Parks. Knoxville: University of Tennessee Press, 2008.

The Great Battlefield of Shiloh: History, Memory, and the Establishment of a Civil War National Military Park. Knoxville: University of Tennessee Press, 2004.

Soule, George H. *Prosperity Decade: From War to Depression, 1917–1929*. Armonk, NY: M. E. Sharp, 1947.

Stewart, Catherine A. *Long Past Slavery: Representing Race in the Federal Writers' Project*. Chapel Hill: University of North Carolina Press, 2016.

Stewart, Mart A. *"What Nature Suffers to Groe": Life, Labor, and Landscape on the Georgia Coast, 1680–1920*. Athens: University of Georgia Press, 1996.

Stokes, Barbara J. *Myrtle Beach: A History, 1900–1980*. Columbia: University of South Carolina Press, 2007.

Stone, Edward Durell. *The Evolution of an Architect*. New York: Horizon Press, 1962.

Stoney, Louisa Cheves, ed. and comp. *A Day on Cooper River*. 2nd ed. Columbia: R. L. Bryan Company, 1932.

Stronge, William B. *The Sunshine Economy: An Economic History of Florida since the Civil War*. Gainesville: University Press of Florida, 2008.

Sullivan, C. John. *Waterfowling on the Chesapeake, 1819–1936*. Baltimore: Johns Hopkins University Press, 2003.

Susman, Warren I. *Culture as History: The Transformation of American Society in the Early Twentieth Century*. Washington, DC: Smithsonian Institution Press, 2003.

Swanson, Drew. *Remaking Wormsloe Plantation: The Environmental History of a Lowcountry Landscape*. Athens: University of Georgia Press, 2012.

Swarns, Rachel L. *American Tapestry: The Story of the Black, White, and Multiracial Ancestors of Michelle Obama*. New York: Amistad, 2012.

Taylor, William A. *Cavalier and Yankee: The Old South and American National Character*. New York: George Braziller, 1961.

Tebeau, Charleton W. *A History of Florida*. Coral Gables: University of Miami Press, 1971.

Tindall, George Brown. *The Emergence of the New South, 1913–1945*. Baton Rouge: Louisiana State University Press, 1967.

South Carolina Negroes, 1877–1900. Columbia: University of South Carolina Press, 1952.

Tolber, James A. *Who Owns the Wildlife?: The Political Economy of Game Conservation in Nineteenth-Century America*. Westport, CT: Greenwood Press, 1981.

Trachtenberg, Alan. *The Incorporation of America: Culture and Society in the Gilded Age*. New York: Hill and Wang, 1982.

Train, Frances Cheston. *A Carolina Plantation Remembered: In Those Days*. Charleston: History Press, 2008.

Tuan, Yi-Fu. *Space and Place: The Perspective of Experience*. Minneapolis: University of Minnesota Press, 1977.

Turner, Lorenzo Dow. *Africanisms in the Gullah Dialect*. Chicago: University of Chicago Press, 1949.

Tuten, James H. *Lowcountry Time and Tide: The Fall of the South Carolina Rice Kingdom*. Columbia: University of South Carolina Press, 2010.

Urofsky, Melvin I. *The Levy Family and Monticello, 1834–1923: Saving Thomas Jefferson's House*. Charlottesville: Thomas Jefferson Foundation, 2001.

Van Deburg, William L. *Slavery and Race in American Popular Culture*. Madison: University of Wisconsin Press, 1984.

Veblen, Thorstein. *Theory of the Leisure Class*. New York: Macmillan, 1899.

Vileisis, Ann. *Discovering the Unknown Landscape: A History of America's Wetlands*. Washington, DC: Island Press, 1997.

Vlach, John Michael. *Back of the Big House: The Architecture of Plantation Slavery*. Chapel Hill: University of North Carolina Press, 1993.

   The Planter's Prospect: Privilege and Slavery in Plantation Paintings. Chapel Hill: University of North Carolina Press, 2002.

Wallace, David Duncan. *The History of South Carolina*. 4 vols. New York: American Historical Society, 1934.

Wallace-Sanders, Kimberly. *Mammy: A Century of Race, Gender, and Southern Memory*. Ann Arbor: University of Michigan Press, 2008.

Waterhouse, Richard. *A New World Gentry: The Making of a Merchant and Planter Class in South Carolina, 1670–1777*. New York: Garland Publishing, 1989.

Waterman, Thomas T. *The Dwellings of Colonial America*. Chapel Hill: University of North Carolina Press, 1950.

Way, Albert G. *Conserving Southern Longleaf: Herbert Stoddard and the Rise of Ecological Land Management*. Athens: University of Georgia Press, 2011.

Wecter, Dixon. *The Saga of American Society; a Record of Social Aspiration, 1607–1937*. New York: C. Scribner's Sons, 1937.

Weir, Robert M. *Colonial South Carolina: A History*. Millwood, NY: KTO Press, 1983.

West, Patricia. *Domesticating History: The Political Origins of America's House Museums*. Washington, DC: Smithsonian Institution Press, 1999.

Weyeneth, Robert W. *Historic Preservation for a Living City: Historic Charleston Foundation, 1947–1997*. Columbia: University of South Carolina Press, 2000.

Whitehead, Ruth Holmes. *Broughton Family Sourcebook*. N.p.: n.p., [1998?]).

Williamson, Joel. *After Slavery: The Negro in South Carolina during Reconstruction, 1861–1877*. Chapel Hill: University of North Carolina Press, 1965.

   The Crucible of Race: Black-White Relations in the American South since Emancipation. New York: Oxford University Press, 1984.

Wilson, Charles Reagan. *Baptized in Blood: The Religion of the Lost Cause, 1865–1920*. Athens: University of Georgia Press, 1980.

Wilson, Richard Guy, Dianne H. Pilgrim, and Richard N. Murray. *The American Renaissance, 1876–1917*. New York: Brooklyn Museum, 1979.

Wilson, Richard Guy, Shaun Eyring, and Kenny Marotta, eds. *Re-creating the American Past: Essays on the Colonial Revival*. Charlottesville: University of Virginia Press, 2006.

Wise, Stephen R., and Lawrence S. Rowland. *The History of Beaufort County, South Carolina*, vol. 2: *Rebellion, Reconstruction, and Redemption, 1861–1893*. Columbia: University of South Carolina Press, 2015.

Wood, Peter H. *Black Majority: Negroes in Colonial South Carolina from 1670 through the Stono Rebellion*. New York: Knopf, 1974.

Woods, Mary N. *Beyond the Architect's Eye: Photographs and the American Built Environment*. Philadelphia: Philadelphia University Press, 2009.

Woodward, C. Vann. *Origins of the New South, 1877–1913*. Baton Rouge: Louisiana State University Press, 1951.

Woofter, T. J. *Black Yeomanry: Life on St. Helena Island*. New York: Henry Holt and Company, 1930.

Yuhl, Stephanie E. *A Golden Haze of Memory: The Making of Historic Charleston*. Chapel Hill: University of North Carolina Press, 2005.

Theses and Dissertations

Betsworth, Jennifer. "'Then Came the Peaceful Invasion of the Northerners': The Impact of Outsiders on Plantation Architecture in Georgetown County, South Carolina." MA thesis, University of South Carolina, 2011.

Brock, Julia. "Land, Labor, and Leisure: Northern Tourism in the Red Hills Region, 1890–1950." PhD dissertation, University of California Santa Barbara, 2012.

Cann, Mary Katherine Davis. "The Morning After: South Carolina in the Jazz Age." PhD dissertation, University of South Carolina, 1984.

Harper, Marilyn M. "'What It Ought to Have Been': Three Case Studies of Early Restoration Work in Virginia." MA thesis, George Washington University, 1989.

Hart, T. Robert, Jr. "The Santee-Cooper Landscape: Culture and Environment in the South Carolina Lowcountry." PhD dissertation, University of Alabama, 2004.

Lockhart, Matthew A. "From Rice Fields to Duck Marshes: Sport Hunters and Environmental Change on the South Carolina Coast, 1890–1950." PhD dissertation, University of South Carolina, 2017.

Smith, Shelley Elizabeth. "The Plantations of South Carolina: Transmission and Transformation in Provincial Culture." PhD dissertation, Columbia University, 1999.

Wayne, Lucy Bowles. "Burning Brick: A Study of a Lowcountry Industry." PhD dissertation, University of Florida, 1992.

Index

For EU product safety concerns, contact us at Calle de José Abascal, 56–1°, 28003 Madrid, Spain or eugpsr@cambridge.org.

www.ingramcontent.com/pod-product-compliance
Ingram Content Group UK Ltd.
Pitfield, Milton Keynes, MK11 3LW, UK
UKHW010855090126
466816UK00012B/248